Science and Soccer

OTHER TITLES FROM E & FN SPON

Coaching Children in Sport
M. Lee

Drugs in Sport (Second edition)
D. Mottram

Foods, Nutrition and Sports Performance
C. Williams and J.R. Devlin

Kinanthropometry and Exercise Physiology Laboratory Manual
R. Eston and T. Reilly

Physiology of Sports
T. Reilly, N. Secher, P. Snell and C. Williams

Science and Football II
T. Reilly, J. Clarys and A. Stibbe

Sociology of Leisure
C. Critcher, P. Bramham and A. Tomlinson

Journal of Sports Sciences

Leisure Studies

For more information about these and other titles please contact:
The Promotion Department, E & FN SPON, 2–6 Boundary Row, London SE1 8HN.
Telephone 0171 865 0066, Fax 0171 522 9623

Science and Soccer

Edited by

Thomas Reilly

Professor of Sports Science
Liverpool John Moores University
Liverpool, UK

E & FN SPON
An Imprint of Chapman & Hall

London · Glasgow · Weinheim · New York · Tokyo · Melbourne · Madras

Published by E & FN Spon, an imprint of Chapman & Hall, 2–6 Boundary Row, London SE1 8HN, UK

Chapman & Hall, 2–6 Boundary Row, London SE1 8HN, UK

Blackie Academic & Professional, Wester Cleddens Road, Bishopbriggs, Glasgow G64 2NZ, UK

Chapman & Hall GmbH, Pappelallee 3, 69469 Weinheim, Germany

Chapman & Hall USA, 115 Fifth Avenue, New York, NY 10003, USA

Chapman & Hall Japan, ITP-Japan, Kyowa Building, 3F, 2-2-1 Hirakawacho, Chiyoda-ku, Tokyo 102, Japan

Chapman & Hall Australia, 102 Dodds Street, South Melbourne, Victoria 3205, Australia

Chapman & Hall India, R. Seshadri, 32 Second Main Road, CIT East, Madras 600 035, India

First edition 1996

© 1996 E & FN Spon

Typeset in 10/12 Times by WestKey Limited, Falmouth, Cornwall

Printed in Great Britain by T.J. Press (Padstow) Ltd

ISBN 0 419 18880 0

Apart from any fair dealing for the purposes of research or private study, or criticism or review, as permitted under the UK Copyright Designs and Patents Act, 1988, this publication may not be reproduced, stored, or transmitted, in any form or by any means, without the prior permission in writing of the publishers, or in the case of reprographic reproduction only in accordance with the terms of the licences issued by the Copyright Licensing Agency in the UK, or in accordance with the terms of licences issued by the appropriate Reproduction Rights Organization outside the UK. Enquiries concerning reproduction outside the terms stated here should be sent to the publishers at the London address printed on this page.

The publisher makes no representation, express or implied, with regard to the accuracy of the information contained in this book and cannot accept any legal responsibility or liability for any errors or omissions that may be made.

A catalogue record for this book is available from the British Library

Library of Congress Catalog Card Number: 95-74655

∞ Printed on permanent acid-free text paper, manufactured in accordance with ANSI/NISO Z39.48–1992 and ANSI/NISO Z39.48–1984 (Permanence of Paper).

RC
1220
S57
S24
1996

Contents

Contributors

Jens Bangsbo
August Krogh Institute, University of Copenhagen, Denmark

Dick Bate
Formerly, Football Association of Malaysia, Malaysia

Andy Borrie
Centre for Sport and Exercise Sciences, School of Human Sciences, Liverpool John Moores University, UK

Malcolm Cook
Dewsbury College, West Yorks, and Director, Pro Sport UK Ltd

Ian M. Franks
School of Human Kinetics, University of British Columbia, Vancouver, Canada

David Gilbourne
Centre for Sport and Exercise Sciences, School of Human Sciences, Liverpool John Moores University, UK

Tracey Howe
Centre for Habilitation and Muscle Performance, School of Nursing Studies, University of Manchester, UK

Mike Hughes
Centre for Notational Analysis, Cardiff Institute, UK

Adrian Lees
Centre for Sport and Exercise Sciences, School of Human Sciences, Liverpool John Moores University, UK

Tim McGarry
School of Human Kinetics, University of British Columbia, Vancouver, Canada

Don MacLaren
Centre for Sport and Exercise Sciences, School of Human Sciences, Liverpool John Moores University, UK

John Minten
Centre for Sport and Exercise Sciences, School of Human Sciences, Liverpool John Moores University, UK

Benny Josef Peiser
Centre for Sport and Exercise Sciences, School of Human Sciences, Liverpool John Moores University, UK

Thomas Reilly
Centre for Sport and Exercise Sciences, School of Human Sciences, Liverpool John Moores University, UK

Frank Sanderson
Centre for Sport and Exercise Sciences, School of Human Sciences, Liverpool John Moores University, UK

Tony Shelton
Centre for Sport and Exercise Sciences, School of Human Sciences, Liverpool John Moores University, UK

Introduction to science and soccer

1

Thomas Reilly

Introduction

Football is the world's most popular form of sport, being played in every nation without exception. The most widespread code is association football or soccer. The sport has a rich history though it was formalized as we know it today by the establishment of the Football Association in 1863. The game soon spread to continental European countries and later to South America and the other continents. The world's governing body, the Federation of the International Football Association (FIFA), was set up in 1904 and the first Olympic soccer competition was held four years later. The United Kingdom won the final 2–0, defeating Denmark, another nation playing a leading role in the popularization of the game. Uruguay played host to the first World Cup tournament in 1930. This competition is held every four years and is arguably the tournament with the most fanatical hold on its spectators and TV audiences. Only six nations have won the tournament – Uruguay, Argentina, Brazil, Germany, England and Italy. Whilst they may represent the top teams at elite level, the popularity of the game is reflected in the millions who participate in soccer at lower levels of play.

Science and Soccer. Edited by Thomas Reilly. Published in 1996 by E & FN Spon, London. ISBN 0 419 18880 0.

1.1 DEVELOPMENT OF SPORTS SCIENCE

In recent years there has been a remarkable expansion of sports science. The subject area is now recognized both as an academic discipline and a valid area of professional practice. Sports science is well respected within its parent disciplines, for example biomechanics, biochemistry, physiology, psychology, sociology and so on. A new maturity became apparent as the sports sciences were increasingly applied to address problems in particular sports rather than to sports in general. One of these specific applications has been to soccer.

The applications of science to soccer predated the formal acceptance of sports science as an area of study in university programmes. South American national teams used specialists in psychology, nutrition and physiology in the preparation of squads for the major international tournaments from the early 1970s. The comprehensive systems of scientific support accessible to Eastern European athletes since the 1970s dwarfed the commitments of Western countries to top-level sport. The gulf was notably wide with respect to British soccer, where the sports scientist was more often than not shunned or at best frostily welcomed.

In the 1980s it became apparent that the football industry and professionals in the game could no longer rely on the traditional methods of previous decades. Coaches and trainers were more open to contemporary scientific approaches to preparing for competition. Methods of management science were applied to organizing the big soccer clubs and the training of players could be formulated on a systematic basis. In general the clubs that moved with the times were rewarded with success by gaining an advantage over those that did not change.

It has taken some years for the knowledge accumulating within sports science to be translated into a form usable by practitioners. Efforts have been made to compile scientific knowledge and expertise and make them more widely available to the football world. The production of this textbook is but one step in that direction.

1.2 SCIENCE AND FOOTBALL

The First World Congress of Science and Football, held at Liverpool in 1987, represented a milestone in the application of science to football. The Congress embraced all the football codes, but a definite attempt was made to establish common threads between them. The broad aim was to bring together those scientists whose research work was directly related to football and practitioners of football interested in obtaining current information about its scientific aspects. Practitioners included players, trainers, coaches, managers and administrators. The list of Congress themes (Table 1.1) demonstrates the scope of topics that were communicated. The Congress is held every four years under

Table 1.1 Congress themes at the First World Congress of Science and Football

•Clothing and footwear	•Structuring football skills and practices
•Football surfaces	•Physiology of training and match-play
•Biomechanics of kicking	•Nutritional factors in football
•Computer-aided match analysis	•Playing in heat or cold
•Team management	•Football at altitude
•Group dynamics in match-play	•Coaching the problem player
•Decision-making by referees	•The injury-prone player
•Soccer violence	•Post-injury fitness testing
•Pre-match stress and performance	•Strain in adolescent footballers

the auspices of the International Council of Sports Science and Physical Education and the World Commission for Sports Biomechanics. The meeting at Liverpool was followed by the Second World Congress on Science and Football at Eindhoven in the Netherlands (1991), the Third at Cardiff (Wales) in 1995; the fourth event is scheduled for Sydney, Australia, in 1999.

Many national governing bodies of soccer set up their own system of scientific support. Mostly, this was implemented through their sports medicine programmes. An example was the Football Association's National Training and Rehabilitation Centre at Lilleshall in the early 1980s. This reflected the perceived potential of sports science as a component of sports medicine. This applied also to the science input to the world's ruling body, FIFA, which historically had been through the medium of its Medical Committee.

A consensus statement concerned with food and nutrition as they applied to soccer was approved at FIFA headquarters in 1994. The event marked another milestone in the progress of scientific information related to the game. A parallel within the European Federation (UEFA) was the launching in 1989 of the *Journal of Science and Football*, the official publication of the European Society of Team Physicians in Football.

1.3 ACADEMIC PROGRAMMES IN SCIENCE AND FOOTBALL

The first academic programmes in sports science were studied in the United Kingdom in 1975. The background to and development of these undergraduate courses have been described elsewhere (Reilly, 1992). The disciplines included in the pioneering programmes were biology, biochemistry, physiology, biomechanics, mathematics, psychology and sociology. Contemporary programmes may incorporate economics, recreation, coaching and computer science but the major thrust of scientific method is maintained.

Whilst the professional preparation of coaches in some countries includes substantial components of sports science, the emphasis is firmly on coaching competence rather than intellectual skills. The first formal academic programme in Science and Football was offered at Diploma level at Liverpool

Table 1.2 Course programme for Diploma in Science and Football (soccer)

Core programme
Football skills
Assessment of football performance
Football violence
Ergonomics of football
Mental training for football
Research project
Option programme
Information technology and football
Environmental physiology and football
Electives
Sports science modules *including*:
Sports mechanics
Sports nutrition
Psychology of sport and exercise
Cardiorespiratory physiology
Exercise and health
Applied soccer research
Provision for leisure

John Moores University in 1991. The syllabus was dedicated to scientific subjects applied to soccer. The modules of study include elements of choice, as shown in Table 1.2.

Formal academic activity is not restricted to undergraduate, or postgraduate courses. The University of Leicester set up its research unit in the 1980s to focus on sociological aspects of soccer. It was funded by the Football Trust and made a major contribution to the study of the 'football hooligan' phenomenon throughout the decade. The Centre of Excellence for Science and Football at Liverpool John Moores University supports an active programme of postgraduate research. Most notable achievements elsewhere have included the award of a DSc degree for a thesis on physiological investigations directly related to soccer play (Bangsbo, 1993).

1.4 THE FIELD OF STUDY

Clearly there are many aspects of science and football and plenty of subject areas which have benefited from scientific knowledge and know-how. These include the natural and physical sciences, the disciplines allied to medicine and the social sciences.

An ergonomics model of the application of science to the game itself is illustrated in Figure 1.1 (Reilly, 1991). It shows how the role of the scientist is

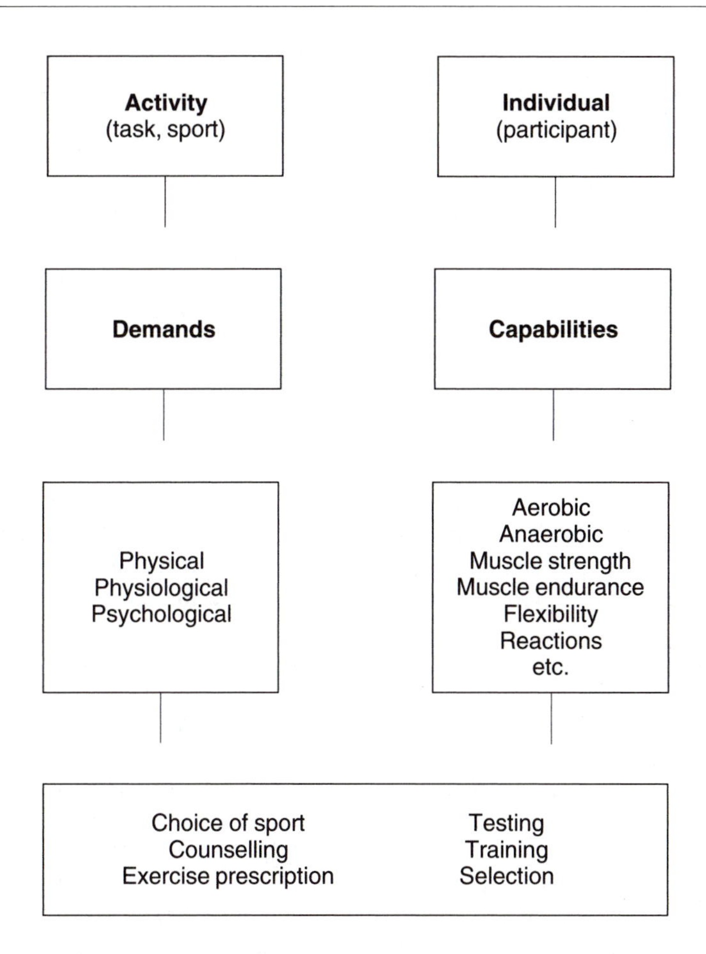

Figure 1.1 An ergonomics model of sports participation. (Redrawn from Reilly, 1991.)

to match the characteristics of individuals to the demands of the game. This is a complex problem in team sports where eventual success is determined by how the collection of individuals form an effective unit. There are implications for fitness testing, training, and selection. The study of the organization of the entire group is also highly relevant.

Similarly, the prediction of performance is more difficult by far in soccer than in individual sports. In competition, success may be determined by choice of tactics of either team. There are also elements of chance that determine the outcome of critical events and tilt the balance of the contest. This makes even the most complex of game theories hard to relate to the outcome of a particular match. Nevertheless, match analysis can be approached from a scientific perspective.

The physical sciences provide insights into the nature and appropriateness of artificial pitches. There have also been applications to the design of shoes and evaluations of the need for protective equipment. Principles of biomechanics are relevant in considering prevention of soccer injuries. The physical sciences also embrace agronomy, the cultivation of grasses and the maintenance of playing conditions in cold and wet weather.

The widest field of application of sports science to soccer is probably apparent in the behavioural disciplines. The many opportunities for investigation include the study of crowds and their control, the management of large groups and the organization of personnel, the stresses on playing officials and on management of the clubs.

1.5 SOCCER AS ART

Followers of soccer frequently criticize the game as lacking creativity and flair. Some critics may go so far as to blame use of scientific methods by soccer teams for lack of entertainment. Underlying these points is the fact that soccer at top level has an obligation to entertain the viewing public but financial rewards to the players depend on their securing victory. Consequently fear of failure to win may motivate players to err on the side of caution and emphasize defence rather than attack. The negative emphasis on preventing the opposition from playing to its strength may leave the 'fans' disenchanted.

The coach and trainer may use scientific information to avoid errors and to maximize the chances of preparing the team well. The style of play and choice of tactics are judgements made by the coach on the basis of the best available information about his own team, the opposition and the playing conditions. The scientific support may be utilized to guide the right course for the practitioner and so in no sense is science taking over control of the game.

The professional soccer player is comparable with the actor in that hours of practice or rehearsal underpin the preparation for public performance. The expertise of the player or actor is judged largely on a subjective basis by a critical audience or attendance of the public event.

That soccer itself is an art rather than a science is exemplified by the craft of great players like Johann Cruyff or Brazil's Romario, the guile of Diego Maradona or Franco Baresi. The game is aleatory and is partly determined by chance. This uncertainty of outcome is part of its appeal.

A scientific approach towards preparation for play can nevertheless enhance the enjoyment of both players and spectators. It can achieve this by enabling the team to play to its potential. This realization of possibilities can apply to the recreational player participating for pleasure, or the professional playing for material reward. It can apply to the parents gaining satisfaction from watching talented offspring at play or to the home supporter whose zeal may border on prejudice. It is this microcosm that is subjected to scientific scrutiny in the chapters that follow.

REFERENCES

Bangsbo, J. (1993) The physiology of soccer – with special reference to intense intermittent exercise. DSc thesis, August Krogh Institute, University of Copenhagen.

Reilly, T. (1991) Physical fitness – for whom and for what? in *Sport for All* (eds P. Oja and R. Telama), Elsevier Science, Amsterdam, pp. 81–8.

Reilly, T. (1992) *Strategic Directions for Sports Science Research in the United Kingdom*, The Sports Council, London.

PART ONE

Biology and soccer

Functional anatomy 2

Tracey Howe

Introduction

Soccer practitioners require many attributes to become successful players. These include cardiovascular fitness, muscle strength, endurance, flexibility, agility, coordination, skill and tactical knowledge. Few players possess 'natural ability' in all areas, indeed the vast majority of players undergo training programmes, in some or all attributes, to improve their ability on the field. An understanding of basic anatomy and physiology and a knowledge of muscle actions during soccer skills such as running, kicking, jumping, heading and throwing will be useful to the player, coach, trainer and medical staff. This knowledge may be employed in the design of training programmes to enhance the performance of soccer skills, in injury prevention and diagnosis and rehabilitation programmes.

A comprehensive description of human anatomy is provided by texts such as Tortora and Grabowski (1993) and Williams and Warwick (1980). Essential background for applications of anatomy to a soccer context is given in this chapter.

2.1 ARTICULATIONS

Joints (**articulations**) refer to the junction between bones which allows flexibility of the body segments. Bones are held together by strong connective tissues (**ligaments**) which stabilize the joint by restricting movement in certain

Science and Soccer. Edited by Thomas Reilly. Published in 1996 by E & FN Spon, London. ISBN 0 419 18880 0.

directions depending on their position. Muscles also help to stabilize joints but when they contract they produce movement of the joint. Some joints are very stable as the bones fit closely together and the surrounding ligaments are very strong, e.g. the joint between the sacrum and the pelvis (sacro-iliac joint). Other joints are very mobile as the bones are not in as close proximity to each other, e.g. the shoulder joint. These joints with great mobility are more prone to dislocation.

Most joints are freely movable (**synovial joints**). A fibrous sleeve surrounds the joint; ligaments also form part of this **articular capsule**. The joint surface of the bones possesses **articular cartilage** which is smooth to allow movement to occur. There is a gap between the articulating surfaces of the bones, the **synovial cavity**, which is filled with **synovial fluid**. This fluid lubricates the joint and provides nourishment to the articular cartilage (Figure 2.1).

2.2 SKELETAL MUSCLE

Muscles constitute approximately 40–50 % of total body weight and the percentage is even greater in soccer players. The main function of skeletal (or striated) muscle is to control the movement of body segments, by a series of patterns of contractions and relaxations, which are under conscious (voluntary) control.

Muscle has four main characteristics: excitability, contractility, extensibility

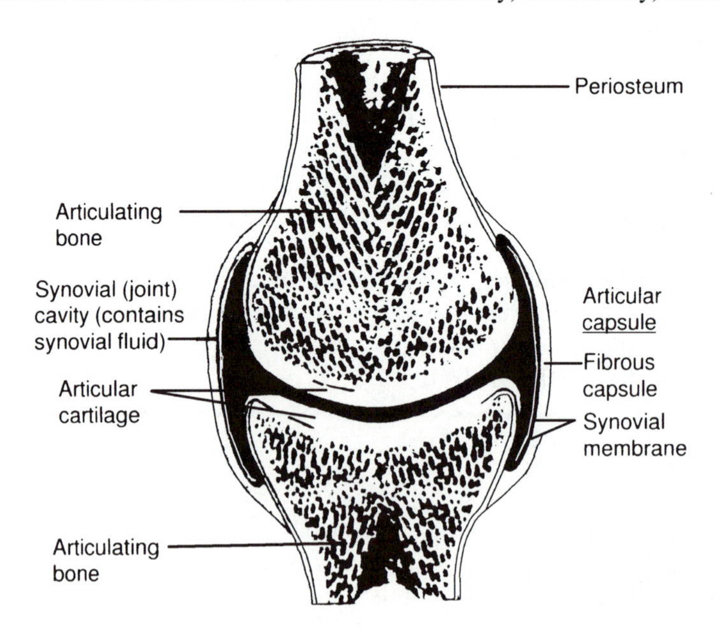

Figure 2.1 The structure of a typical synovial joint. (Reproduced with permission from Tortora and Grabowski, 1993.)

and elasticity. **Excitability** is the ability to respond to certain stimuli, neurotransmitters, by producing action potentials. **Contractility** is the ability of a muscle to contract and generate tension in response to action potentials from the motor nerve. **Extensibility** is the ability of a muscle to extend without causing damage to the tissue. **Elasticity** is the ability of a muscle to return to its original size and shape after shortening or being stretched.

2.2.1 Gross structure

Skeletal muscle is comprised of long thin muscle fibres which are grouped together to form bundles of various sizes (**fasciculi**). The spaces between muscle fibres are filled by connective tissue (**endomysium**) and each fascicle is surrounded by a connective tissue sheath (**perimysium**). Finally, the outermost protective layer of muscle is the **epimysium**. All three layers of connective tissue extend beyond the muscle fibres to form tendons that attach to other muscles or to bone (Figure 2.2).

2.2.2 Ultrastructure

The contractile elements of muscle fibres are the myofibrils which are arranged in parallel in small compartments called **sarcomeres**. The two contractile proteins are **actin** and **myosin** and it is the arrangement of these protein chains that gives skeletal muscle its striated appearance under the microscope.

2.2.3 Muscle contraction: the 'sliding-filament theory'

Skeletal muscles are innervated by motor nerves. Impulses travel from the brain via motor nerves as electrical signals, **action potentials**. The arrival of these action potentials at the muscle (motor end-plate) causes the release of the neurotransmitter acetylcholine. This generates an action potential which travels down the **transverse tubules**, which surround the myofilaments, and causes an increase in Ca^{2+} in the sarcoplasm.

The myosin cross-bridges pull the thin actin filaments towards the centre of the sarcomere. Although the lengths of the thick and thin filaments do not alter, the sarcomere becomes shortened as the actin and myosin filaments overlap. This causes shortening of the muscle fibre and eventually the whole muscle. This is called the **sliding-filament theory** of muscle contraction.

2.2.4 Muscle fibre types

Muscle fibres have been classified as type I, slow oxidative (SO), type IIb, fast glycolytic (FG), and type IIa, fast oxidative glycolytic (FOG). Type I fibres have a high oxidative capacity and are fatigue-resistant. Type IIb fibres have a low oxidative capacity and a low fatigue resistance. Type IIa fibres possess intermediate properties. Most skeletal muscles are a mixture of the three types of fibres; however, the relative proportions of each varies according to the regular action of the muscle. Muscles that function in a predominantly ballistic manner, e.g. gastrocnemius to plantarflex the ankle joint during jumping, contain a high proportion of fast type II fibres. Conversely, muscles that have a predominantly postural function, e.g. soleus, which plantarflexes the ankle during walking, contain a high proportion of slow type I fibres.

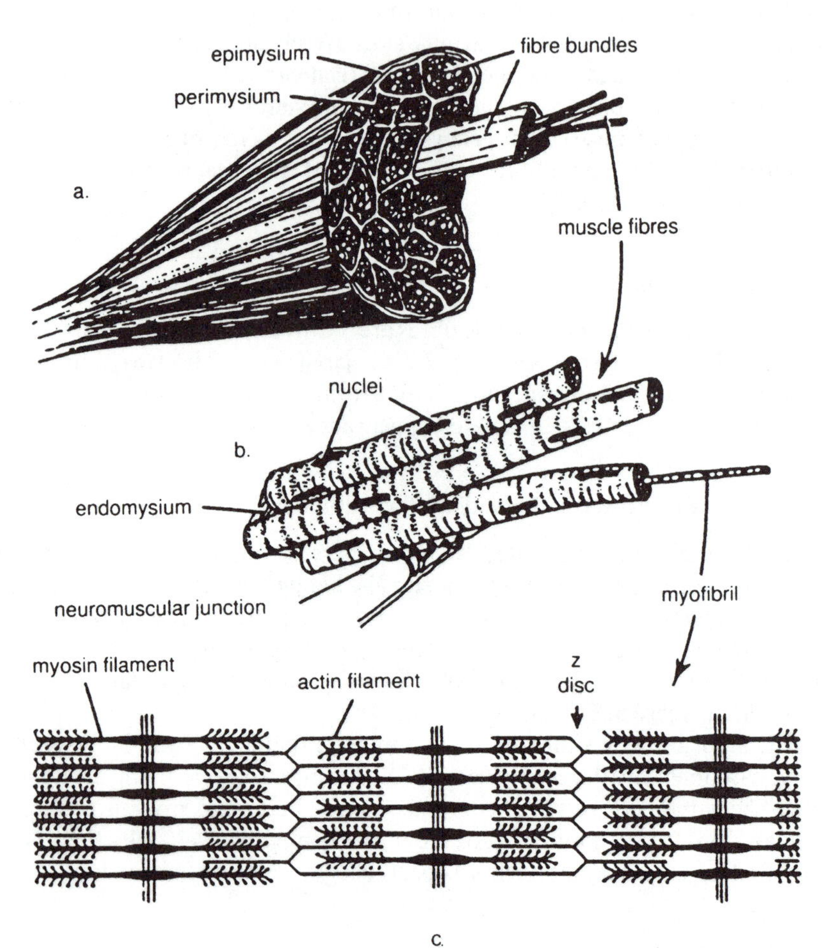

Figure 2.2 The structure of skeletal muscle as revealed by the light microscope. (*a*) Whole muscle in transverse section showing organization of connective tissue and muscle fibres. (*b*) Individual muscle fibres revealing characteristic cross-striations and nuclei. (*c*) Ultra structure of muscle fibril showing individual protein filaments. (Reproduced with permission from T. Reilly and D. Jeeps, *Sports Fitness and Sports Injuries*, (published by Mosby-Wolfe, an Imprint of Times Mirror International Publishers Ltd, London, UK, 1981.)

The proportion of fibre types within a muscle is not constant. It is affected by the ageing process, injury, denervation and specific training methods, either by voluntary contractions or electrical stimulation. This property of skeletal muscle tissue is known as **plasticity**. The muscles of soccer players are composed of a mixture of the fibre types (Bangsbo, 1993). This is due to the nature of the game, with periods of relative inactivity interspersed with periods of high-intensity activity.

2.2.5 The motor unit

The stimulus for a muscle fibre to contract is delivered by a motor neurone. A motor neurone and all the muscle fibres that it innervates is termed a **motor unit**. The constituent fibres of individual motor units are found to be distributed throughout the muscle and are homogeneous, possessing identical structural, metabolic and contractile properties. The number of muscle fibres innervated by a single motor neurone varies from muscle to muscle. Those muscles that control precise movements, such as in the eye, have few muscle fibres per motor unit. Conversely muscles such as the quadriceps which are responsible for powerful gross movements of body segments have approximately 2000 muscle fibres per motor unit. When a motor neurone is stimulated all the muscle fibres it innervates contract simultaneously. Motor units are progressively recruited and 'decruited' in order of size so that force may be modulated in a step-wise manner. Small fatigue-resistant motor units (type I) are recruited most frequently, providing control at the low forces required to perform everyday activities (walking, standing from sitting) and maintain posture. Larger motor units (type II) are recruited when strong and fast contractions are required (running, jumping and kicking a ball).

2.2.6 Movement

Muscles span joints and it is the contraction, or shortening, of the muscle belly that produces movement at a joint. Common terms for types of movement produced about a joint are **flexion** (usually a reduction of the joint angle) and **extension** (usually increasing the joint angle), **abduction** (movement away from the body), **adduction** (movement towards the body), **medial rotation** (towards the body) and **lateral rotation** (away from the body). Common movements at the ankle joint are **dorsiflexion** (movement of the foot towards the shin), **plantarflexion** (pointing of the foot downwards), **inversion** (turning the foot inwards) and **eversion** (turning the foot outwards). The type of movement produced when muscles contract depends on the attachments of the muscle and the architecture of the joint itself.

Muscles initiating or maintaining a movement are termed **prime movers** or **agonists**, e.g. the quadriceps extending the knee joint during kicking. Muscles that oppose this movement, or initiate and maintain its converse, are termed **antagonists**, e.g. the hamstrings opposing knee extension during kicking. When prime movers and antagonists contract together as **fixators** they stabilize a joint, e.g. the quadriceps and hamstrings to stabilize the knee joint during a block tackle.

The contraction of a prime mover that acts across a single joint may produce the desired movement. Many muscles cross more than one joint, e.g. rectus femoris is an extensor of the knee and a flexor of the hip, and many joints are multiaxial. The unrestrained contraction of such muscles may produce additional and unnecessary movements. The restriction of unwanted movements occurs by the contraction of a partial antagonist muscle acting as a **synergist**. An example of this is powerful flexion of the fingers by the long flexors, e.g. during a throw-in. Unrestricted contraction of these muscles would also produce flexion of the wrist joint and would thus reduce function. Function is improved by contraction of the long extensors acting as synergists to prevent wrist flexion.

2.2.7 Types of muscle action

During an **isometric** action the muscle does not shorten and therefore does not perform any external work as no distance is moved about the joint, e.g. sustaining a squatting position. Both slow and fast twitch fibres are equally involved in the development of isometric tension.

Dynamic actions occur when a muscle attempts to overcome a resistance. These actions will be either **concentric** or **eccentric**. During a concentric action, e.g. the action of the quadriceps during jumping, the muscular tension rises and the muscle shortens. This approximates the attachments of the muscle, which extends the knee joint and the body is propelled upwards. During an eccentric action, e.g. quadriceps during landing from a jump, the distance between the origin and insertion of the muscle is increased and flexion of the knee is increased. The tension developed within the muscle regulates the speed of motion.

Isokinetic movements incorporate dynamic contractions through a range of movement at a constant angular limb velocity. These movements are not natural movements but are produced using accommodating resistance devices which load the muscle with maximal force at every point throughout the range of movement. Such computer controlled dynamometers, e.g. Lido, Kin-Com, Bio-dex, Cybex, are used to test muscle performance and for training purposes during rehabilitation programmes.

2.2.8 Flexibility and strength

Flexibility refers to the range of movement about a joint. The flexibility of a joint is dependent upon the extensibility and elasticity of the structures surrounding it. These structures include muscles, the fibrous joint capsule and ligaments. Flexibility reduces the risk of injury sustained during overstretching, e.g. attempting to reach the ball when intercepting a pass. Conversely, hypermobility may be disadvantageous as joints become unstable and prone to injury.

Increases in the force-generating capacity of a muscle (strength) are accompanied by an increase in the cross-sectional area and a decrease in muscle length. Shortening of a muscle leads to a reduction in joint flexibility and therefore increases the risk of injury. It is essential that strength training regimes are accompanied by flexibility programmes.

When muscle strength is tested the dominant limb is usually slightly stronger than the non-dominant limb. This is not as apparent in muscles of the lower limb as in the upper limb and is less so in highly trained individuals. Often a ratio of strength exists between agonists and antagonist muscle groups, e.g. the strength of the hamstrings is only 60% that of the quadriceps. Knowledge of such ratios is important when designing strength training programmes and when diagnosing the cause of, and attempting to prevent, injuries.

2.3 ANATOMY (Figure 2.3)

2.3.1 The muscles of the back

The erector spinae refer to a group of muscles that surround the spine and run from the sacrum to the head. These muscles act to extend, flex laterally and rotate the trunk and the neck.

2.3.2 The muscles of the abdominal wall

The abdominals (rectus abdominis, and the internal and external obliques) run from the ribs to the pelvis. They act to flex, laterally flex and rotate the trunk, e.g. during sit-ups.

2.3.3 The gluteal muscles

The gluteal muscles (gluteus maximus, gluteus minimus and gluteus medius) all abduct and laterally rotate the hip joint. Gluteus maximus is also a strong extensor of the hip.

2.3.4 Main flexors of the thigh

These muscles (psoas and iliacus) arise in the abdomen from the pelvis and the front of the spine and insert on the front of the femur. When the lower limb is fixed these muscles will flex the trunk on the thigh, e.g. during a sit-up exercise.

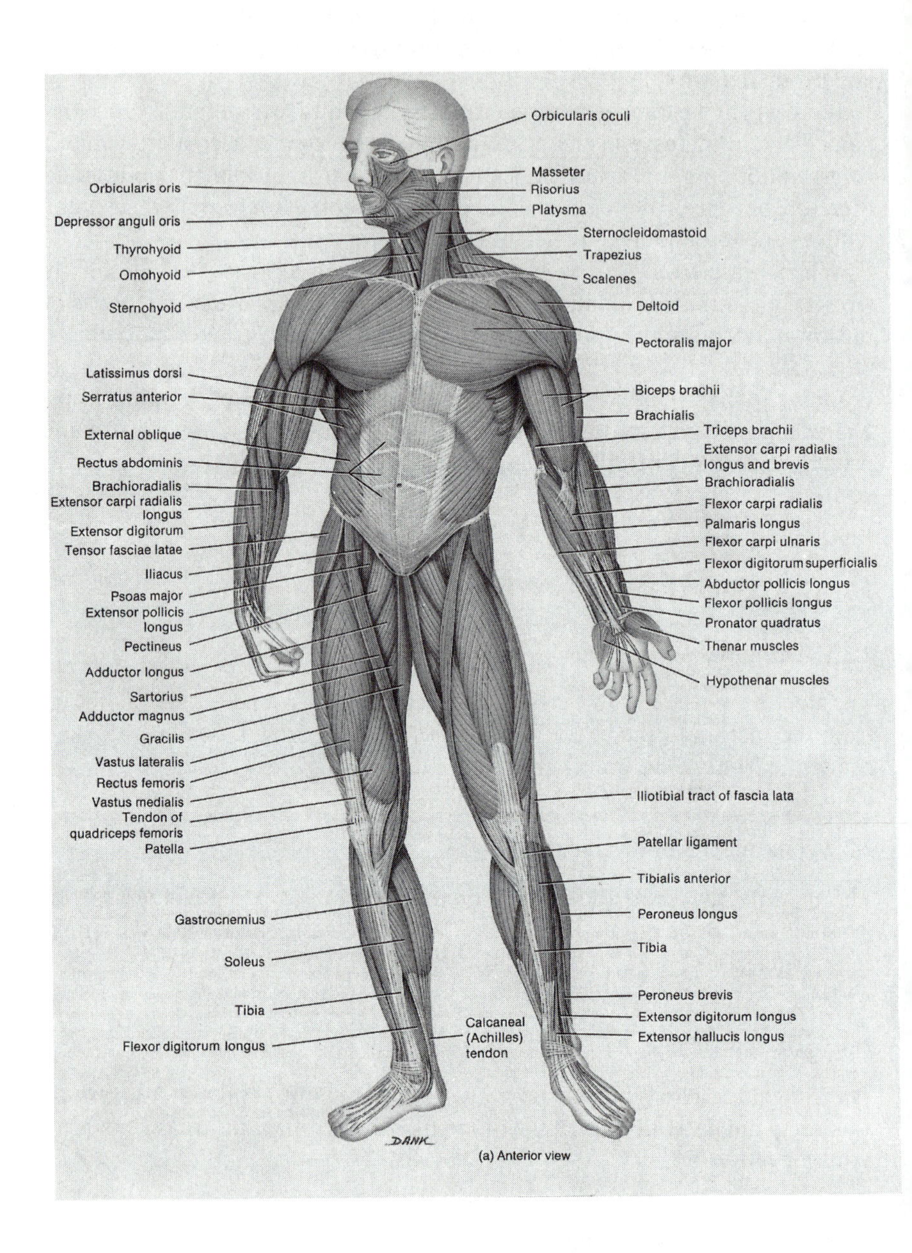

Figure 2.3 Anterior and posterior views showing the principal superficial skeletal muscles. (Reproduced with permission from Tortora and Grabowski, 1993.)

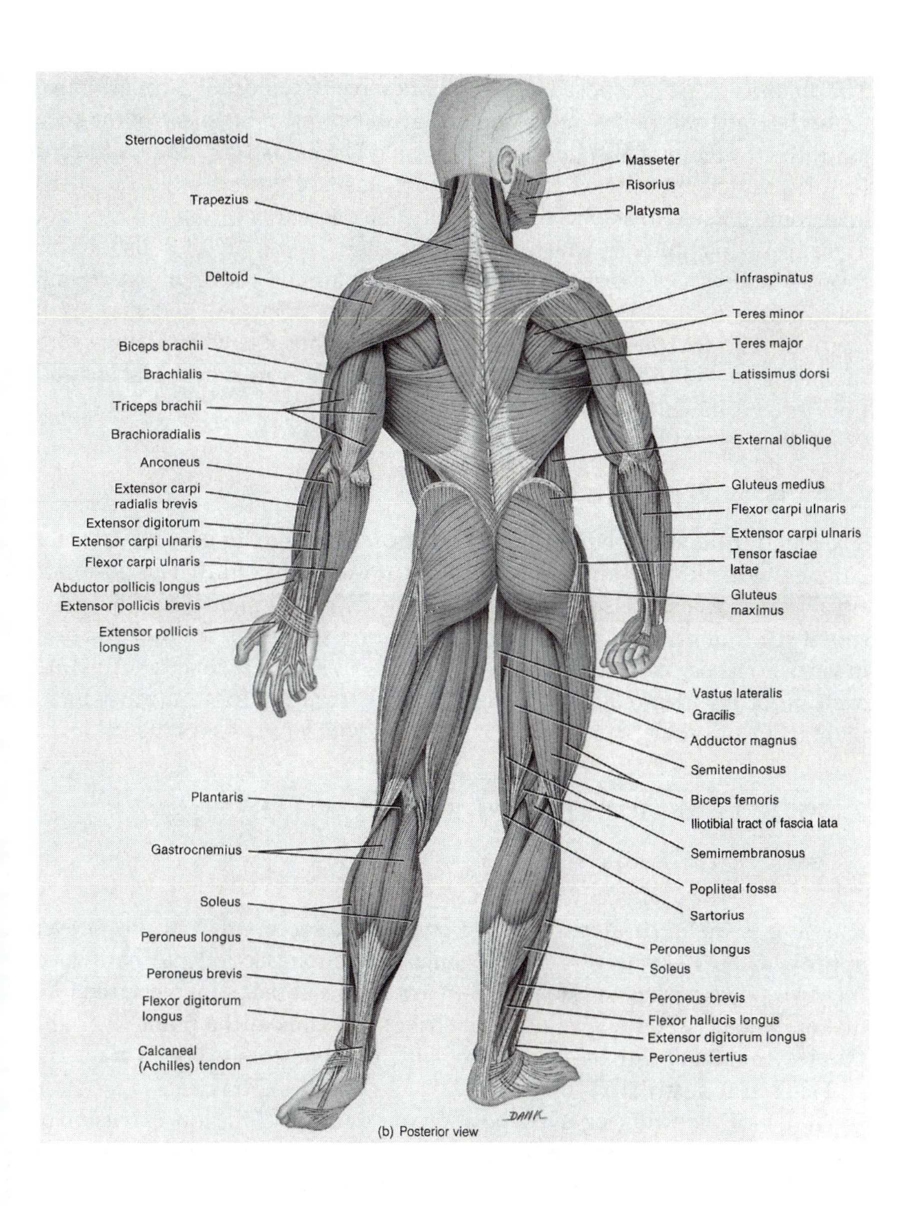

(b) Posterior view

Figure 2.3 continued

2.3.5 Muscles of the thigh

The quadriceps are composed of four muscles (rectus femoris, vastus medialis, vastus lateralis and vastus intermedius) that are powerful extensors of the knee joint. Rectus femoris also flexes the hip joint. The hamstring muscles (biceps femoris, semimembranosus and semitendinosus) are powerful flexors of the knee joint. They also extend the hip joint. The adductors (groin muscles) are a group of five muscles which adduct the thigh; some muscles also act as rotators and weak flexors of the thigh. They bring the abducted lower limb back into a neutral position, e.g. when running sideways, and may even continue this movement to cross one leg over the other, e.g. when kicking with the inside of the foot. These muscles are often tight in soccer players and therefore groin strains are a common injury.

2.3.6 Muscles of the lower leg

The anterior tibials (tibialis anterior, extensor digitorum longus and extensor hallucis) dorsiflex the ankle (pull the foot towards the shin). The posterior tibials (gastrocnemius, soleus and tibialis posterior) all plantarflex the ankle (point the foot downwards). Tibialis posterior also inverts the foot (turns it inwards). The peronei (peroneus longus, peroneus brevis, peroneus tertius) are a group of muscles that act to evert the foot (turn the foot outwards) and plantarflex the ankle, except peroneus tertius, which dorsiflexes the ankle.

2.4 MUSCLE ACTIONS DURING SOCCER SKILLS

2.4.1 Running

Running is an integral part of soccer. Indeed soccer players may cover approximately 10 km during a single game. The running action may be divided into two stages, **swing** and **support**. Support begins at the point when the foot makes contact with the ground (**foot strike**) and ends at the point when the foot leaves contact with the ground (**toe-off**). The swing phase begins at toe-off and ends at foot strike.

At toe-off the swing leg is in a position of extension of the hip, extension of the knee and plantarflexion of the ankle. The gluteals and hamstrings are still acting to extend the hip and the gastrocnemius to plantarflex the ankle giving a good push off. The actions of psoas and iliacus flex the hip, the hamstrings flex the knee and the anterior tibials dorsiflex the ankle. The hip continues to flex and the ankle to dorsiflex to bring the leg forwards in front of the support leg; the adductors act to prevent the thigh from swinging outwards. The quadriceps then begin to extend the knee in preparation for foot strike.

When foot strike occurs the hip is in flexion, the knee is in slight flexion and the ankle is normally dorsiflexed and slightly inverted. At this point the weight of the body must be controlled as it hits the ground. The gluteals contract to extend the hip, the quadriceps and hamstrings contract to stabilize the knee joint and the adductors to stabilize the hip. The anterior tibials work eccentrically and the gastrocnemius concentrically to control the foot as it strikes the ground. The momentum of the body carries it forwards over the ankle joint which acts as a rocker as the foot becomes flat to the ground and then toe-off occurs.

As the speed of running increases longer strides are taken. In this instance the swing phase involves greater knee flexion and hip extension (the heel almost touching the buttock) and greater hip flexion in the later part of the phase.

When running with a ball much shorter strides are taken as the player must be ready to change direction and speed. At the toe-off phase the leg may not be as extended. Heel strike may not be as pronounced, instead the foot may land in a more neutral position or be plantarflexed.

The muscles of the arms and trunk also play an important role during running. They act to maintain balance and to counterbalance the rotation of the body when the pelvis rotates.

2.4.2 Kicking a ball

There are many different types of kick in soccer, e.g. running kick, volley and push pass (Pronk, 1991). Skilled players can also impose spin on the ball and cause it to dip quickly in flight. In such cases the kicking action is quite complex. For the purposes of this text the kick is simplified into that of movement in one plane. This action may also be divided into four phases: phase one, priming the thigh and leg during backswing; phase two, rotation of the thigh and leg laterally and flexion of the hip; phase three, deceleration of the thigh and acceleration of the leg; and finally stage four, the follow-through.

During **phase one** the hip of the kicking leg is rapidly extended by the action of the gluteals and the pelvis is rotated backwards. The knee is flexed by the hamstrings and the anterior tibials dorsiflex the ankle. These actions are limited by the hip flexors and the adductors which often become overstretched in many players. The harder the subsequent kick the further the stretch on these muscles. During **phase two** the psoas and iliacus contract and the hip flexes to move the thigh and leg forwards and the pelvis rotates forwards. **Phase three** involves the hamstrings acting to decelerate the thigh and the quadriceps rapidly extend the knee joint. The position of the ankle joint during ball strike is dependent upon the type of kick performed. In addition, the adductors will contract to pull the leg towards the body. This is especially relevant during a side kick or push pass. **Phase four** begins after the ball has lost contact with the foot, the leg and thigh will follow through due to the momentum of the thigh, leg and foot. This causes a stretch on the muscles opposing these actions, especially the hamstrings as they pass over two joints (De Proft *et al.*, 1988).

The muscles of the non-kicking leg act in a similar fashion to their behaviour during the stance phase of running. However, they act mainly to stabilize the body to provide a stable platform on which the kicking leg may act. This leg is usually abducted and rotated. Again the muscles of the arms and trunk work to maintain poise and balance and to provide a counterbalance to the kicking leg, thus providing more control and speed.

2.4.3 Jumping and heading

Jumping to control the ball in the air is of major importance in soccer. Jumping can occur from a standing position or from a run-up. Take off from a standing jump is usually from both feet and from one leg when using a run-up. When performing a standing jump the player will sink down into a position of flexion. The trunk, hips and knees will flex and the ankle will dorsiflex under the action of body weight and gravity but controlled by the agonists to these movements acting eccentrically (erector spinae, gluteals, hamstrings, quadriceps and plantarflexors). The elbows will flex and the shoulders will be extended. In this position the body is almost spring-like; the prime movers of the jumping action are on a stretch, storing potential energy ready to be released at the appropriate moment. When the jump itself begins the prime movers act to launch the body weight up in the air. This is achieved by rapid and powerful contractions of the erector spinae, gluteals, hamstrings, quadriceps and plantarflexors to produce extension of the trunk, hips, knees and plantarflexion of the ankles. The arms are also moved rapidly forwards and upwards by flexion of the shoulders and extension of the elbows. When the spine becomes extended during the jumping action a severe stretch may be placed on the abdominal muscles and the hip flexors and injury to these muscles may occur.

Landing from a jump is just as important as the jump itself, as the weight of the body must be controlled as it hits the ground. Basically it is a reverse of the jumping action. However, this time the muscles of jumping act eccentrically to control joint movement and decelerate the action, thereby increasing shock absorption and decreasing the risk of injury.

The primary aim of most jumps in soccer is to head the ball, but heading may also occur from a standing position. As a player jumps the neck becomes extended partly from the effects of gravity and partly due to the action of the erector spinae muscles. As a player attempts to make contact with the ball they will aim their head at it. This may involve a combination of movements. Flexion of the neck is the most powerful action but this may be combined with rotation or lateral flexion to direct the ball.

2.4.4 Throwing a football

Throw-ins are usually taken from a short run-up and a two-footed stance. With both feet level the erector spinae, gluteals and the hamstrings contract to extend the spine and the hips. The dorsiflexors act eccentrically to allow the ankles to move into a small degree of plantarflexion without losing balance. The ball is held in both hands and the two arms are held up above the head. The shoulders are moved into full flexion and the elbows also are now fully flexed. This creates full stretch on the antagonist groups and potential energy is stored. As the throw begins these now become prime movers which contract from a stretched position. The elbows become extended, the shoulders become more extended. The contraction of the abdominals and psoas and iliacus causes the spine and hips to flex. Dorsiflexion of the ankles is controlled by the eccentric action of gastrocnemius and soleus.

Summary

An understanding of the importance of joint flexibility and muscle strength combined with a basic knowledge of anatomy and physiology and muscle actions during soccer skills will aid the coach, trainer and medical staff in the design of appropriate training programmes for soccer players.

REFERENCES

Bangsbo, J. (1993) The physiology of soccer – with special reference to intense intermittent exercise. DSc thesis, August Krogh Institute, University of Copenhagen, pp. 79–82.

De Proft, E., Clarys, J.P., Bollens, E. *et al.* (1988) Muscle activity in the soccer kick, in *Science and Football* (eds T. Reilly, A. Lees, K. Davids and W.J. Murphy), E. & F.N. Spon, London, pp. 434–40.

Pronk, N.P. (1991) The soccer push pass. *National Strength and Conditioning Journal*, **13**(2), 6–81.

Tortora, G.T. and Grabowski, S.R. (1993) *Principles of Anatomy and Physiology*, 7th edn, Harper Collins, New York, pp. 216–17, 237–58.

Williams, P.L. and Warwick, R. (eds) (1980) *Gray's Anatomy*, 36th edn, Churchill Livingstone, Edinburgh, pp. 420–39, 506–621.

Fitness assessment

3

Thomas Reilly

Introduction

A prerequisite for playing soccer is possession of the skills and fitness to do so. Fitness refers to a range of individual characteristics and in a game like soccer is a composite of many attributes and competencies. In consequence, fitness for soccer is said to be multivariate and also specific to the sport. It comprises physical, physiological, psychomotor and psychological factors. Such qualities are needed in contesting and retaining possession of the ball, maintaining a high work-rate for 90 minutes of play, reacting quickly and appropriately as opportunities arise and regulating mental attributes before and during match-play.

The fitness requirements for football depend on the level of performance, positional role and styles of play. They vary also with age groups, between men and women, and at different stages of the playing season. Re-acquiring desirable fitness levels is especially important after injury, prior to returning to competitive play. Otherwise the individual is vulnerable to re-injury if uncorrected weaknesses, in muscle strength for example, are carried into a game.

The foundation for performance in soccer is represented by the array of skills and tactical sense of individual players. Coaching staff must nurture tactical know-how alongside improvements in fitness profiles. The success of the team depends on how individuals are blended into an effective playing unit. When teams roughly equal in ability meet, the one with the higher overall fitness level will have the advantage of being more able to cope with a fast pace of play.

Science and Soccer. Edited by Thomas Reilly. Published in 1996 by E & FN Spon, London. ISBN 0 419 18880 0.

Coaches, trainers and sports scientists acknowledge that preparation for competitive match-play calls for a systematic approach. This includes consideration of fitness levels of individual players as well as overall throughout the team. Attention to fitness profiles is relevant not just in the build-up towards key matches and tournaments but also throughout the competitive league season. Fitness profiling is achieved by means of a battery of tests. The test items may either be part of a comprehensive physiological assessment or be dedicated to performance in soccer. The fitness profiles have some value in allowing comparisons between individuals and with global standards; individual weaknesses may be identified and remedial training prescribed. Repeated fitness assessment is of further value in that changes in fitness profiles within individuals and through the team as a whole can be measured. In this way the effects of training, overtraining or detraining can be evaluated.

Fitness tests commonly employed in sports science laboratories are described in this chapter. The application to soccer is highlighted as are some of their limitations. Field tests are included as a complement to laboratory based tests, and in some circumstances may be an alternative. Descriptive profiles are provided to examine the variability between individuals, especially at the elite level of soccer play.

3.1 ANTHROPOMETRY

Members of top soccer teams tend to have an average age of 25–27 years with a standard deviation of about 2 years. Players of teenage years do feature in top club teams, for example Ryan Giggs in Manchester United's England Premier League winning team of 1993 and 1994. There are examples also of others playing in the World Cup finals at a similar age, such as Pele of Brazil in 1958 and Whiteside of Northern Ireland in 1982. These players tended to be part of a very experienced squad and also reach the pinnacle of their own playing careers at a later age. The majority of professional players are in their twenties and traditionally there was a reluctance of managers at top club level to retain the professional services of players once they were into their thirties. A loss of motivation to train may have contributed also to an earlier than necessary retirement from playing professional soccer. Active athletes can maintain fitness levels well into their thirties before physiological functions begin to show signs of deterioration.

Nowadays professional players do seem prepared to stay in the game for longer than was traditional. This is probably due to the commercial attractions of maintaining one's playing career as long as possible. Development in orthopaedic procedures for repair of tissue damage, that in previous decades might have halted a player's career, may also contribute to the trend of professionals staying active for longer than previously. It is noticeable how many players into their thirties were key members of the teams in the 1994 World Cup finals.

Goalkeepers seem to have longer playing careers than outfield players and it is not unusual to have players at top level who are well into their thirties. Indeed there are examples of players who have represented their countries at major international events in their forties. These include Dino Zoff (Italy), Peter Shilton (England) and Pat Jennings (Northern Ireland). This may be related to the special requirements of the position, players maturing in tactical judgement with experience in the game. It may be related also to a lower incidence of chronic injuries and degenerative trauma in goalkeeping compared to outfield positions.

Data on height and body mass of soccer teams suggest that players vary widely in body size (Table 3.1). Lack of height is not in itself a bar to success in soccer, though it might determine the choice of playing position. Being tall is an advantage for the goalkeeper, for centre-backs and for a forward player used as a 'target' for winning possession of the ball with the head. In contrast the players deployed in midfield, in full-back and on the wings tend to be smaller in size than those in other positional roles.

Average values have but a limited use for comparative purposes when the variability is large. A coach may modify his team configuration and playing

Table 3.1 Mean (± s.e.) height and weight of soccer teams in a sample of reports in the literature

	Height (cm)	*Body mass (kg)*
English League, First Division (White *et al.*, 1988)	180.4 ± 1.7	76.7 ± 1.5
English League, First Division (Reilly, 1979)	176.0 ± 1.1	73.2 ± 1.5
Tottenham Hotspur (Reilly, 1979)	178.5 ± 1.3	77.5 ± 1.3
Aberdeen FC (Williams *et al.*, 1973)	174.6 ± 0.9	69.4 ± 2.1
Dallas Tornado (Raven *et al.*, 1976)	176.3 ± 1.2	75.7 ± 1.9
South Australian representatives (Withers *et al.*, 1977)	178.1 ± 3.6	75.2 ± 2.2
Italian professionals (Faina *et al.*, 1988)	177.2 ± 0.9	74.4 ± 1.1
Ujpesti Dozja, Budapest (Apor, 1988)	176.5 ± 1.7	70.5 ± 1.3
Honved, Budapest (Apor, 1988)	177.6 ± 1.1	73.5 ± 1.6
Danish national squad (Bangsbo and Mizuno, 1988)	183.0 ± ?	77.0 ±
Finnish national squad (Rahkila and Luhtanen, 1991)	180.4 ± 0.8	76.0 ± 1.3

style to accommodate individuals without the expected physical attributes of conventional playing roles, provided they compensate by superior skills and motivation. The average body size may also represent ethnic or racial influences. Many top European and South American teams contain players of different racial backgrounds and this can make interpretation of anthropometric profiles more difficult.

A particular body size may encourage acquisition of certain skills and force a gravitation towards a specific playing position. This is likely to occur before maturity so that the individual will tend to favour one positional role before playing at senior level. Under-age soccer is organized according to chronological rather than biological age. Advantages bestowed by body size in adolescent and youth soccer may disappear as the late maturers catch up and the gap to the early maturers is narrowed.

Physique represents body shape rather than body size and its measurement is referred to as somatotyping. Somatotype is represented on three dimensions – endomorphy, mesomorphy and ectomorphy. Endomorphy reflects roundness and adiposity, mesomorphy indicates muscularity whilst ectomorphy suggests a tendency towards linearity. The somatotype is calculated from a number of limb girth, bone diameter and skinfold thickness measures, in conjunction with height and body mass (Eston and Reilly, 1995). Typical somatotype values for soccer players are 3–5–3, reflecting a trend towards mesomorphy (Reilly, 1990). The muscular make-up would be of benefit in game contexts such as tackling, shielding the ball, contesting possession, turning, accelerating, kicking and so on. Thus the muscular development is more pronounced in the thigh and calf compared to the upper body in top soccer players.

Body composition is an important aspect of fitness for soccer as superfluous adipose tissue acts as dead weight in activities where body mass must be lifted repeatedly against gravity. This applies to locomotion during play and in jumping for the ball. The most commonly used model of body composition divides the body into two compartments – fat and fat-free mass. An alternative is to estimate muscle mass from anthropometric measures using the equation of Martin *et al.* (1990). Such estimates confirm the tendency towards a muscular make-up among soccer players.

Generally the amount of fat in the adult male in his mid-twenties is about 16.5% of body weight. A comparable figure for the adult female is 26%. Lowest values for percentage body fat among athletic groups are found in distance runners, with mean values as low as 4–7% in men. Figures for soccer players are higher than this and reports for average team values have ranged from 9 to 16% (Reilly, 1990). Higher values are found in goalkeepers than in outfield players, probably because of the higher metabolic loading imposed by match-play and training on outfield players. Soccer players accumulate body fat in the off-season and lose weight more during pre-season training time than in other periods. They may also put on weight when they are recovering from injury and unable to train strenuously, unless they modify their intake of food. Thus the habitual activity of players at the time of measurement, their diet and the stage of the competitive season should be considered when body composition is evaluated.

The method of estimating or measuring adiposity or percentage body fat should also be considered when interpreting observations. Body fat (or body adiposity) is determined indirectly in live subjects. It may be estimated from chemical measures such as total body water or total body potassium. These methods are not accessible for routine use with soccer players and facilities are not generally available to sports science support groups. Similarly, medical imaging techniques are impractical with athletic groups and the radiographic dose with computerized tomography makes it unsuitable for repeated application. Magnetic resonance imaging (MRI) overcomes this problem but its expense makes it unrealistic for routine use. Portable devices, such as bioelectric impedance analysis and infrared interactance, have not been sufficiently validated for universal use. The scientifically accepted reference method is underwater weighing but it is doubtful whether this is *actually* true at present for all purposes; the assumptions made with regard to body density are not transferable to all highly trained athletes. Besides, it cannot be prescribed for testing squads of soccer players, particularly in field conditions.

The most accessible method for obtaining data on body composition is assessment of skinfold thickness by means of calipers. The measurements must be made by an appropriately trained individual, identifying the correct sites for measurement of skinfolds and using the proper equipment.

Five anatomical sites are recommended – biceps, triceps, subscapula and suprailiac – as used by Durnin and Womersley (1974) – and anterior thigh as a fifth site (Clarys *et al.*, 1987). The five skinfolds should be summed and the resultant value used as an index of subcutaneous adiposity since the skinfold thickness data themselves provide indications of changes in body composition. A percentage body fat value can be calculated in order to provide a target if weight control is desirable.

3.2 MUSCLE FUNCTION

3.2.1 Muscle strength

Various tests of muscle strength and power have been employed for assessment of soccer players. These have ranged from performance tests and measurement of isometric strength to contemporary dynamic measures using computer-linked isokinetic equipment. Tests of anaerobic power output have also evolved as well as short-term performance on the force platform, which have relevance to soccer play.

Strength in the lower limbs is of obvious concern in soccer: the quadriceps, hamstrings and triceps surae groups must generate high forces for jumping, kicking, tackling, turning and changing pace. The ability to sustain forceful contractions is also important in maintaining balance and control. Isometric strength is possibly important in maintaining a player's balance on a slippery pitch and also in contributing to ball control. For a goalkeeper almost all the body's muscle groups are important for executing his skills. For outfield players the lower part of the trunk, the hip flexors and the plantarflexors and dorsiflexors of the ankle are used most. Upper body strength is employed in throw-ins and the strength of the neck flexors could be important in forcefully heading the ball. At least a moderate level of upper body strength should prove helpful in preventing being knocked off the ball.

Soccer players are generally found to be only a little above average in isometric muscle strength. This may reflect inadequate attention to resistance training in their habitual programme. Besides, isometric strength may not truly reflect the ability to exert force in dynamic conditions. It may also be a poor predictor of muscle performance in the game (Reilly, 1994).

Some studies have indicated a relationship between dynamic muscle performance in laboratory and field contexts. Asami and Togari (1968) reported a significant correlation between knee extension power and ball speed in instep kicking, both increasing with experience in the game. Cabri and co-workers (1988) also reported a significant relation between leg strength, measured as peak torque during an isokinetic movement, and kick performance indicated by the distance the ball travelled. The relationship was significant for both eccentric and concentric contractions of hip and knee joints in flexion and extension.

The relation between leg strength and kick performance implies that strength training could be effective in improving the kicking performance of soccer players. Given a certain level of technique, it seems that strength training added to the normal soccer training improves both muscular strength and kick performance (De Proft *et al.*, 1988). Soccer players have greater fast speed capabilities than normal (Oberg *et al.*, 1986) and this may be an important determinant of technique in kicking the ball.

The relationship between dynamic muscle strength of the knee extensors and kick performance may be dependent on the level of skill already acquired. Trolle *et al.* (1993) measured isokinetic strength of the leg extensors in skilled soccer players at angular velocities between 0 and 4.18 rad/s. No relationship was found between these measures and ball velocity recorded during a standardized indoor soccer kick. Ball velocity was unchanged after 12 weeks of strength training. Motor control may override muscular strength in well-trained soccer players' performance, whilst the choice of criterion for kick performance may also be a factor.

The shoulder and trunk muscles are engaged in throwing in the ball from the sidelines and a long throw into the opponents' penalty area can be a rich source of scoring opportunities. The throw-in distance of soccer players has been reported to be related to pull-over strength and trunk flexion strength.

Training methods, using a medicine ball, increased strength measures but without a corresponding increase in throw distance (Togari and Asami, 1972). This demonstrates a degree of specificity in the throwing skill and suggests that individual players should be pre-selected to take tactical long throws.

Since fitness requirements tend to vary with positional roles, muscle strength values may depend on the player's position. Goalkeepers and defenders were found by Oberg *et al.* (1984) to have higher knee extension torque at 0.52 rad/s than midfield players and forwards. The result was attributable to differences in body size since correction for body surface area removed the positional effect. A similar observation was recorded by Togari *et al.* (1988) in their tests on Japanese Soccer League players. The goalkeepers were significantly stronger than forwards at slow (1.05 rad/s) speeds of movement, midfield players being intermediate. The differences tended to disappear when the angular velocity was raised to 3.14 rad/s.

It is now common to monitor the muscle strength of soccer players using isokinetic apparatus such as Cybex, Kin-Com, Bio-Dex or Lido (Figure 3.1) systems. These machines offer facilities for determining torque-velocity curves in isokinetic movements and joint-angle curves in a series of isometric contractions. The more complex systems allow for measurement of muscle actions in eccentric as well as concentric and isometric modes. In eccentric actions the

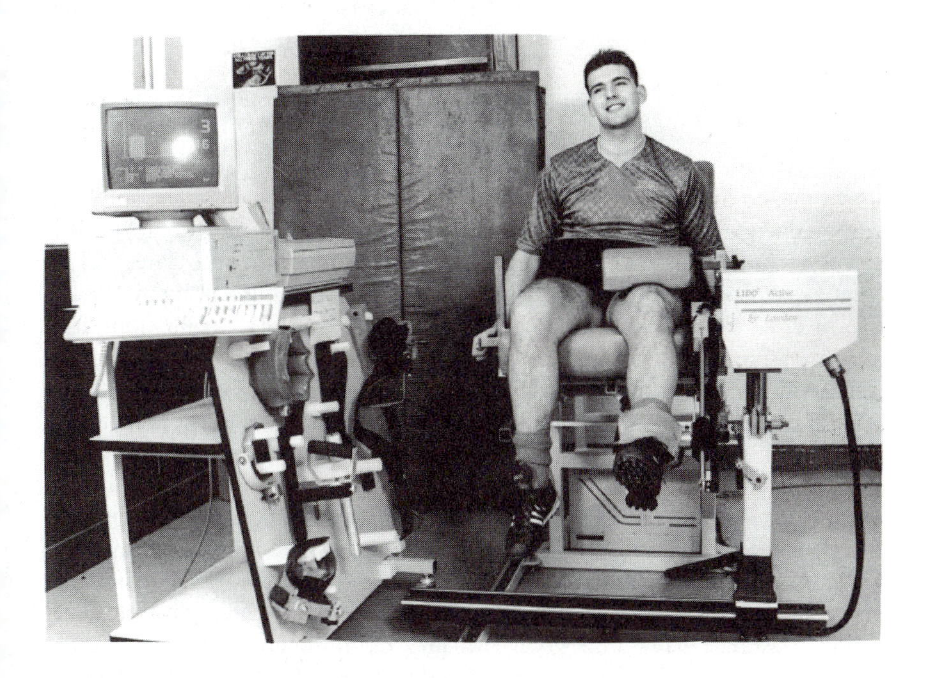

Figure 3.1 Isokinetic dynamometer (Lido) for measurement of muscular function in isometric, concentric or eccentric modes.

limb musculature resists a force exerted by the machine: it is lengthened in the process and hence produces an eccentric contraction.

Isokinetic tests of soccer players have been concentrated almost exclusively on lower limb muscle groups and on concentric contractions. Whilst knee extension strength in concentric contractions is correlated with kick performance, an even higher correlation has been reported for knee flexion strength in an eccentric contraction (Cabri *et al.*, 1988). The possession of strong hamstrings, particularly in eccentric modes, is an important requirement for playing soccer.

The balance between hamstrings and quadriceps strength may predispose towards injury in soccer players. At slow speeds and under isometric conditions, a knee flexor–extensor ratio of 60–65% is recommended (Oberg *et al.*, 1986). This ratio is increased at the higher angular velocities of commercially available apparatus, although the reliability of measurement is reduced at fast speeds. Isokinetic strength testing allows comparison of left and right legs to identify any muscle imbalance, the weaker side being the one most liable to injury. Test profiles are also important in monitoring regains in muscle strength using the uninjured side as reference. These comparisons to identify asymmetry or weakness within an individual player may be more important than comparison between teams and between team members. This reservation applies especially to comparisons with data from other laboratories using alternative test protocols.

3.2.2 Anaerobic power

Soccer players are frequently required to produce high power output and sometimes to maintain it with only a brief respite for recovery. The splitting of high energy intramuscular phosphagens contributes along with anaerobic glycolysis to the maximal power a player can develop. These substrates (ATP, creatine phosphate and glycogen) may be used for combustion by muscle at the onset of exercise and result in a high anaerobic work production. The maximum power output can be calculated from performance on the stair-run test of Margaria and colleagues (1966). Measurement is made of the time taken for the player to run between two steps on the stairs, the vertical distance between which is known.

The maximal ability to generate muscular power can be measured as a response to jumping on a force platform. This requires expensive and complex equipment which is not available for routine assessments. Power output in vertical jumping can be calculated, knowing the subject's body mass, the vertical distance through which body mass is moved and the flight time. The vertical distance itself is a good measure of muscular performance, i.e. mechanical work done. This can be recorded using a digital system attached to the subject's waist and based on the extension of a cord which is pulled as the subject jumps vertically (Figure 3.2). This method has now replaced the

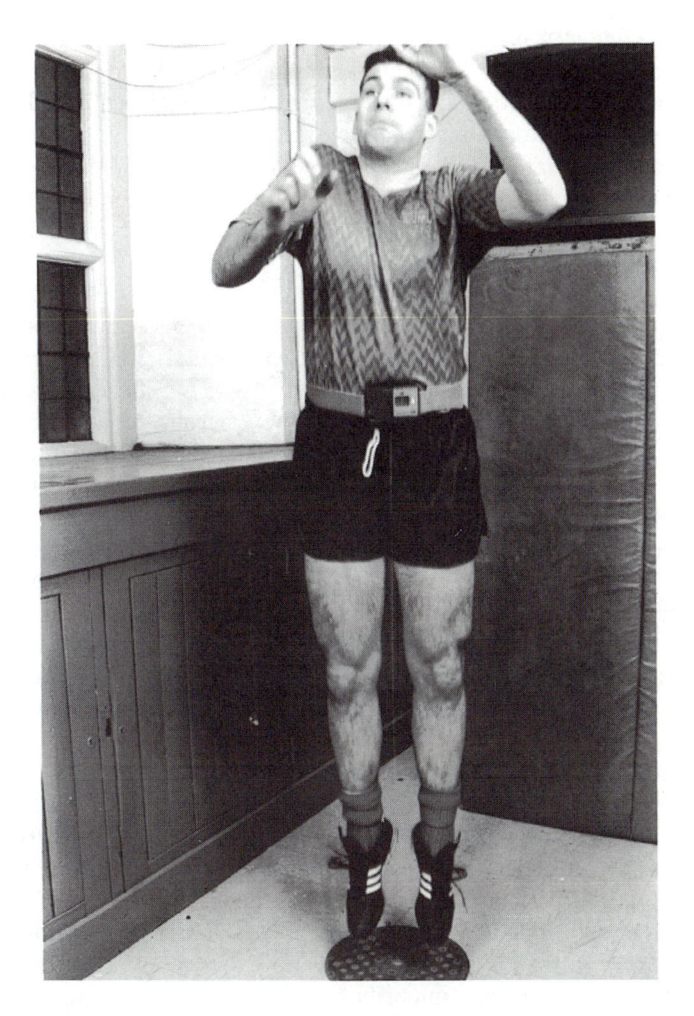

Figure 3.2 Measurement of vertical jumping ability: power output is calculated from the height of jump and body mass if flight time is measured.

classical Sargent jump technique for measuring vertical jump. It is preferable to the standing broad jump which is influenced by leg length and which does not permit calculation of power output.

Performance of soccer players in such tests of jumping ability tends to show up influences of positional role. Superior performances of England League players, for both standing broad jump and vertical jump, were found among goalkeepers and centre-backs and in forwards operating as target men. Midfielders had relatively low scores in both tests. The performances of two centre-backs and one striker in the vertical jump were similar to results reported for international high jumpers (Reilly, 1979).

Bosco *et al.* (1983) described another method for measuring mechanical power output in jumping. It requires jumping repeatedly for a given period, usually 60 s, the higher time and jumping frequency being recorded. The jumps are performed on a touch-sensitive mat which is connected to a timer. Power output can be estimated knowing the subject's body mass and the time between contacts on the mat. Performance at various parts of the 1-minute test can be compared, the tolerance to fatigue as the test progresses being indicative of the anaerobic glycolytic capacity.

As soccer players must be prepared to repeat fast bursts of activity supported by anaerobic glycolysis, the high anaerobic capacity should be important to play well. The Wingate Test which entails 30 s all-out effort on a cycle ergometer has been widely adopted as a test of 'anaerobic capacity'. The test duration of 30 s is too short to tax the anaerobic capacity completely. Another major limitation of such anaerobic capacity profiles is the mode of exercise. Measurement of power production and anaerobic capacity on a treadmill is more appropriate for soccer players.

Power output may be measured whilst the player runs as fast as possible on a 'self-powered' treadmill (Figure 3.3). The speed of the belt is determined by the effort of the subject. The horizontal forces produced can be determined from a load cell attached to the individual by means of a harness worn around the waist (Lakomy, 1984). Repeated bursts of exercise, such as 6 s in duration, may be performed and power profiles determined at different recovery periods.

An anaerobic capacity test of potential relevance to soccer players is that used by Medbo *et al.* (1988). It employs measurement of the maximum accumulated oxygen deficit. It is based on the close relation between the observed oxygen deficit during intense exercise and the anaerobic energy production. It is thought that the energy demand during exercise of higher intensities than the maximal aerobic power is under-estimated by this method (Bangsbo, 1993).

Ability to perform high-intensity exercise in soccer training contexts is usually recorded by coaching staff from time trials over short distances. Performance in such tests tends to be unreliable, due to lack of motivation or suspicion by some players. There is also difficulty in obtaining reproducible test conditions. Problems are partly overcome by use of the 'anaerobic capacity' test of Cunningham and Faulkner (1969). The subject runs to voluntary exhaustion at a treadmill speed of 18 km/h and an incline of 20°. The test has been more widely used in North America than in Europe and so far has not been applied widely to assessment of soccer players.

3.2.3 Muscle fibre types

Muscle performance characteristics of soccer players in many respects are determined by their distribution of fibre types (Figure 3.4). Muscle fibre types

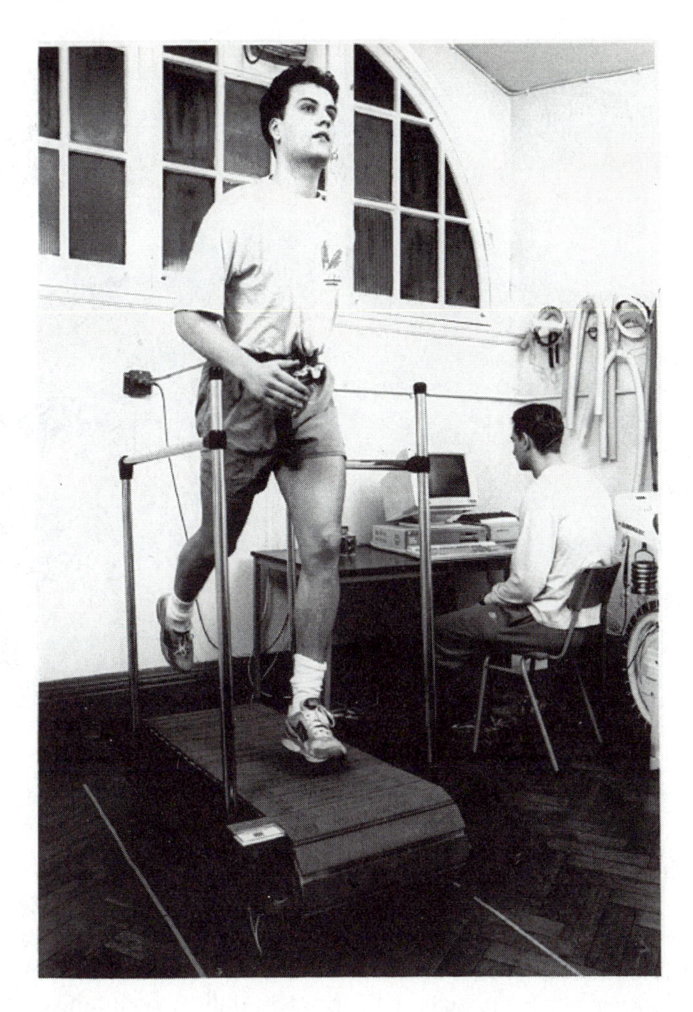

Figure 3.3 Experimental set-up for measuring power production whilst running.

are categorized as fast twitch (FT) or slow twitch (ST) based on the speed of response when stimulated. An alternative classification is based on the histochemical characteristics of the different motor units. This classification distinguishes between slow oxidative (SO or type I), fast glycolytic (FG or type IIb), and fast oxidative glycolytic (FOG or type IIa). The functions of the different fibre types and their recruitment during exercise have been reviewed by Williams (1990).

Soccer play demands an ability to sustain physical effort, albeit discontinuously, over 90 minutes, some of which is at high intensity. As the activity profile is compatible with both slow twitch and fast twitch muscle fibre characteristics, a balanced combination of muscle fibre types would be

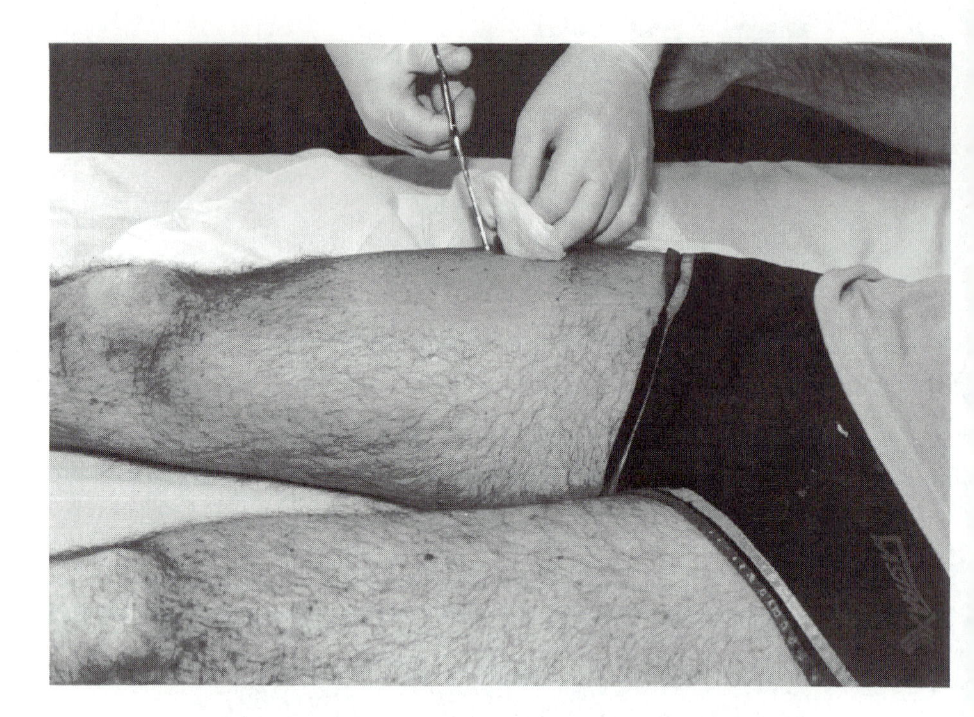

Figure 3.4 Muscle biopsy being taken from vastus lateralis: the technique shown here is known as 'conchotome'.

expected in top players. The muscle fibres in the vastus lateralis of Swedish professional club players was found to be 59.8 (± s.d. = 10.6)% fast twitch. The percentage fast twitch area was 65.6 (± 10.6)%, depicting a FT/ST mean fibre area of 1.28 (± 0.22) (Jacobs *et al.*, 1982). These figures suggest that the fibre types of soccer players are closer to sprinters than to endurance athletes in make-up. However, a large variability was observed within the squad, the number of FT fibres ranging from 40.8 to 79.1%. It would be expected that the fibre types of the goalkeeper and central defenders, in which an anaerobic profile of physical performance dominates, would be biased towards FT fibres.

Examination of the oxidative enzymes of soccer players' gastrocnemius muscles provides a different perspective (Bangsbo and Mizuno, 1988). The relative occurrence of ST (type I), FTa (type IIa) and FTb (type IIb) fibres in four Danish professionals was 55.9 (range 48–63.6), 39.8 (33–46.5) and 4.4 (3.0–5.5)% respectively. A reduction in fibre area with 3 weeks of detraining was observed only in the FTa fibres and the decrease was small (7%). The number of capillaries around the fibres decreased with detraining only in the ST fibres. At the time of full training mitochondrial activities of oxidative enzymes were similar to those for cross-country skiers in the case of 3-

hydroacyl coenzyme A dehydrogenase (HAD). Values for citrate synthase were intermediate between middle-distance runners and non-athletes.

Later studies from the same laboratory showed a fibre type distribution of 48.5 (34.6–61.0)% ST, 44.1 (32.0–65.4)% FTa and 7.4 (0.3–18.7)% FTb in the gastrocnemius of eight Danish professional club members. The mean number of capillaries around each fibre type was 4.67, 4.73 and 4.39 per fibre, respectively (Bangsbo and Lindquist, 1992).

The biochemical properties of fibre types may be affected by the nature and intensity of training. Andersen *et al.* (1992) studied nine national level Danish players during a 3-month strength training programme for the quadriceps muscles. A decrease in type IIa fibres from 34.9 to 26.5% was linked to a corresponding significant increase in type IIb fibres. An inverse correlation was noted between percentage type I fibres and the number of knee extensions performed over 50 s in an all-out test. The results demonstrated that well-trained soccer players can increase short-term muscle performance by strength training but that changes other than those in fibre type are responsible for the progress created by the training.

Eight Finnish second division players were examined by Smaros (1980). Biopsies taken from the vastus lateralis muscle showed an average fibre type distribution of 47% FT and 53% ST. Moreover, muscle biopsies taken at the end of a game showed that the reduction in glycogen stores occurred mainly in the ST fibres. Ryushi *et al.* (1979) reported fibre type percentages for Japanese university players that were very close to the values found in the Finnish study. They reported no relation between percentage fibre area and isometric strength but the maximal power of the knee extensors per kilogramme body mass was highly correlated ($r = 0.734$) with the percentage fibre area.

Kuzon *et al.* (1990) compared the skeletal muscle characteristics of the vastus lateralis of 11 top Canadian junior (mean age 18.1 ± 1.3 years) players to those of age-matched controls. Percentage values for the controls were 51.4 ± 12.5% type I (ST), 29.5 ± 7.3% type IIa (FTa) and 19.1 ± 13.1% type IIb (FTb), similar figures being reported for the soccer players. The footballers had the greater capillary supply, characterized by a significantly greater number of capillaries around each muscle fibre (5.7 ± 0.9 vs 4.9 ± 0.4), a significantly larger capillary density (282.7 ± 42 vs 220 ± 38.1/mm) and a significantly higher capillary to fibre ratio (2.2 ± 0.6 vs 1.7 ± 0.1). The findings were taken to indicate that soccer may induce simultaneous adaptations in both muscle fibre types.

Any inferences about fitness levels, muscle fibre types and elite soccer play must be tentative. Fibre type distributions could reflect positional role and so a team may be relatively heterogeneous in muscle composition. Parente and colleagues (1992) showed that there was a higher percentage of type I fibres (67%) among the midfield players than in defenders (44%) and forwards (38%) that they studied. The defenders had more type IIb fibres (49%) than the forwards (40%) or the midfield players (17%). These characteristics correspond to the demands imposed by the playing role. Fibre type distributions also vary between skeletal muscles. The gastrocnemius has an important function in locomotion whereas the quadriceps comprise the important muscle group in powerful kicking. The observations on soccer players may therefore reflect a relatively higher FT proportion in vastus lateralis than in the gastrocnemius compared to the ratio noted in other athletic populations.

3.3 AEROBIC FITNESS

The aerobic system is the main source of energy provision during soccer match-play (Bangsbo, 1993). This is indicated both by measurements of physiological responses during games and by the metabolic characteristics of soccer players' muscles. The upper limit of the body's ability to consume oxygen is indicated by the maximum oxygen uptake or VO_2max. The VO_2max represents an integrated physiological function with contributions from lungs, heart, blood and active muscles.

Lung function is measured with single-breath spirometry. Measures obtained for vital capacity and forced expiratory volume are made from a maximal exhalation after a maximal inspiration. Lung function values tend to be higher in soccer players than predicted from anthropometric measures (Reilly, 1979), but this is characteristic of athletes in general. Pulmonary function is not normally a limiting factor in maximal aerobic performance and the main use of single-breath spirometry lies in screening for any impairment or lung obstruction.

The oxygen transport system is influenced by the O_2 carrying capacity of the blood. Along with the maximal cardiac output, this determines the amount of oxygen delivered to the active muscle cells. This is important in soccer because of the large contribution of the aerobic system towards energy production. The oxygen carrying capacity is determined by the concentration of haemoglobin in blood, which affects the binding of O_2 in red blood cells, and the blood volume. Thus total body haemoglobin is highly correlated with the maximal oxygen uptake. Blood volume and total body haemoglobin tend to be about 20% higher in endurance trained athletes than in non-athletes. Haemoglobin concentration and haematocrit (the percentage of blood volume occupied by red blood cells) of soccer players generally fall within the normal range. Blood tests tend to have most value in screening for anaemia or deficient iron stores in cases of players whose exercise performances fall below expectations.

The amount of blood delivered to the active muscles during strenuous exercise depends on the cardiac output. This is a function of the stroke volume and the heart rate. The maximal heart rate is not increased as a result of training and is not itself an indicator of fitness. The heart responds to strenuous training by becoming larger and more effective as a pump. The chambers (particularly the left ventricle) increase in volume from a repetitive overload stimulus such as endurance running whilst the walls of the heart thicken and may grow in strength as a result of a pressure stimulus. Hypertrophy of cardiac muscle is reflected in a greater stroke volume and a larger left ventricular size enables more blood to fill the chamber before the heart contracts. Both are manifested in a lower heart rate at rest and this is apparent in observations on well-trained professional soccer players. Resting heart rates of 48 (± 1) beats/min have been reported for English League players. The slower than normal heart rate allows extended relaxation time during diastole for the pressure to drop below the normal level of about 80 mmHg. The pulse pressure, the difference between systolic and diastolic pressures, with a value of 50 mmHg for the English League players, was superior to the normal 40 mmHg (Reilly, 1979).

The $\dot{V}O_2max$ may be affected by pulmonary ventilation ($\dot{V}E$), pulmonary diffusion, the O_2 carrying capacity of the blood, the cardiac output and the arteriovenous difference in O_2 concentration. It is measured in a progressive exercise test to voluntary exhaustion (Figure 3.5). A motorized treadmill provides the most appropriate mode of exercise for testing soccer players. Expired air is analysed for its O_2 and CO_2 content and the minute ventilation ($\dot{V}E$) is also measured. The attainment of maximal oxygen uptake ($\dot{V}O_2max$) is indicated by a plateau in $\dot{V}O_2$ near exhaustion, a rise in the respiratory exchange ratio above 1.1 ($\dot{V}CO_2/\dot{V}O_2$), the elevation of heart rate to its age-predicted maximum or the blood lactate concentration reflecting anaerobic metabolism.

The average values of $\dot{V}O_2max$ for top-level soccer players tend to be high, supporting the belief that there is a large contribution from aerobic power to playing the game. Nevertheless values do not reach the same levels as in specialist endurance sports such as cross-country running and skiing, distance running or orienteering where values frequently exceed 80 ml kg^{-1} min^{-1}. Values for elite players lie in the region 55–70 ml kg^{-1} min^{-1}, the higher figures tending to be found at the top level of soccer and when players are at peak fitness.

Whilst $\dot{V}O_2max$ values may be influenced by differences in standards of play and training regimes, the stage of the season should also be considered. The $\dot{V}O_2max$ of professional soccer players does improve significantly in the pre-season period when there is an emphasis on aerobic training (Reilly, 1979). Further emphasis on improving the $\dot{V}O_2max$ adds little to the quality of play. When two teams of equal skill meet, the one with superior aerobic fitness would have the edge, being able to play the game at a faster pace throughout. Apor (1988) provided data on Hungarian players which showed a high rank-order correlation between mean $\dot{V}O_2max$ of the team and finishing position in the Hungarian First Division Championship. Mean $\dot{V}O_2max$ for the first, second, third and fourth teams were 66.6, 64.3, 63.3 and 58.1 ml kg^{-1}

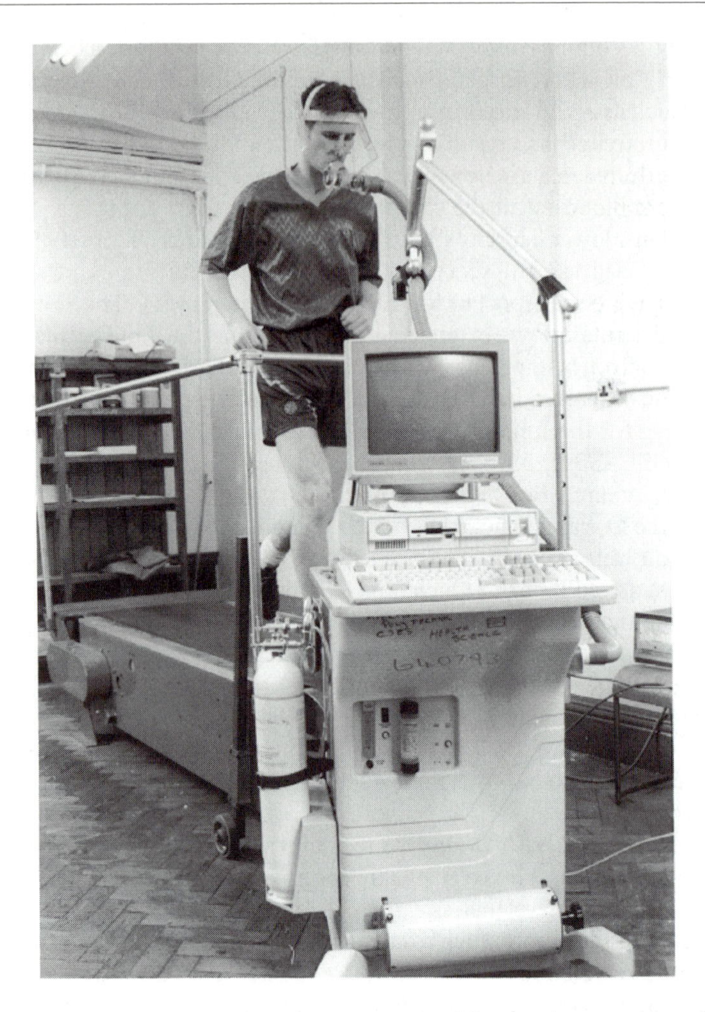

Figure 3.5 Maximal oxygen uptake measurement whilst the player runs to exhaustion on a motor-driven treadmill.

min^{-1}, respectively. Common factors such as stability in the team, avoidance of injury and so on help to maintain both $\dot{V}O_2max$ and team performance independently.

The $\dot{V}O_2max$ varies with positional role, when such roles can be clearly differentiated. When English League players were subdivided into positions according to 4-3-3 and 4-2-2 configurations, the midfielders had significantly higher aerobic power values than those in the other positions. The central defenders had significantly lower relative values than the other outfield players while the full-backs and strikers had values that were intermediate (Reilly, 1979). The significant correlation between $\dot{V}O_2max$ and distance covered in a game ($r = 0.67$) underlines the need for a high work-rate in midfield players who link between defence and attack. The goalkeepers had lower values than the centre-backs, an observation confirmed by other researchers and reinforced by the highest values for adiposity among goalkeepers. Four goalkeepers in the German national team had values of 56.2 (± 1.2) ml kg^{-1} min^{-1} compared to 62.0 (± 4.5) ml kg^{-1} min^{-1} for the squad as a whole (Hollmann *et al.*, 1981). The $\dot{V}O_2max$ of 19 professional players in the First Division of the Portuguese League was 59.6 (± 7.7) ml kg^{-1} min^{-1}: the average values for goalkeepers and central defenders were below, whilst midfield players and forwards were above 60 ml kg^{-1} min^{-1} (Puga *et al.*, 1993).

Whilst the $\dot{V}O_2max$ indicates the maximal ability to consume oxygen in strenuous exercise, it is not possible to sustain exercise for very long at an intensity that elicits $\dot{V}O_2max$. The upper level at which exercise can be sustained for a prolonged period is thought to be indicated by the so-called 'anaerobic threshold': this is usually expressed as the work-rate corresponding to a blood lactate concentration of 4 mM, the onset of accumulation of lactate in the blood (OBLA) or as a deflection in the relation between ventilation and oxygen consumption with incremental exercise (the ventilatory threshold). The inflection point in blood lactate response to incremental exercise represented 83.9% $\dot{V}O_2max$ in the 31 top Finnish players studied by Rahkila and Luhtanen (1991). The $\dot{V}O_2max$ corresponding to a blood lactate concentration of 3 mM was about 80% of $\dot{V}O_2max$ for both a continuous and an interval test on Danish players on a treadmill (Bangsbo and Lindquist, 1992). This reference lactate level for the continuous test was significantly correlated with distance covered in a game. The ventilatory threshold has been measured at 77% $\dot{V}O_2max$ in English League First Division players (White *et al.*, 1988). These values are close to the usual work intensity in running a marathon. The intermittent nature of soccer means that frequently players operate at above this intensity although the average fractional utilization of $\dot{V}O_2max$ is deemed to be 75–80% $\dot{V}O_2max$ (Reilly, 1990).

3.4 FIELD TESTS

Soccer is a practical activity and so its coaches are continually on the look-out for appropriate tests which allow them to assess fitness of players in field conditions. A convenient practical test for estimating the $\dot{V}O_2max$ of soccer players is the multi-stage shuttle run (Leger *et al.*, 1988). The speed is dictated by a rhythm on a tape recorder, the individual response to each running intensity being monitored. This test is now accepted as a valid method of indirectly estimating the maximal oxygen uptake. It is suitable for testing of football squads as it has a reasonable fidelity to movements in the game.

The ability to recover quickly from strenuous exercise may be important in soccer, which involves intermittent efforts interspersed with short rests. The Harvard Step Test designed initially for college men and later used in testing of military conscripts provides a fitness index which is based on the recovery of pulse rates over 3.5 min after a standard work-rate. The test has a long history of use in the fitness assessment of soccer players (Reilly, 1990). It has lost favour in recent years due to the availability of alternatives such as the 20 m shuttle run and the greater specificity associated with these other tests.

Bangsbo (1994) described various running tests specifically designed for soccer players. They included a sprint test performed seven times over a slalom course of about 35 m with 25 s rest between sprints. The duration of each sprint is recorded and a fatigue index is obtained by comparing the fastest and slowest sprints. Blood lactate concentrations (between 9 and 14 mmol l^{-1}) attest to the large anaerobic involvement in the activity.

Aerobic mechanisms are stressed in an intermittent field test conducted in a space equal to the penalty area and over a running course incorporating ziz-zag, backwards, sideways and forwards running (Figure 3.6). The relevance of the intermittent endurance test was shown by the correlation between test performance and distance covered by the player in competitive matches.

Whilst elements of soccer skills, such as dribbling and passing, have been incorporated into field tests, the assessment of soccer skill has received only little attention. Reilly and Holmes (1983) validated a series of skills tests which included a wall-volley test, a shooting test, a slalom dribble test and a straight dribble test. The tests were designed primarily as an aid in identifying and screening for young talent.

3.5 AGILITY AND FLEXIBILITY

Agility refers to the capability to change the direction of the body abruptly. The ability to turn quickly, dodge and side-step calls for good motor co-

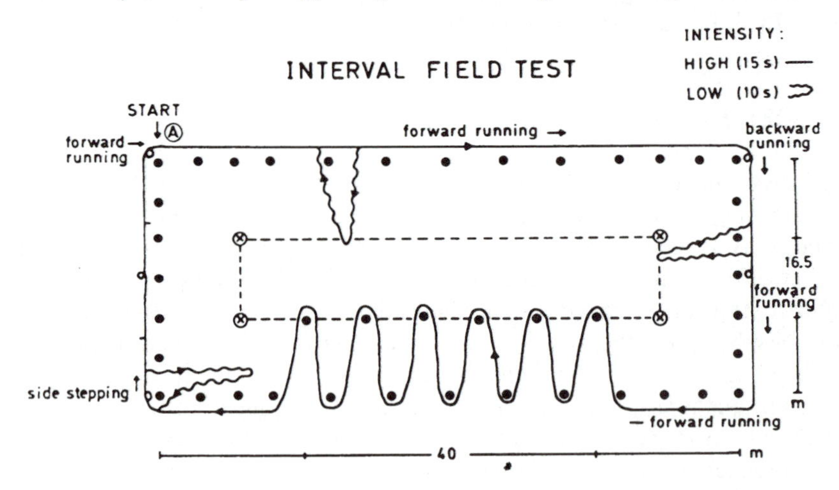

Figure 3.6 Drawing of field test for footballers. (Data from Bangsbo, 1993.)

ordination and is reflected in a standardized agility run test. Dallas Tornado players were found to have average times on the Illinois Agility Run above the 99.95 percentile for the test norms (Raven *et al.*, 1976). The test distinguished the soccer players as a group from the normal population better than any field test used for strength, power and flexibility. This is understandable, since soccer players have to be capable of dodging and weaving past opponents.

Joint flexibility is an important safety factor in soccer. Testing for inflexibility can be of benefit in screening for injury predisposition. Factor analysis of a number of fitness tests on English games players showed that flexibility in a range of movements in the hip joint afforded protection against injury (Reilly and Stirling, 1993).

Muscle tightness, particularly in the hamstring and adductor groups, has been linked with increased risk of muscle injury in Swedish professionals (Ekstrand, 1982). Two-thirds of the players had flexibility values poorer than non-players. This may be an adaptation, but it could also reflect a lack of attention to flexibility practices in training. Poorer range of motion has also been noted at the ankle joint in Japanese (Haltori and Ohta, 1986) and English League (Reilly, 1979) players, although the goalkeepers were exceptions among the English professionals. The Japanese players were less flexible than a reference group in inversion, eversion, plantarflexion and dorsiflexion. This may reflect an adaptive response of soft tissue around the ankle which improves stability at the joint.

3.6 PSYCHOPHYSIOLOGICAL CHARACTERISTICS

Simple reaction time gives a measure of how quickly a subject can respond to a stimulus in the immediate environment. This ability is predominantly due to heredity but deteriorates with age. It might be important in soccer where players have to respond immediately to environmental stimuli. Reaction times of English League soccer players to a visual stimulus were found to be faster than the normal values but similar to reaction times of track and field athletes (Reilly, 1979).

This is in agreement with results for a range of sportsmen and the observation of many investigators that athletes are superior to non-athletes in this measure (Reilly, 1979). No significant differences were found between the first team and the reserve team squads nor between goalkeepers and outfield soccer players. The apparently rapid responses of goalkeepers in competitive conditions can be attributed to their trained ability to anticipate the attackers' play from antecedent cues. This ability to predict opponents' manoeuvres correctly and to select the appropriate response rapidly from a large array of possible ones is a recognized hallmark of the skilled performer. The highly skilled player has the ability to read and interpret complex situations quickly and to initiate decisive action. The faster the simple reaction time of the individual,

Table 3.2 Muscle fibre composition of soccer players from various sources[a]

Players	*n*	*Muscle*	*Fibre type*		$\dot{V}O_2max$ ($ml\ kg^{-1}min^{-1}$)	*Reference*
			FT	*ST*		
Malmo FC	19	Vastus lateralis	59.8 ± 10.6	40.2	—	Jacobs *et al.* (1982)
Danish professionals	4	Gastrocnemius (IIa, IIb)	44	55.9	66.2	Bangsbo and Mizuno (1988)
Danish professionals	8	Gastrocnemius	51.5 (IIa, IIb)	48.5	60.4 ± 3.1	Bangsbo and Lindquist (1992)
Finnish Second Division	8	Vastus lateralis (IIa, IIb)	53	47 ± 13.3	63.6 ± 6.6	Smaros (1980)
Spanish and Italian semi-professionals	12	Quadriceps (undefined)	61.2 (type IIa, IIb)	38.8 (type I, IIc)	—	Montanari *et al.* (1990)
Japanese university players	12	Vastus lateralis	55.4	44.6	—	Ryushi *et al.* (1979)
Canadian national junior team	11	Vastus lateralis	47.1 (IIa, IIb)	52.9 ± 18.8	—	Kuzon *et al.* (1990)
Italian players		Quadriceps				Parente *et al.* (1992)
(unspecified)		(undefined)	56	44		
defenders	10	″	33	67		
midfield	10	″	62	38		
forwards	10	″	(IIa, IIb)	(type I)		

[a] Values of $\dot{V}O_2$max (± s.d.) are also included where they were reported or could be calculated.

the quicker will be the responses to complex situations. It will also give him an advantage in initiating abrupt movements. Thus a fast simple reaction time denotes a general athletic ability whilst fast responses in a complex or choice reaction time test specific to soccer characterizes game-related decision-making.

In games, the player has to move the entire body quickly, rather than one segment. Whole body reaction time (WRT) of soccer players was studied by Togari and Takahashi (1977). No differences were found in simple WRT between regular and substitute players but the regular players had the faster choice WRTs. No differences were observed between any of the various playing positions, although goalkeepers were generally faster to react in choice WRT. This superiority is likely to be largely a product of training specific to that position.

Fast diving movements are particularly relevant to goalkeeping skills. Suzuki *et al.* (1988) showed that skilled goalkeepers could dive faster than lesser skilled counterparts. The skilled players could generate a faster take-off velocity and turn better to meet the ball compared to those less skilled. Fast reactions in physical activity may be split between reaction time and movement time. Limb movement time should be an important attribute of the successful soccer player. This measure is independent of reaction time and is quicker in athletic subjects than in non-athletes. It does not, however, appear to have been studied in soccer players (except when linked with force production at high angular velocities), so that there is as yet no indication as to whether it discriminates between different levels of playing proficiency or different positional roles in the game. Movement time in a throwing action has been shown to influence the distance of the throw. The build up of speed in the final phase of a throw-in movement is especially influential (Kullath and Schwirtz, 1988).

During a game the player must recognize an array of complex stimuli in a wide range of vision. Peripheral vision would therefore seem to be an important prerequisite for success in soccer. Singer (1972) cited norms of 84.5° for athletes compared with 71.25° for non-athletes. Visual acuity and depth perception are additional aspects of vision important for soccer. The need for excellent dynamic visual acuity would apply particularly to the goalkeeper because of the relationship between this function and ball catching. Despite the obvious relevance of visual characteristics for success in soccer play, they do not seem to have been studied in great detail. In recent years more attention has been paid to tracking eye movements in simulated games contexts.

The field dependence/independence construct (Witkin *et al.*, 1962) regards perceptual ability as a continuum whose ends represent two contrasting modes of perceiving. This refers to the perception one has of oneself in relation to the surrounding field. Field dependency might be an advantage in performance of team sports in which the performer must relate the skill to the environment. This involves the ability to discriminate among complex visual stimuli. The hypothesis did not receive support from results of the experiments by Barrell and Trippe (1975). They found that 30 English League First Division players were not significantly different from a control group in field dependency. The construct, whilst relevant to many daily tasks, may not be sufficiently sensitive for application to the perceptual faculties of top soccer players.

Summary

Fitness for soccer cannot be determined by a single parameter since the game demands a large ensemble of physical, physiological and psychological capabilities. Fitness profiles are likely to vary within the playing season and the emphasis placed on different components of fitness will change with the stage of the season. Variables linked with fitness are influenced not just by training regimes but also by the stimulus provided by competing regularly.

Successful play at top level in contemporary football depends on how individuals are knitted together into a competent unit and so the combination of physiological characteristics may vary from player to player. Nevertheless it is possible to generalize on physiological characteristics of specialists in this sport. Anthropometric factors can determine the positional role most appropriate for the player. Body adiposity is highest among players pre-season but players at major international tournaments (with the exception of the goalkeeper) tend to have very little surplus fat. The physique of players generally shows muscular development, reflected in a high mesomorphy and low ectomorphy somatotype profile. Muscular strength, particularly in fast isokinetic movements, does seem to favour game-related performance. Anaerobic power of soccer players tends towards the profile of sprinters for the goalkeeper and central defenders whereas anaerobic capacity would seem to be more important in the other positions. Imbalanced muscular development (inappropriate flexor–extensor ratios, unilateral weakness, muscle tightness) may predispose towards injury. Leg muscle composition is not extreme, the fibre type distribution favouring fast movements but demonstrating histochemical properties of aerobically trained athletes. This is complemented by the moderately large heart volumes and aerobic power of players. Values for $\dot{V}O_2max$ above 60 ml kg^{-1} min^{-1} would seem desirable for outfield players. Sensory physiological mechanisms are also relevant considerations in the make-up of soccer players. It is likely that central factors in deciding the timing of game-related movements, supported by sufficiently well-developed muscular strength, motor coordination and oxygen transport mechanisms to implement the decisions, are the keys that open up opportunities for success in soccer play.

REFERENCES

Andersen, J.L., Bangsbo, J., Klitgaard, H. and Saltin, B. (1992) Changes in short-term performance and muscle fibre-type composition by strength training of elite soccer players. *Journal of Sports Sciences*, **10,** 162–3.

Apor, P. (1988) Successful formulae for fitness training, in *Science and Football* (eds T. Reilly, A. Lees, K. Davids and W.J. Murphy), E. & F.N. Spon, London, pp 95–107.

Asami, T. and Togari, H. (1968) Studies on the kicking ability in soccer. *Research Journal of Physical Education*, **12,** 267–72.

Bangsbo, J. (1993) The physiology of soccer – with special reference to intense intermittent exercise. DSc thesis, August Krogh Institute, University of Copenhagen.

Bangsbo, J. (1994) *Fitness Training in Football – a Scientific Approach*, HO & Storm, Bagsvaerd.

Bangsbo, J. and Lindquist, F. (1992) Comparison of various exercise tests with endurance performance during soccer in professional players. *International Journal of Sports Medicine*, **13,** 125–32.

Bangsbo, J. and Mizuno, M. (1988) Morphological and metabolic alterations in soccer players with detraining and retraining and their relation to performance, in *Science and Football* (eds T. Reilly, A. Lees, K. Davids and W.J. Murphy), E. & F.N. Spon, London, pp. 114–24.

Barrell, G.V. and Trippe, H. (1975) Field dependence and physical ability. *Perceptual and Motor Skills*, **41,** 216–18.

Bosco, C.P., Luhtanen, P. and Komi, P. (1983) A simple method for measurement of mechanical power in jumping. *European Journal of Applied Physiology*, **50,** 273–82.

Cabri, J., De Proft, E., Dufour, W. and Clarys, J.P. (1988) The relation between muscular strength and kick performance, in *Science and Football* (eds T. Reilly, A. Lees, K. Davids and W.J. Murphy), E. & F.N. Spon, London, pp. 186–93.

Clarys, J.P., Martin, A.D., Drinkwater, D.T. and Marfell-Jones, M.J. (1987) The skinfold: myth and reality. *Journal of Sports Sciences*, **5,** 3–33.

Cunningham, D.A. and Faulkner, J.A. (1969) The influence of training on aerobic and anaerobic metabolism during a short exhaustive run. *Medicine and Science in Sports*, **1,** 65–9.

De Proft, E., Cabri, J., Dufour, W. and Clarys, J.P. (1988) Strength training and kick performance in soccer players, in *Science and Football* (eds T. Reilly, A. Lees, K. Davids and W.J. Murphy), E. & F.N. Spon, London, pp. 108–13.

Durnin, J.V.G.A. and Womersley, J. (1974) Body fat assessed from total body density and its estimation from skinfold thickness: measurements on 481 men and women aged 17 to 72 years. *British Journal of Nutrition*, **32,** 77–97.

Ekstrand, J. (1982) Soccer injuries and their prevention. Doctoral thesis, Linkoping University.

Eston, R.G. and Reilly, T. (eds) (1995) *A Laboratory Manual for Kinanthropometry*, E. & F.N. Spon, London.

Faina, M., Gallozzi, C., Lupo, S. *et al.* (1988) Definition of the physiological profile of the soccer player, in *Science and Football* (eds T. Reilly, A. Lees, K. Davids and W.J. Murphy), E. & F.N. Spon, London, pp. 158–63.

Haltori, K. and Ohta, S. (1986) Ankle joint flexibility in college soccer players. *Journal of Human Ergology*, **15,** 85–9.

Hollmann, W., Liesen, H., Mader, A. *et al.* (1981) Zur Hochstund Dauerleistungsfahigkeit der deutschen Fussball-Spitzenspieler. *Deutsch Zeitschrift für Sportmedizin*, **32,** 113–20.

Jacobs, I., Westlin, N., Karlsson, J. *et al.* (1982) Muscle glycogen and diet in elite soccer players *European Journal of Applied Physiology*, **48,** 297–302.

Kullath, E. and Schwirtz, A. (1988) Biochemical analysis of the throw-in, in *Science and Football* (eds T. Reilly, A. Lees, K. Davids and W.J. Murphy), E. & F.N. Spon, London, pp. 460–7.

Kuzon, W.M. Jr, Rosenblatt, J.D., Huebel, S.C. *et al.* (1990) Skeletal muscle fibre type, fibre size and capillary supply in elite soccer players. *International Journal of Sports Medicine*, **II,** 99–102.

Lakomy, H. (1984) An ergometer for measuring the power generated during sprinting. *Journal of Physiology*, **354,** 33P.

Leger, L.A., Mercier, D., Gadoury, C. and Lambert, J. (1988) The multistage 20 metre shuttle run test for aerobic fitness. *Journal of Sports Sciences*, **6,** 93–101.

Margaria, R., Aghemo, P. and Rovelli, E. (1966) Measurement of muscular power (anaerobic) in man. *Journal of Applied Physiology*, **21,** 1661–4.

Martin, A.D., Spenst, L.F., Drinkwater, D.T. and Clarys, J.P. (1990) Anthropometric estimates of muscle mass in men. *Medicine and Science in Sports and Exercise*, **22,** 729–33.

Medbo, J.I., Mohn, A., Tabata, I., Bahr, R. and Sejersted, O. (1988) Anaerobic capacity determined by the maximal accumulated oxygen deficit. *Journal of Applied Physiology*, **64,** 50–60.

Montanari, G., Vecchiet, L. and Recoy Campo, G.L. (1990) Structural adaptations to the muscle of soccer players, in *Sports Medicine Applied to Football* (ed. G. Santilli), CONI, Rome, pp. 169–78.

Oberg, B., Eskstrand, J., Moller, M. and Gillquist, J. (1984) Muscle strength and flexibility in different positions of soccer players. *International Journal of Sports Medicine*, **5,** 213–16.

Oberg, B., Moller, M., Gillquist, J. and Ekstrand, J. (1986) Isokinetic torque levels in soccer players. *International Journal of Sports Medicine*, **7,** 50–3.

Parente, C., Montagnari, S., De Nicola, A. and Tajana, G.F. (1992) Anthropometric and morphological characteristics of soccer players according to positional role. *Journal of Sports Sciences*, **10,** 155.

Puga, N., Ramos, J., Agostinho, J. *et al.* (1993) Physical profile of a First Division Portuguese professional football team, in *Science and Football II* (eds T. Reilly, J. Clarys and A. Stibbe), E. & F.N. Spon, London, pp. 40–2.

Rahkila, P. and Luhtanen, P. (1991) Physical fitness profile of Finnish national soccer teams candidates. *Science and Football*, **5,** 30–3.

Raven, P., Gettman, L., Pollock, M. and Cooper, K. (1976) A physiological evaluation of professional soccer players. *British Journal of Sports Medicine*, **109,** 209–16.

Reilly, T. (1979) *What Research Tells the Coach about Soccer*, American Alliance for Health, Physical Education, Recreation and Dance, Washington DC.

Reilly, T. (1990) Football, in *Physiology of Sports* (eds T. Reilly, N. Secher, P. Snell and C. Williams), E. & F.N. Spon, London, pp. 371–425.

Reilly, T. (1994) Physiological profile of the player, in *Football (Soccer)* (ed. B. Ekblom), Blackwell Scientific, London, pp. 78–94.

Reilly, T. and Holmes, M. (1983) A preliminary analysis of selected soccer skills. *Physical Education Review*, **6,** 64–71.

Reilly, T. and Stirling, A. (1993) Flexibility, warm-up and injuries in mature games players, in *Kinanthropometry IV* (eds W. Duquet and J.A.P. Day). E. & F.N. Spon, London, pp. 119–23.

Ryushi, T., Asami, T. and Togari, H. (1979) The effect of muscle fibre composition on the maximal power and the maximal isometric strength of the leg extensor muscle. *Proceedings of the Department of Physical Education,* (College of General Education, University of Tokyo), **13,** 11–15.

Singer, R.N. (1972) *Coaching, Athletics and Psychology*, McGraw-Hill, New York.

Smaros, G. (1980) Energy usage during a football match, in Proceedings of the 1st International Congress of Sports Medicine Applied to Football (ed. L. Vecchiet), vol. II, D. Guanillo, Rome, pp. 795–801.

Suzuki, S., Togari, H., Isokawa, M. *et al.* (1988) Analysis of the goalkeeper's diving motion, in *Science and Football* (eds T. Reilly, A. Lees, K. Davids and W.J. Murphy), E. & F.N. Spon, London, pp. 468–75.

Togari, H. and Asami, T. (1972) A study of throw-in training in soccer. *Proceedings of the Department of Physical Education* (College of General Education, University of Tokyo), **6,** 33–8.

Togari, H. and Takahashi, K. (1977) Study of 'whole-body reaction' in soccer players. *Proceedings of the Department of Physical Education* (College of General Education, University of Tokyo), **11,** 35–41.

Togari, H., Ohashi, J. and Ohgushi, T. (1988) Isokinetic muscle strength of soccer players, in *Science and Football* (eds T. Reilly, A. Lees, K. Davids and W.J. Murphy), E. & F.N. Spon, London, pp. 181–5.

Trolle, M., Aagard, P., Simonsen, E.B., Bangsbo, J. and Klausen, K. (1993) Effects of strength training on kicking performance in soccer, in *Science and Football II* (eds T. Reilly, J. Clarys and A. Stibbe), E. & F.N. Spon, London, pp. 95–7.

White, J.E., Emery, T.M., Kane, J.E. *et al.* (1988) Pre-season fitness profiles of professional soccer players, in *Science and Football* (eds T. Reilly, A. Lees, K. Davids and W.J. Murphy), E. & F.N. Spon, London, pp. 164–71.

Williams, C. (1990) Metabolic aspects of exercise, in *Physiology of Sports* (ed. T. Reilly, N. Secher, P. Snell and C. Williams), E. & F.N. Spon, London, pp. 3–67.

Williams, C., Reid, R.M. and Coutts, L. (1973) Observations on the aerobic power of university rugby players and professional soccer players. *British Journal of Sports Medicine*, **7,** 390–1.

Withers, R.T., Roberts, R.G.D. and Davies, G.J. (1977) The maximum aerobic power, anaerobic power and body composition of South Australian male representatives in athletics, basketball, field hockey and soccer. *Journal of Sports Medicine and Physical Fitness*, **17,** 391–400.

Witkin, H.B., Dyk, R.B., Faterson, H.F. *et al.* (1962) *Psychological Differentiation*, John Wiley, New York.

Physiology of training

4

Jens Bangsbo

Introduction

Soccer players need a high level of fitness to cope with the physical demands of a game (see Chapter 3) and to allow for their technical skills to be utilized throughout a match. Therefore, fitness training is an important part of the overall training programme.

Common to all types of fitness training in soccer is that the exercise performed should resemble match-play as closely as possible. This is one of the main reasons why the majority of fitness training should be performed with a ball. Other advantages of conducting training as a drill or game are that the players develop technical and tactical skills under conditions similar to those encountered during a match, and that this form of training usually provides greater motivation for the players compared to training without a ball.

4.1 COMPONENTS OF FITNESS TRAINING

Fitness training has to be multifactorial in order to cover the different aspects of physical performance in soccer. Thus, training can be divided into a number of components based on the different types of physical demands during a match (see Figure 4.1). The terms aerobic and anaerobic training are based on the energy pathway that dominates during the activity periods of the training session. Aerobic and anaerobic training represent exercise intensities below and above the maximum oxygen uptake, respectively. However, during a

Science and Soccer. Edited by Thomas Reilly. Published in 1996 by E & FN Spon, London. ISBN 0 419 18880 0.

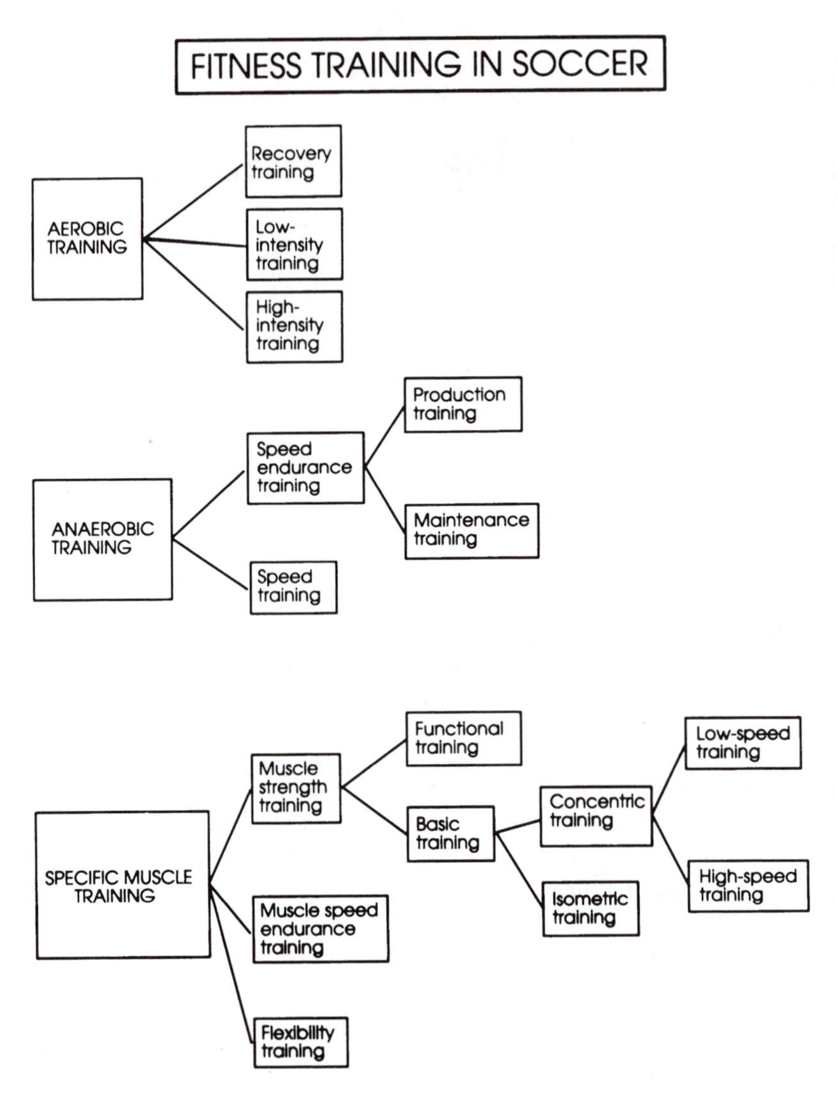

Figure 4.1 Components of fitness training in soccer.

training game, the exercise intensity for a player varies continuously, and some overlap exists between the two categories of training. Figure 4.2 shows examples of exercise intensities during games and drills within aerobic and anaerobic training.

In this chapter the separate components within fitness training are briefly described. They include aerobic, anaerobic and specific muscle training. For further discussion of the practical aspects of fitness training in soccer and for additional suggestions on activities which can be used in soccer, the reader is

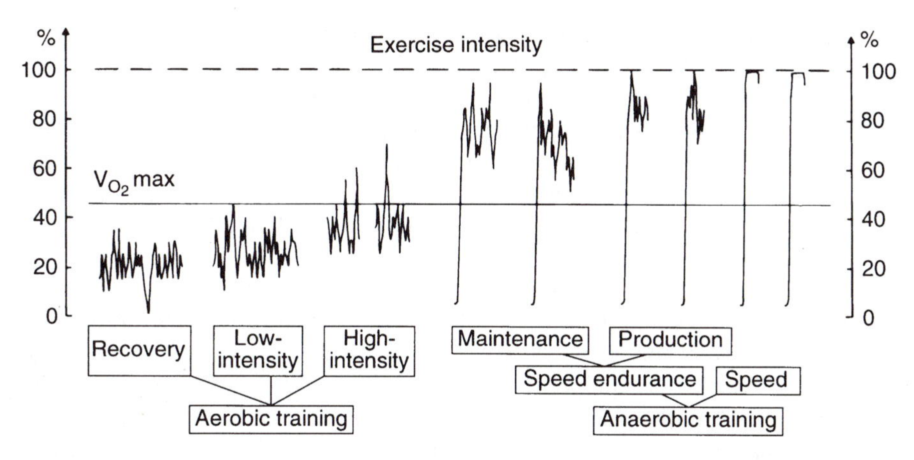

Figure 4.2 Examples of the exercise intensities of a player during games, within aerobic and anerobic training, expressed in relation to maximal intensity (100%). The exercise intensities eliciting maximum oxygen uptake and maximal exercise intensity of the player are represented by the lower and upper horizontal lines, respectively.

referred to *Fitness Training in Football – a Scientific Approach* (Bangsbo, 1994a).

4.1.1 Aerobic training

It has been demonstrated that the total distance covered by high-intensity exercise during a match is related to the standard of soccer, with top-class players covering the most distance. It is therefore important that players are capable of exercising at high intensities for prolonged periods of time. Thus, a player needs a relatively high maximum oxygen uptake ($\dot{V}O_2max$). In addition, a player should have a well-developed endurance capacity, as the average oxygen uptake during a soccer match can be as high as 70% of $\dot{V}O_2max$ (Bangsbo, 1994b). An increase in $\dot{V}O_2max$ and the ability to sustain exercise for a prolonged period can be obtained by aerobic training.

Aerobic training causes changes in central factors such as the heart and blood volume, which result in a higher maximum oxygen uptake. Peripheral adaptations also occur with this type of training. The training leads to a proliferation of capillaries and an elevation of the content of mitochondrial enzymes, as well as the activity of lactate dehydrogenase 1–2 (LDH_{1-2}) isozymes. Furthermore, the mitochondrial volume and the capacity of one of the shuttle systems for NADH are elevated. These changes cause marked alterations in muscle metabolism. The overall effects are an enhanced oxidation of lipids and sparing of glycogen, as well as a lowered lactate production, both at a given and at the same relative work-rate.

The optimal way to train the central versus the peripheral factors is not the same. Maximum oxygen uptake is most effectively elevated by exercise intensities of 80–100% of $\dot{V}O_2$max (about 40% of maximal intensity, see Figure 4.2). For a muscle adaptation to occur, an extended period of training appears to be essential, and therefore, the mean intensity should be below 80% of $\dot{V}O_2$max. This does not imply that high-intensity training does not elevate the number of capillaries and mitochondrial volume in the muscles engaged in the training, but the duration of this type of training is often too short to obtain optimal adaptations at a local level.

The dissociation between changes in $\dot{V}O_2$max and muscle adaptation by means of training and detraining is illustrated by results from two studies on top-class players (Bangsbo, 1994b). Fourteen players were tested before and after a 5 week pre-season period. As a result of the training $\dot{V}O_2$max was only slightly higher, whereas the muscle oxidative enzymes and the number of capillaries per muscle fibre were increased by 10 and 15%, respectively. In accordance with the local adaptations, the blood lactate concentration during submaximal running was significantly lower after the period of pre-season training. In another study players abstained from training for 3 weeks. It was found that $\dot{V}O_2$max was unaltered, whereas performance in a field test was lowered by 8%, and there was a reduction in oxidative enzymes of 20–30% (Figure 4.3).

The recovery processes from intense exercise are related both to the oxidative potential and to the number of capillaries in the muscles. Thus, aerobic training not only improves endurance performance of a soccer player, but also appears to influence a player's ability to repeatedly perform maximal efforts.

The overall aim of aerobic training in soccer is to increase the work-rate

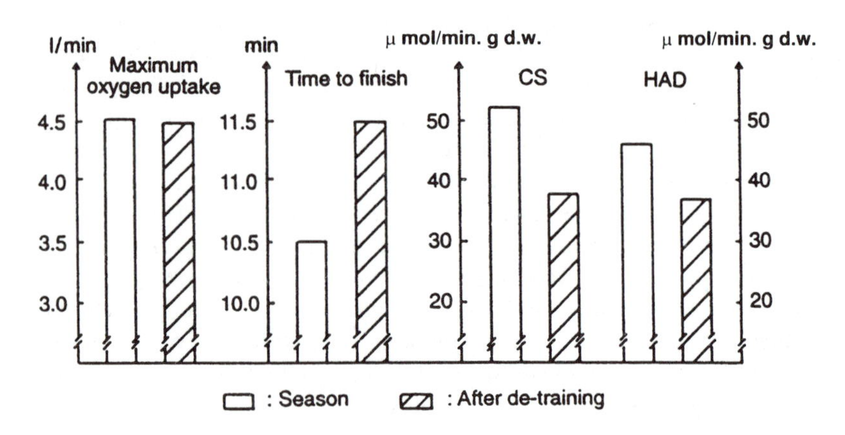

Figure 4.3 Maximal oxygen uptake, performance in a field test, activities of the muscle oxidative enzymes citrate synthase (CS) and b-hydroxy-CoA-dehydrogenase (HAD; involved in fat oxidation) of Danish top-class soccer players during the season and after three weeks of holiday.

during a match, and to minimize a decrease in technical performance as well as lapses in concentration induced by fatigue towards the end of a game. The specific aims of aerobic training in soccer are as follows.

- To improve the capacity of the cardiovascular system to transport oxygen. Thus, a larger percentage of the energy required for intense exercise can be supplied aerobically, allowing a player to work at a higher exercise intensity for prolonged periods of time.
- To improve the capacity of muscles specifically used in soccer to utilize oxygen and to oxidize fat during prolonged periods of exercise. Thereby, the limited store of muscle glycogen is spared and a player can exercise at a higher intensity towards the end of a game.
- To improve the ability to recover after a period of high-intensity exercise. As a result, a player requires less time to recover before being able to perform in a subsequent period of high-intensity exercise.

Components of aerobic training

Aerobic training can be divided into three overlapping components: **recovery training, aerobic low-intensity training (Aerobic$_{LI}$)** and **aerobic high-intensity training (Aerobic$_{HI}$)** (see Figures 4.1 and 4.2). Table 4.1 illustrates the principles behind the various categories of aerobic training, which take into account that the training may be performed as a game, and thus, the heart rate (HR) of a player will frequently alternate during the training.

During recovery training the players perform light physical activities, such as jogging and low-intensity games. This type of training may be carried out the day after a match or the day after a hard training session to help a player return to a normal physical state. Recovery training may also be used to avoid the players getting into a condition known as 'overtraining' in periods involving frequent training sessions (maybe even twice a day) and a busy competitive schedule of matches.

The purpose of Aerobic$_{LI}$ training is to elevate the capillarization and the

Table 4.1 Principles of aerobic training

	Heart rate				*Oxygen uptake*	
	% of HRmax		*Beats/min*		*% of $\dot{V}O_2max$*	
	Mean	*Range*	*Mean*[a]	*Range*[a]	*Mean*	*Range*
Recovery training	65	40–80	130	80–160	55	20–70
Low-intensity training	80	65–90	160	130–180	70	55–85
High-intensity	90	80–100	180	160–200	85	70–100

[a] If HR_{max} is 200 beats/min.

oxidative potential in the muscle (peripheral factors). Thus, the functional significance is an optimization of the substrate utilization and thereby an improvement in endurance capacity. The main aim of $Aerobic_{HI}$ training is to improve central factors such as the pump capacity of the heart which is closely related to VO_2max. These improvements increase a player's capability to exercise repeatedly at high intensities for prolonged periods of time during a match.

4.1.2 Anaerobic training

During a match a player frequently performs activities that require rapid development of force, such as sprinting or quickly changing direction. Furthermore, findings of high blood lactate concentrations in top-class players during match-play indicate that the lactate-producing energy system (glycolysis) is highly stimulated during periods of a game (see p. 76). Therefore, the capacity to repeatedly perform high-intensity exercise should be specifically trained. This can be achieved through anaerobic training.

Anaerobic training results in an increase in the activity of creatine kinase and glycolytic enzymes. Such an increase implies that a certain change in an activator results in a higher rate of energy production of the anaerobic pathways. Intense training does not appear to influence the total creatine phosphate (CP) pool, but it allows the muscle glycogen concentration to be elevated, which is of importance for performance during repeated high-intensity exercise. The capacity of the muscles to release and neutralize H^+ (buffer capacity) also appears to be increased after a period of anaerobic training. This will lead to a lower reduction in pH for a similar amount of lactate produced during high-intensity exercise. Therefore, the inhibitory effects of H^+ within the muscle cell are smaller, which may be one of the reasons for a better performance in high-intensity tests after a period of anaerobic training.

The overall aim of anaerobic training in soccer is to increase a player's potential to perform high-intensity exercise during a game. The specific aims of anaerobic training in soccer are summarized below.

- To improve the ability to act quickly and to produce power rapidly. Thus, a player reduces the time required to react and elevates performance of a sprint during a game.
- To improve the capacity to produce power and energy continuously via the anaerobic energy-producing pathways. Thereby, a player elevates the ability to perform high-intensity exercise for longer periods of time during a game.
- To improve the ability to recover after a period of high-intensity exercise. As a result, a player requires less time before being able to perform maximally in a subsequent period of exercise, and is therefore able to perform high-intensity exercise more frequently during a game.

Components of anaerobic training

Anaerobic training can be divided into **speed training** and **speed endurance training** (see Figure 4.1). The aim of speed training is to improve a player's ability to act quickly in situations where speed is essential. Speed endurance training can be separated into two categories: **production training** and **maintenance training**. The purpose of production training is to improve the ability to perform maximally for a relatively short period of time, whereas the aim of maintenance training is to increase the ability to sustain exercise at a high intensity. Table 4.2 illustrates the principles of the various categories of anaerobic training.

Anaerobic training must be performed according to an interval principle. During speed training the players should perform maximally for a short period of time (< 10 s). The periods between the exercise bouts should be long enough for the muscles to recover to near resting conditions, so as to enable a player to perform maximally in a subsequent exercise bout. In soccer, speed is not merely dependent on physical factors. It also involves rapid decision-making which must then be translated into quick movements. Therefore, speed training should mainly be performed with a ball. Speed drills can be designed to promote a player's ability to sense and predict situations, and the ability to decide on the opponents' responses in advance.

By speed endurance training the creatine kinase and glycolytic pathways are highly stimulated. The exercise intensity should be almost maximal to elicit major adaptations in the enzymes associated with anaerobic metabolism. In **production training** the duration of the exercise bouts should be relatively short (20–40 s), and the rest periods in between the exercise bouts should be comparatively long (2–4 min) in order to maintain a very high intensity during the exercise periods throughout an interval training session. In **maintenance**

Table 4.2 Principles of anaerobic training

		Duration			
		Exercise (s)	*Rest*	*Intensity*	*Number of repetitions*
Speed training		2–10	> 5 times the exercise duration	Maximal	2–10
Speed endurance training	Production	20–40	> 5 times the exercise duration	Almost maximal	2–10
	Maintenance	30–90	Equal to or less than exercise duration	Almost maximal	2–10

training the exercise periods should be 30–90 s and the duration of the rest periods should approximately equal the exercise periods, to allow players to become progressively fatigued.

The adaptations caused by speed endurance training are mostly localized to the exercising muscles. Thus, it is important that a player performs movements in a manner similar to during match-play. This can be obtained with high-intensity games or drills with a ball. Figure 4.4 illustrates a game within the maintenance category of speed endurance training and it also shows HR and blood lactate values for a player during the game.

Speed endurance training is both physically and mentally demanding for the players. Therefore, it is recommended that this type of training is only used with top-class players. When there is a limited amount of time available for training, time can be better utilized on other forms of training.

4.1.3 Specific muscle training

Specific muscle training involves training of muscles in isolated movements. The aim of this type of training is to increase performance of a muscle to a higher level than can be attained just by playing soccer. Specific muscle training can be divided into **muscle strength**, **muscle speed endurance** and **flexibility** training (Figure 4.4). The effect of this form of training is specific to the muscle groups that are engaged, and the adaptation within the muscle is limited to the kind of training performed.

A brief description of muscle strength training is given below. Further information about strength training as well as an overview of muscle endurance and flexibility training for soccer players can be obtained elsewhere (Klausen, 1990; Bangsbo, 1994a).

Area of field: One-third of a soccer field

Number of players: Ten

Organization: A team consists of five players. The teams perform man-to-man marking. Six small 'goals' (pairs of cones) are placed at various positions on the field.

Description: Ordinary soccer play. The players must try to play the ball through the goals to a team-mate.

Rule: The players are not allowed to run through the goals.

Intervals: Exercise periods of 1.5 min and rest periods of 0.5 min.

Scoring: A point is scored for passing the ball through one of the goals to a team-mate.

Comments: The players should be motivated to continuously exercise at a high intensity. If one player cannot cope with the marking of an opponent the intensity of the other players can be affected. It is therefore important to have players of equal ability marking each other. In order to avoid delays extra balls should be placed around the field. The exercise demands can be controlled by changing the number or the width of the goals.

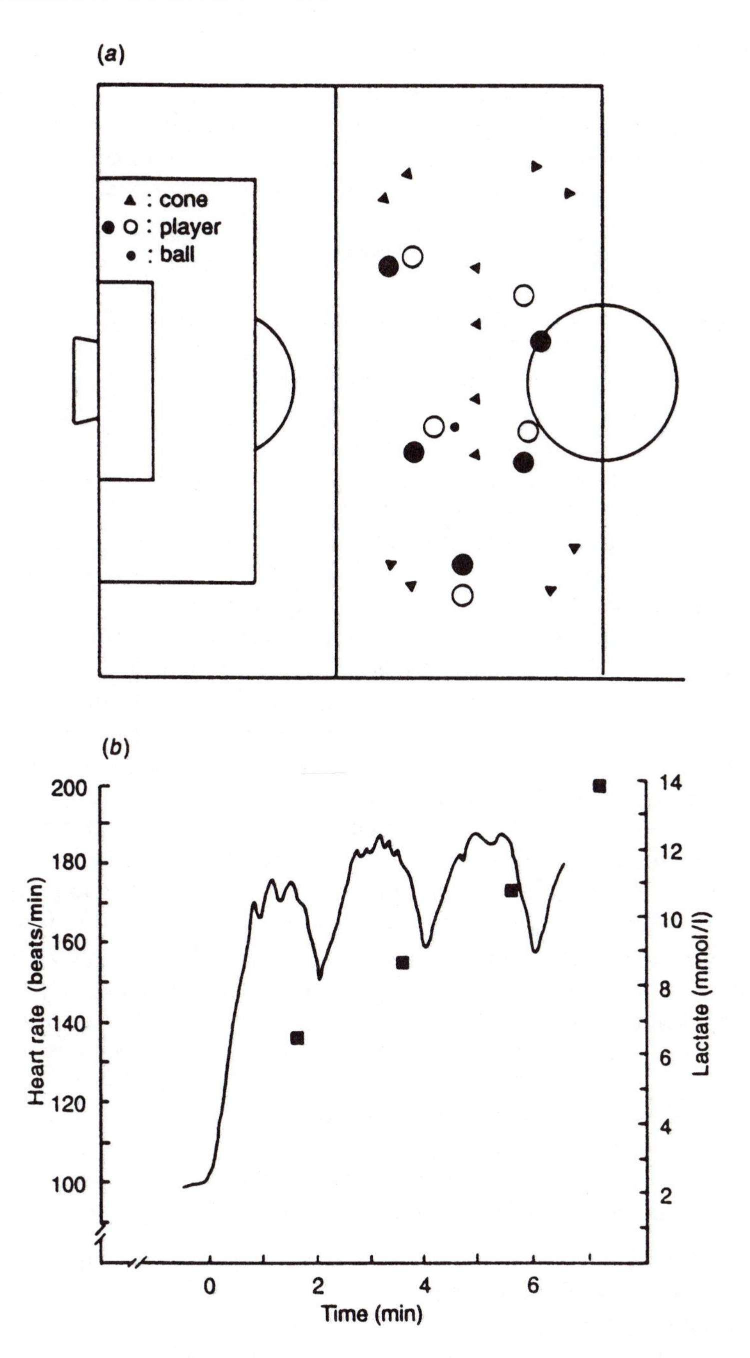

Figure 4.4 Speed endurance game (maintenance training). The game is described in the boxed text and the playing field is illustrated in (*a*). (*b*) Blood lactate and heart rate of one player performing the game.

Strength training

Many activities in soccer are forceful and explosive, e.g. tackling, jumping, kicking, turning and changing pace. The power output during such activities is related to the strength of the muscles involved in the movements. Thus, it is beneficial for a soccer player to have a high level of muscular strength, which can be obtained by strength training.

Strength training can result in hypertrophy of the muscle, partly through an enlargement of muscle fibres. In addition, training with high resistance can change the fibre type distribution in the direction of more fast twitch fibres (Andersen *et al.*, 1994). There is also a neuromotor effect of strength training and part of the increase in muscle strength can be attributed to changes in the nervous system. Improvements in muscular strength during isolated movements seem closely related to training speeds. However, significant increases in force development at very high speeds (10–18 rad/s) have also been observed with slow-speed high-resistance training (Aagaard *et al.*, 1993).

Various studies have focused on the effectiveness of strength training for soccer players. Significant increases in strength of the leg muscles only lead to a minor or no improvement in kick performance (De Proft *et al.*, 1988; Aagaard *et al.*, 1993). This illustrates that a player's ability to produce force during a soccer activity is not solely dependent on the strength of the muscles involved in the movement, but is also influenced by the player's ability to coordinate the muscle actions. Therefore, to utilize improvements in muscle strength effectively during match-play, it is important to combine strength training with technical training to enhance the synchronization of force development between the agonist and antagonist muscles in movements specific to soccer.

One essential function of the muscles is to protect and stabilize joints of the skeletal system. Hence, strength training is of importance also in preventing injuries as well as re-occurrence of injuries. A prolonged period of inactivity, e.g. during recovery from an injury, will considerably weaken the muscles. Thus, before a player returns to soccer training after an injury, a period of strength training is needed. The length of time required to regain strength depends on the duration of the inactivity period but generally several months are needed. For a group of players observed two years after a knee operation, it was found that the average strength of the quadriceps muscle of the injured leg was only 75% of the strength in the other leg (Ekstrand, 1982).

The overall aim of muscle strength training in soccer is to develop a player's muscular make-up. The specific aims of muscle strength training in soccer are:

- to increase muscle power output during explosive activities in a soccer match such as tackling, jumping and accelerating;
- to prevent injuries;
- to regain strength after an injury.

Table 4.3 Principles of muscle strength training

		Workload	*Number of repetitions*	*Rest between repetitions (s)*	*Number of sets*
Concentric	Low-speed	5RM[a]	5	2–5	2–4
	High-speed	50% of 5RM	15	1–3	2–4
Isometric		85–100% of max maintained for 5–15 s	5–10	5–15	2–4

[a] RM, repetition maximum.

Components of strength training

Strength training can be divided into **functional strength training** and **basic strength training** (see Figure 4.1). In functional strength training, movements related to soccer are used. The training can consist of games in which typical soccer movements are performed under conditions that are physically more stressful than normal. During basic strength training muscle groups are trained in isolated movements. For this training different types of conventional strength training machines and free weights can be used, but the body weight may also be used as resistance. Strength training should be carried out in a manner that resembles activities and movements specific to soccer. Based on the separate muscle actions the basic strength training can be divided into isometric, concentric and eccentric muscle strength training (see Figure 4.1, Table 4.3). Several principles can be used in concentric strength training. Table 4.3 illustrates a principle which is based on determinations of five repetition maximum (5RM) and which allows for muscle groups to be trained at both slow and fast speeds.

Common to the two types of strength training is that the exercise should be performed with a maximum effort. After each repetition a player should rest a few seconds to allow for a higher force production in the subsequent muscle contraction. The number of repetitions in a set should not exceed 15. During each training session two to four sets should be performed with each muscle group, and rest periods between sets should be longer than one minute. During this time the players can exercise with other muscle groups.

4.2 PLANNING FITNESS TRAINING

The time course of adaptations in the various tissues should be taken into account when planning fitness training. A change in heart size is rather slow, and there is a need for training over a long period of time (years) to improve the pump capacity of the heart significantly. Blood volume changes more

Table 4.4 Priority of fitness training through a year

		Off-season				*Season*							
Aerobic training	Low-intensity	3344	4445	5555	4433	4343	4343	4343	4343	4343	4343	4343	4343
	High-intensity	2223	3234	4445	4555	5545	5545	5545	5545	5545	5545	5545	5444
Anaerobic training	Speed endurance	1111	1111	2234	4555	4353	4353	4353	4353	4353	4353	4353	3453
	Speed	1111	1111	2234	4555	5555	5555	5555	5555	5555	5555	5555	5554
Muscle strength training	Basic	3334	5555	5543	2222	2222	2222	2222	2222	2222	2222	2222	2222
	Functional	2222	3333	3344	4343	4343	4343	4343	4343	4343	4343	4343	4322
Muscle speed training		1111	1112	3333	3333	3333	3333	3333	3333	3333	3333	3333	3333
Flexibility training		3232	3434	4444	4444	4444	4444	4444	4444	4444	4444	4444	4444

Each single number represents a week. For practical purposes each month is given four weeks. The values represent the following priorities: 1, very low priority; 2, low priority; 3, moderate priority; 4, high priority; 5, very high priority.

quickly than the heart size, but this adaptation is optimal first after a dimensional development of the cardiovascular system has occurred. The content of oxidative enzymes in a tissue and the degree of capillarization of skeletal muscle change more rapidly than the volume of a tissue, e.g. the heart, but months of regular training are needed to obtain considerable increases in muscle capillaries and oxidative enzymes. On the other hand, a reduction in these parameters can occur with a time constant of weeks (see Figure 4.3). The changes in glycolytic enzymes are rapid and they can be markedly elevated within a month of appropriate training.

When planning fitness training in soccer the year may be divided into a season and an off-season. Table 4.4 shows how the different types of fitness training can be structured through the two periods of the year. The higher the number given, the more important is the form of training. The scheme is based on an eight-month season and a four-month off-season of which the last eight weeks before the season are spent in the club. The scheme is only a guideline, since there are differences in duration of the season and off-season from country to country, and some countries have a mid-season break. It should also be emphasized that there may be deviations in the priority of the various aspects of fitness training due to specific demands of a team.

During the off-season the players should regularly perform sessions with $\text{Aerobic}_{\text{LI}}$ training, since the oxidative enzyme capacity is rapidly lost with inactivity and it takes months to restore the enzyme levels. The training will reduce the decrease in fitness level, which always occurs on cessation of normal training and competition. A gradual transition between the various phases of the off-season also keeps the risk of injuries low and leaves time for other types of football training, such as tactical and technical training.

In the first part of the off-season it is reasonable to emphasize basic strength training, since the adaptations from this type of training can be easily maintained. As the start of the season approaches, the amount of basic strength training should be reduced and more time should be allocated to functional strength training and playing soccer.

During the last six weeks or so of the off-season, the players should frequently perform sessions of $\text{Aerobic}_{\text{HI}}$ training, speed training and, for elite players, also speed endurance training. Such training should be supplemented by regular matches at a high competitive level.

During a season, $\text{Aerobic}_{\text{HI}}$ training should be given a high priority (see Table 4.4). Speed training and, for top-class players, speed endurance training should also be performed regularly. Endurance capacity may be maintained by frequently including prolonged training sessions with only short rest periods. The extent of strength training during the season should be determined by the total training time available.

Summary

With appropriate training, performance of a player during a match can be increased and the risk of injury can be reduced. In order to design an efficient training programme it is important to be aware of the different components of fitness training in soccer. Aerobic training increases the ability to exercise at an overall higher intensity during a match, and minimizes a decrease in technical performance induced by fatigue towards the end of a game. Anaerobic training elevates a player's potential to perform high-intensity exercise during a game. Muscle strength training, combined with technical training, improves a player's power output during explosive activities in a match.

Fitness training should mainly be performed with a ball. This ensures that the specific muscles used within soccer are trained. Equally important is that players should develop their technical skills under conditions similar to those encountered during competition.

REFERENCES

Aagaard, P., Trolle, M., Simonsen, E.B. *et al.* (1993) High speed knee extension capacity of soccer players after different kinds of strength training, in *Science and Football II* (eds T. Reilly, J. Clarys and A. Stibbe), E. & F.N. Spon, London, pp. 92–4.

Andersen, J.L., Klitgaard, H., Bangsbo, J. and Saltin, B. (1994) Myosin heavy chain isoforms in single fibres from m. vastus lateralis of soccer players: effects of strength-training. *Acta Physiologica Scandinavica* **150,** 21–6.

Bangsbo, J. (1994a) *Fitness Training in Football – a Scientific Approach*, HO & Storm, Bagsvaerd.

Bangsbo, J. (1994b) The physiology of soccer – with special reference to intense intermittent exercise. *Acta Physiologica Scandinavica*, **151,** Suppl. 619.

De Proft, E., Cabri, J., Dufour, W. and Clarys, J.P. (1988) Strength training and kick performance in soccer players, in *Science and Football* (eds T. Reilly, A. Lees, K. Davids and W.J. Murphy). E. & F.N. Spon, London, pp. 108–14.

Ekstrand, J. (1982) Soccer injuries and their prevention. Thesis, Linköping University Medical Dissertation 130.

Klausen, K. (1990) Strength and weight-training, in *Physiology of Sports* (eds T. Reilly, N. Secher, P. Snell and C. Williams). E. & F.N. Spon, London, pp. 41–67.

Motion analysis and physiological demands

5

Thomas Reilly

Introduction

The physiological demands of soccer play are indicated by the exercise intensities at which the many different activities during match-play are performed. There are implications not only for fitness assessment and selection of players but also for their training regimes. Since the training and competitive schedules of players comprise their occupational roles, there are consequences too for their habitual activities, daily energy requirements and energy expenditures. Finally, there are repercussions for the prevention of injuries as far as is possible.

The exercise intensity during competitive soccer can be indicated by the overall distance covered. This represents a global measure of work-rate which can be broken down into the discrete actions of an individual player for a whole game. The actions or activities can be classified according to type, intensity (or quality), duration (or distance) and frequency. The activity may be juxtaposed on a time-base so that the average exercise-to-rest ratios can be calculated. These ratios can then be used in physiological studies designed to represent the demands of soccer and also in conditioning elements of the soccer players' training programmes. These work-rate profiles can be complemented by monitoring physiological responses where possible.

Various aspects of the exercise intensities in soccer are examined in

Science and Soccer. Edited by Thomas Reilly. Published in 1996 by E & FN Spon, London. ISBN 0 419 18880 0.

this chapter. Work-rate during play and factors influencing work-rate profiles are considered prior to a review of physiological responses to playing. These are restricted to heart rate and metabolic measures. The compatibility between the demands of play, training stimuli and fitness measures is addressed.

5.1 MOTION ANALYSIS

In the early applications of motion analysis to professional soccer, it was presumed that work-rate could be expressed as distance covered in a game, since this determines the energy expenditure. Activities were coded according to intensity of movements, the main categories being walking, jogging, cruising, sprinting, whilst other game-related activities such as backing, playing the ball and so on were investigated. The observer utilized a learnt map of pitch markings in conjunction with visual cues around the pitch boundaries and spoke into a tape recorder. The method of monitoring activity was checked for reliability, objectivity and validity (Reilly and Thomas, 1976), and is still considered to be the most appropriate way of monitoring one player per game (Reilly, 1990, 1993, 1994a).

Coded commentary of activities on to a tape recorder by a trained observer has been correlated with measurements taken from video recordings. The latter method entails establishing stride characteristics for each subject according to the various exercise intensities. The data can then be translated into distances or velocities of movements. With careful checks on procedures the two methods are in good agreement (Reilly, 1994a). Video recording has been employed in studies of referees during matches where heart rate can be measured throughout the game (Catterall *et al.*, 1993). Stride frequencies can be counted on playback of the video; these can be expressed as distance for each discrete event, provided the stride length for each activity is determined separately for the individual concerned.

An alternative appropriate for data collection is to set the activity profile alongside a time-base. This permits establishment of exercise-to-rest ratios; these can be useful both in designing training drills and in interpreting physiological stresses. This approach is nowadays straightforward, once video systems are linked with computerized methods of handling the observations.

A comprehensive review of methods of motion analysis in soccer has been published elsewhere (Reilly, 1994a). The various methods used have incorporated cine-film of samples of individuals to eventually encompass the whole team, overhead cine-views of the pitch for computer-linked analysis of movements, and synchronized cameras for calculation of activities using trigonometry. Hand-notation methods of recording activities on paper have

Table 5.1 Mean distance covered per game according to various sources[a]

Source	*n*	*Distance covered (m)*	*Method*
English	40	4834	Hand notation
Finnish	7	7100	TV cameras (2)
English	40	8680	Tape-recorder
Japanese	2	9845	Trigonometry (2 cameras)
Swedish	10	9800	Hand notation
Japanese	—	9971	Trigonometry
Belgian	7	10 245	Cine-film
Danish	14	10 800	Video (24 cameras)
Swedish	9	10 900	Cine-film
Czech	1	11 500	Undisclosed
Australian	20	11 527	Videotape
Japanese	50	11 529	Trigonometry

[a] Sources of the data are cited in Reilly, 1994b.

also been used whilst computerized notation analysis, currently utilized for analysing patterns of play, has great potential for producing work-rate information. Whatever method is adopted must comply with quality control specifications.

A summary of overall work-rates reported in the literature (Table 5.1) indicates that outfield players should be able to cover 8–12 km during the course of a match. This is done more or less continuously. The overall distance covered during a game is only a crude measure of work-rate due to the frequent changes in activities. These amount to approaching 1000 different activities in a game, or a break in the level or type of activity once every 6 s (Reilly and Thomas, 1976). The changes embrace alterations in pace and direction of movement, execution of game skills and tracking opponents' movements.

The overall distance covered by outfield players during a match consists of 24% walking, 36% jogging, 20% cruising submaximally (striding), 11% sprinting, 7% moving backwards and 2% moving in possession of the ball. Masked within the broad categories are sideways and diagonal movements. These figures (Figure 5.1) are representative of contemporary play in the English top divisions and seemingly are indicative also of other major national leagues in Europe and at top level in Japan (Reilly, 1994b).

The categories of cruising and sprinting can be combined to represent high-intensity activity in soccer. The ratio of low-intensity to high-intensity exercise is then found to be about 2.2 to 1 in terms of distance covered (Reilly and Thomas, 1976). In terms of time, this ratio is about 7 to 1, denoting a predominantly aerobic outlay of energy. On average each outfield player has a short rest pause of only 3 s every 2 min, though rest breaks are longer and occur more frequently than this at lower levels of play where players are more

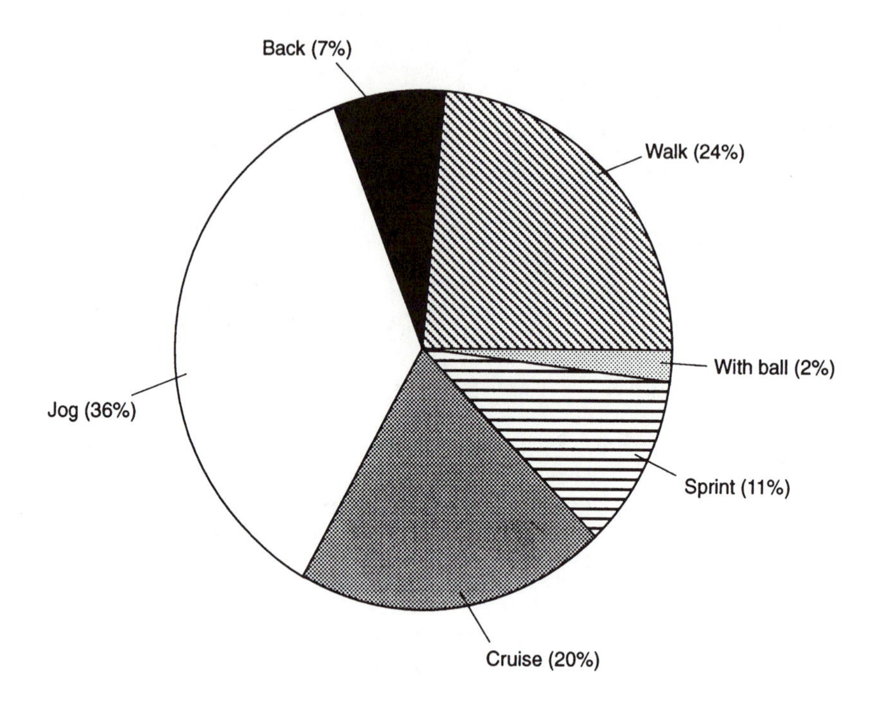

Figure 5.1 Relative distances covered in different categories of activity for outfield players during soccer match-play. (Data from Reilly and Thomas, 1976.)

reluctant to run to support a colleague in possession of the ball. Generally, less than 2% of the total distance covered by top players is with the ball. The vast majority of actions are 'off the ball', either in running to contest possession, support team-mates, track opposing players, execute decoy runs, counter runs by a marking player, jump for the ball or tackle an opponent, or play the ball with one touch only.

Most activity during a game at top level is at a low or submaximal level of exertion but the importance of high-intensity efforts cannot be over-emphasized. Players generally have to run with effort (cruise) or sprint every 30 s but sprint all-out only once every 90 s. The timing of these anaerobic efforts, whether in possession of the ball or without, is crucial since the success of their deployment plays a dominant role in the outcome of the game. Although work-rate profiles are relatively consistent for players from game to game, it is the high-intensity exercise which is the most constant feature (Bangsbo, 1994).

5.2 FACTORS AFFECTING WORK-RATES

The work-rate is determined to a large extent by the positional role of the player. The greatest distances are covered by midfield players who have to act as links between defence and attack. This has been noted in English (Reilly and Thomas, 1976), Swedish (Ekblom, 1986) and Danish (Bangsbo *et al.*, 1991) League matches. In the studies of English League players, the full-backs showed the greatest versatility: although they covered more overall distance than the centre-backs, they covered less distance sprinting. The greatest distances covered sprinting were found in the strikers and midfield players. A characteristic work-rate profile of an English League striker who represented his country in the World Cup Finals of 1990 and 1994 is shown in Figure 5.2.

The greatest overall distance covered by the Danish midfield players was due to more running at low speeds. This denotes an aerobic type of activity profile for the midfield players in particular. A more anaerobic type of profile is found in the centre-back and sweeper or libero. The pace of walking was found to be slower in centre-backs than for any other outfield position. Centre-backs and strikers have to jump more frequently than full-backs or midfield players (Reilly and Thomas, 1976). The frequency of once every 5–6 min denotes that whilst jump endurance may not be as important in soccer as in basketball and volleyball, anaerobic power output and the ability to jump well vertically are requirements for play in central defence and in attack as a 'target player'.

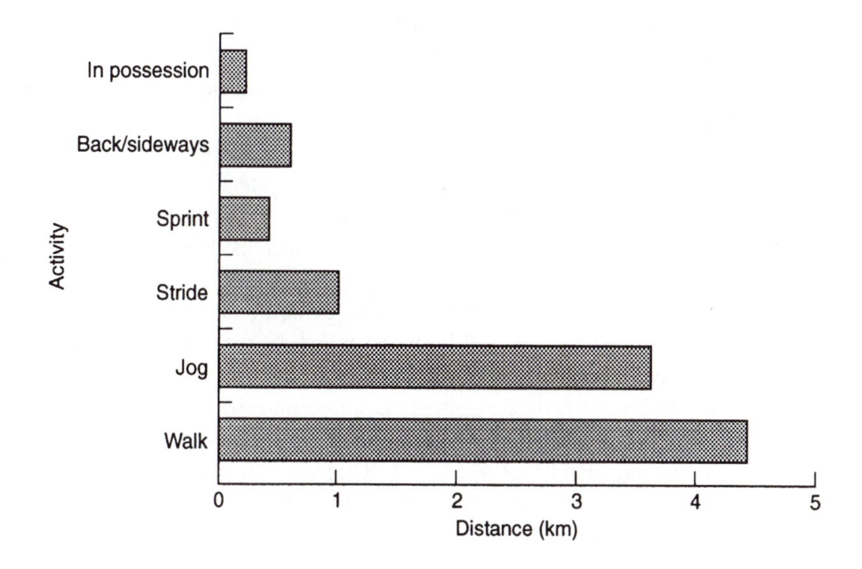

Figure 5.2 Mean distances covered (km) during a soccer match by an international striker playing in an English League match.

The goalkeeper has been observed to cover about 4 km during a match. Time spent standing still is much greater than for outfield players. The work-rate profile emphasizes anaerobic efforts of brief duration when the goalkeeper is involved directly in play. The goalkeeper is engaged in play more than any of the outfield players though the extent of this has been reduced by the rule-changes introduced in 1992 prohibiting back-passes from defenders. This rule has only had a marginal effect on the activities of outfield players.

The ability to sustain prolonged exercise is dependent on a high maximal aerobic power ($\dot{V}O_2max$), but the upper limit at which continuous exercise can be maintained is influenced by the so-called 'anaerobic threshold' and a high fractional utilization of $\dot{V}O_2max$. Soccer play calls for an oxygen uptake corresponding roughly to 75% $\dot{V}O_2max$ (Reilly, 1990), a value likely to be close to the 'anaerobic threshold' of top soccer players. It has been shown that midfield players in the English League have higher $\dot{V}O_2max$ values than players in other outfield positions. The $\dot{V}O_2max$ was found to be correlated significantly with the distance covered in a game, underlining the need for a high work-rate and a high aerobic fitness level, particularly in midfield players (Reilly, 1993). Smaros (1980) confirmed this strong relation between $\dot{V}O_2max$ and distance covered in a game, but noted also that the $\dot{V}O_2max$ influenced the number of sprints the players attempted. Bangsbo and Lindquist (1992) showed that the distance covered was correlated with performance in a continuous field test over 2.16 km with the maximal oxygen uptake, and with the oxygen uptake corresponding to 3 mmol l^{-1} blood lactate level. It seems that work-rate in soccer matches depends on physiological indicators of aerobic fitness as found in distance runners (Jacobs, 1981).

The style of play may influence the work-rates of players. Emphasis on retaining possession, slowing the pace of the game and delaying attacking moves until opportunities to penetrate defensive line-ups are presented place emphasis on speed of movement in such critical phases of the game. Conversely the direct method of play, characteristic of some English clubs, contrasts with the more methodical build-up of offensive plays in European and South American teams. The direct method (known also as 'Route 1'), used by the Republic of Ireland team in the 1988 European Championship and 1990 and 1994 World Cups, raises the pace of the game at all times. The main elements are fast transfer of the ball from defence to attack to create scoring opportunities, use of long passes rather than a sequence of short passes, exploitation of defensive errors, harrying opponents into mistakes when in possession of the ball and midfield players taking turns to support the strikers when on the offensive (Reilly *et al.*, 1991). This style of play has a levelling effect on the work-rate of outfield players since all players are expected to exercise at a high intensity 'off-the-ball'. A similar equalization of aerobic fitness demands applies to the 'total football' style of play as exhibited first by the Netherlands national side in 1974 and characteristic of many top European club sides (for example Milano) today.

Whilst the 'direct method' tends to even out differences in work-rate between playing positions, the Irish style of play evolved to accommodate individual differences. For example, team membership for the 1994 World Cup qualifying matches included some players known for exceptionally high work-rates and one player whose training programme was habitually hampered by chronic injury. This acknowledges that team managers do consider building teams around exceptionally gifted individuals.

Computerized methods for notating movements during play have been developed for describing playing patterns (Hughes, 1988). Variables associated with each player, the position in the field of play and the player's action are entered into this computer by means of a 'concept' keyboard. This comprises a pad with 128 touch-sensitive cells over which a chart illustrating the pitch is superimposed. Information from viewing a video tape of a game is input by attributing codes to each activity on the ball. A comprehensive profile of patterns of play and individual players' contributions to it is built up and illustrated by computer graphics at the end of each game. Comparison of England and Ireland international matches shows the superiority of the latter in fewer lost possessions per touch on the ball (particularly when in possession in defence), a feature of the 'direct method' of play (Reilly *et al.*, 1991). Such observations should be interpreted against the basic background of physiological stress imposed on players by the game strategies adopted.

5.3 FATIGUE

Fatigue can be defined as a decline in performance due to the necessity to continue performing. In soccer it is manifest in a deterioration in work-rate towards the end of a game. Studies which have compared work-rates between first and second halves of matches have provided evidence of the occurrence of fatigue.

Belgian university players were found to cover on average a distance of 444 m more in the first half than in the second half (Van Gool *et al.*, 1988). Bangsbo *et al.* (1991) reported that the distance covered in the first half was 5% greater than in the second. This decrement does not necessarily occur in all players. Reilly and Thomas (1976) noted an inverse relation between aerobic fitness ($\dot{V}O_2max$) and decrement in work-rate. The players with the higher $\dot{V}O_2max$ values, those in midfield and full-back positions, did not exhibit a significant drop in distance covered in the second half. In contrast all the centre-backs and 86% of the strikers had higher figures for the first half, the difference between halves being significant. It does seem that the impact of a high aerobic fitness level is especially evident in the later parts of a match.

The amount of glycogen stored in the thigh muscles pre-match appears to have an important protective function against fatigue. Swedish club players with low glycogen content in the vastus lateralis muscle were found to cover

25% less overall distance than the other players (Saltin, 1973). A more marked effect was noted for running speed; those with low muscle glycogen stores pre-match covered 50% of the total distance walking and 15% at top speed compared to 27% walking and 24% sprinting for players who started with high muscle glycogen concentrations. Attention to diet and maintaining muscle glycogen stores by not training too severely are recommended in the immediate build-up for competition. These considerations would be most important in deciders where drawn matches are extended into 30 minutes extra time.

Youth players of a professional club showed positive responses to consuming a maltodextrin solution during training. The subjective assessments of coaches of their players' performance corroborated the judgements of the players (Miles *et al.*, 1992). Dietary advice given to the senior professionals resulted in an alteration of nutritional support. The distribution of macronutrients in the diet of the players also improved (Reilly, 1994b). Manipulation of energy intake by provision of a high carbohydrate diet improved performance in a running test designed to interpret the activity profile of a soccer match (Bangsbo *et al.*, 1992). Whilst goals may be scored at any time during a game, most are scored towards the end of a game. This is exemplified by data from the Scottish League during an extended period of the 1991–92 season (Figure 5.3). A higher than average scoring rate occurred in the final 10 minutes of play. This cannot be explained simply by a fall in work-rate, as logically this would be balanced out between the two opposing teams. It might be accounted for by the more pronounced deterioration among defenders which gives an advantage to the attackers towards the end of a game. Alternatively it may be linked with 'mental fatigue', lapses in concentration as a consequence of sustained physical effort leading to tactical errors that open up goal-scoring chances. The phenomenon may be a factor inherent in the

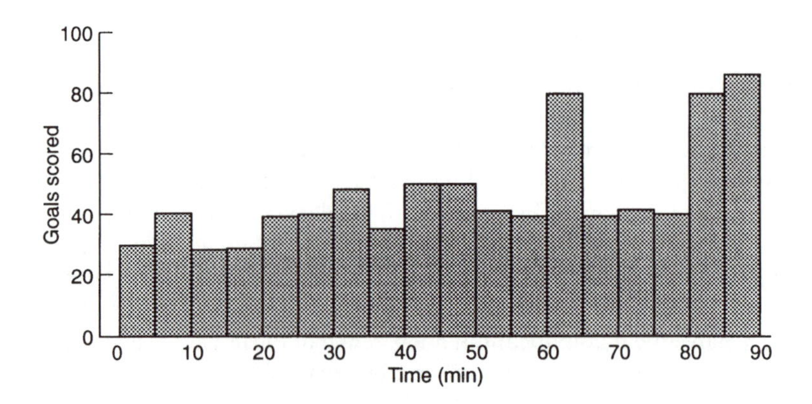

Figure 5.3 Time (min) at which goals were scored in a sample of 283 Scottish League soccer matches played in the 1991–2 season.

game, play becoming more urgent towards the end despite the fall in physical capabilities. Irrespective of the nature of the phenomenon, a team that is physiologically and tactically prepared to last 90 minutes of intense play is likely to be an effective unit.

Environmental conditions may also impose a limit on the exercise intensity that can be maintained for the duration of a soccer game or hasten the onset of fatigue during it. Major soccer tournaments, for example the World Cup finals in Spain in 1982, in Italy in 1990 and in the USA in1994, have been held in hot countries with ambient temperatures above 30 °C. The work-rate is adversely affected when hot conditions are combined with high humidity. Performance is influenced both by the rise in core temperature and dehydration, and sweat production will be ineffective for losing heat when relative humidity is 100%. Cognitive function, akin to the kind of decision-making required during match-play, is better maintained during 90 minutes of continuous exercise when water is supplied intermittently to subjects compared to a control condition (Reilly and Lewis, 1985). Adequate hydration pre-exercise and during the intermissions is important when players have to play in the heat. The opportunity to acclimatize to heat prior to competing in tournaments in hot climates is an essential element in the systematic preparation for such events. This may be realized by astute location of training camps, a good physiological adaptation being realized within 10–14 days of the initial exposure in hot weather or regular and frequent exposures to heat in an environmental chamber.

The major consequences of playing in cold conditions are likely to be associated with liability to injury. This would be pronounced when playing on icy pitches without facilities for underground heating. Muscle performance deteriorates as muscle temperature falls; therefore a good warm-up prior to playing in cold weather and use of appropriate sportswear to maintain warmth and avoid the deterioration in performance synonymous with fatigue would be important. It is also established that injury is more likely to occur in games players if the warm-up routine is inappropriate (Reilly and Stirling, 1993). Therefore, pre-match exercises should engage the muscle groups employed during the game, particularly in executing soccer skills.

The interactions between environmental variables and soccer performer are covered more extensively elsewhere in this volume. A consensus statement of nutritional needs of the soccer player and guidelines for fluid replacement to offset work-rate deterioration are outlined by Ekblom and Williams (1994).

5.4 PHYSIOLOGICAL RESPONSES TO MATCH-PLAY

The relative metabolic loading during soccer play could be indicated if direct measurements were available for both energy expenditure during competition and the maximal aerobic power ($\dot{V}O_2max$). Direct measurements made from collections of expired air in Douglas bags have indicated energy expenditure rates of 22–44 kJ/min (Covell *et al.*, 1965) and 32.3 kJ/min (Yamaoka, 1965). These values are likely to be underestimates due to the restrictions placed on the players by the apparatus and also to the low skills of the subjects used in these investigations. Higher values than these were reported by Seliger (1968a, 1968b) for Czech players, mean figures being 54.8 kJ/min for energy expenditure and 76.0 l/min for minute ventilation. The $\dot{V}O_2$ of 35.5 ml kg^{-1} min^{-1} is in close agreement with figures of 35–38 and 29–30 ml mg^{-1} min^{-1} for two Japanese players (Ogushi *et al.*, 1993). These approaches to data collection are likely to have hampered the activities of the players. A lightweight telemetric system for measuring oxygen concentrations, such as the K2 device (Kawakami *et al.*, 1992), is an improvement. The original devices lacked the facility for recording CO_2 but contemporary designs can measure O_2 and CO_2. An alternative research strategy has been to measure heart rate during match-play and juxtapose the observations on heart rate–$\dot{V}O_2$ regression lines determined during running on a treadmill. The error involved in using this method of estimating energy expenditure is small (Bangsbo, 1994). Allowing for any imperfections in such extrapolations from laboratory to field conditions, the heart rate is a useful indicator of the overall physiological strain during play.

Traditionally long-range radio telemetry has been employed to monitor heart rate data during friendly matches or simulated competitions. In recent years the use of short-range radio telemetry (Sport-Tester) has been adopted. Observations generally confirm that the circulatory strain during match-play is relatively high and does not fluctuate greatly during a game. The variability increases in the second half of play at university level (Figure 5.4), as the player takes more rest periods (Florida-James and Reilly, 1995). Rohde and Espersen (1988) reported that the heart rate was about 77% of the heart rate range (maximal minus resting heart rate) for 66% of the playing time. For the larger part of the remaining time the heart rate was above this level.

The heart rate during soccer varies with the work-rate and so may differ between playing positions and between first and second half. Van Gool *et al.* (1983) reported mean figures of 155 beats/min for a centre-back and for a full-back, 170 beats/min for a midfield player, and 168 and 171 beats/min for two forwards. This pattern was closely related to the distances covered by the players in a match. The same research group reported mean values for a Belgian university team during a friendly match of 169 beats/min in the first half and 165 beat/min in the second half. Again, the physiological responses reflected a drop in the work-rate during the second half. These trends have been confirmed in matches played by English university teams (Florida-James and Reilly, 1995).

The heart rate during soccer has been employed in several reports to estimate the relative metabolic loading during match-play (see Table 5.2). Most estimates are that the exercise intensity during soccer is about 75–80% $\dot{V}O_2max$ (Reilly, 1990). Whilst the limitations of extrapolating from labora-

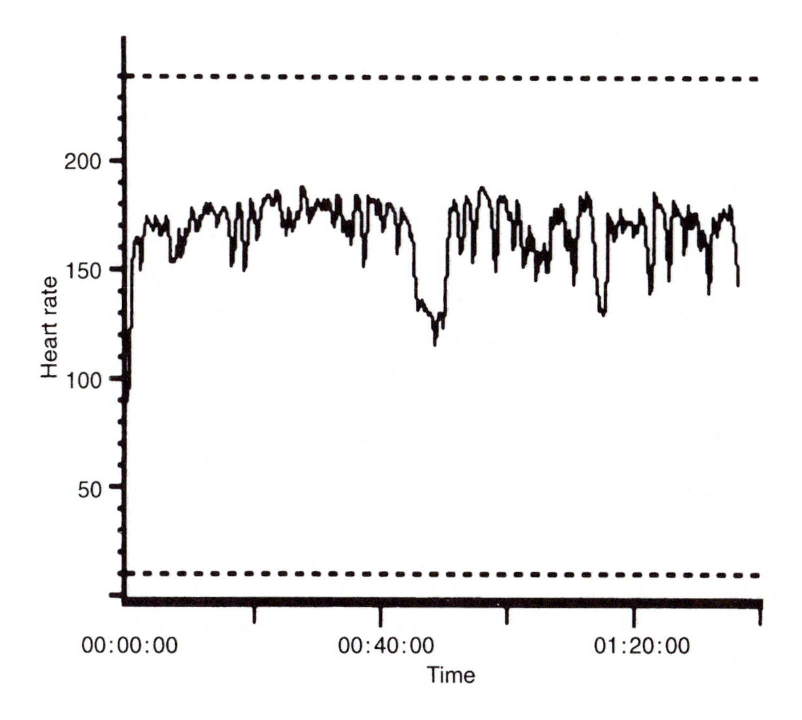

Figure 5.4 Mean heart rate (beats/min) during a whole game for a university player.

tory to field data, using HR–$\dot{V}O_2$ regression lines, suggest that this figure may represent an overestimate, comprehensive calculations indicate this error is not very large (Bangsbo, 1994).

The severity of exercise can also be indicated by measurements of blood lactate concentrations. Progressively higher lactates have been observed in matches from the fourth to the top division in the Swedish League (Ekblom, 1986). Gerisch *et al.* (1988) demonstrated that higher blood lactate levels are associated with man-to-man marking roles compared with a zone-coverage responsibility. Ekblom (1986) claimed that peak values above 12 mmol l^{-1} were

Table 5.2 Mean values for heart rate (beats/min) during soccer

Series	*HR*	*Match-play situation*
Seliger (1968a)	160	Model 10 min game
Seliger (1968b)	165	Model 10 min match
Reilly (1986)	157	Training matches
Ogushi *et al.* (1993)	161	Friendly match (90 min)
Ali and Farrally (1991)	169	Friendly match (90 min)
Florida-James and Reilly (1995)	161	Competitive game (90 min)

Table 5.3 Mean (+ s.d.) blood lactate concentrations (mmol l^{-1}) during soccer

1st Half	*2nd Half*	*Source*
5.1 ± 1.6	3.9 ± 1.6	Rohde and Espersen (1988)
5.6 ± 2.0	4.7 ± 2.2	Gerisch *et al.* (1988)
4.9 —	3.7 —	Bangsbo *et al.* (1991)
4.4 ± 1.2	4.5 ± 2.1	Florida-James and Reilly (1995)

frequently measured at the higher levels of soccer play. Activity could not be sustained continuously under such conditions which reflect the intermittent consequences of anaerobic metabolism during competition. Whilst most studies of blood lactate concentration have shown values of 4–6 mmol^{-1} during play (Table 5.3), such measures are determined by the activity in the 5 minutes prior to obtaining the blood samples. Consequently, higher values are generally noted when observations are made at half-time compared to the end of the match.

Whilst muscle glycogen is the main fuel for exercise during match-play, muscle triglycerides, liver glycogen and blood-borne free fatty acids (FFA) are also used for oxidative metabolism. The increase in FFA is most pronounced in the last 15 min of a game (Bangsbo, 1994). This corresponds to an elevation in catecholamines which increase lipolysis in adipose tissue and raise the concentration of fatty acids in the blood. The plasma concentrations of catecholamines are higher in the second half than in the first half of a game and their effects are enhanced by a rise in growth hormone. Without these contributions from fat as a source of fuel, the available glycogen would be depleted before the end of play. It is likely that there is some contribution to energy production from protein sources, particularly branched chain amino acids, but this is probably small (Wagenmakers *et al.*, 1989).

5.5 GAME-RELATED ACTIVITIES

The distance covered in a game under-represents the energy expended because the extra demands of game skills are not accounted for. These include the frequent accelerations and decelerations, angular runs, changes of direction, jumps to contest possession, tackles, avoiding tackles and the many aspects of direct involvement in play. Some attempts have been made to quantify the additional physiological demands of game skills over and above the physiological cost of locomotion.

Dribbling the ball is an example of a game skill amenable to physiological investigation in a laboratory context. Reilly and Ball (1984) examined physiological responses to dribbling a soccer ball on a treadmill at speeds of 9, 10.5, 12 and 13.5 km/h, each for 5 minutes. A rebound box on the front of the treadmill returned the ball to the player's feet after each touch forward. The

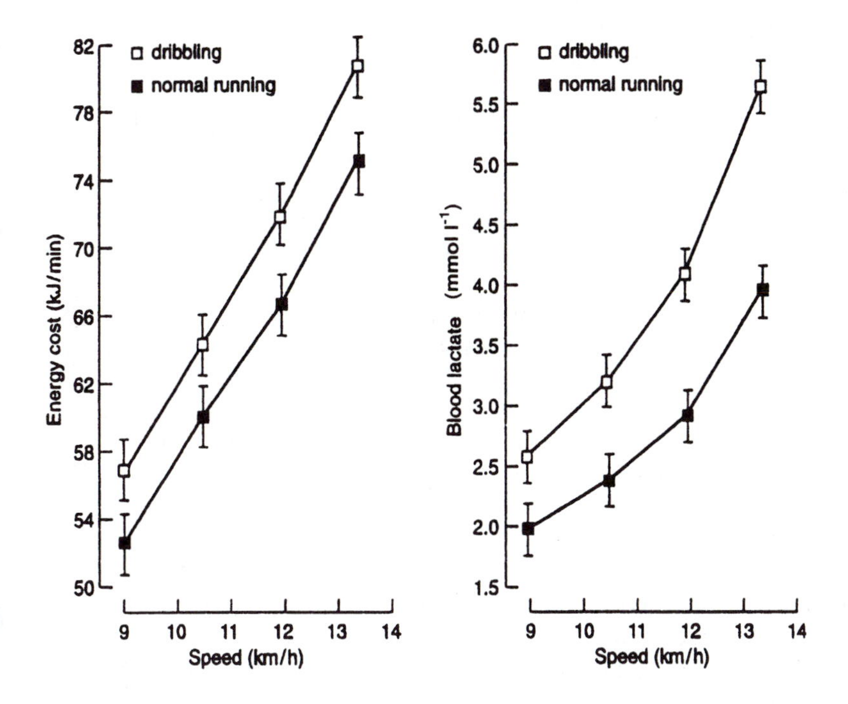

Figure 5.5 Physiological responses to running at different speeds are higher when dribbling a ball than in normal running. (Data from Reilly and Ball, 1984. Reproduced with permission from the American Alliance for Health, Physical Education, Recreation and Dance, Reston, VA 22091.)

procedure allowed precise control over the player's activity while expired air, blood lactate level and perception of effort were measured. The energy cost of dribbling, which entailed one touch of the ball every two to three full stride cycles, was found to increase linearly with the speed of running. The added cost of dribbling was constant at 5.2 kJ/min (Figure 5.5). This value is likely to vary in field conditions according to the closeness of ball control the player exerts.

When dribbling with tight control of the ball, the stride rate increases and the stride length shortens compared with normal running at the same speed; these changes are likely to contribute to the additional energy cost. Increasing or decreasing the stride length beyond that freely chosen by the individual causes the oxygen consumption for a given speed to increase. The energy cost may be further accentuated in matches as the player changes stride characteristics irregularly or feigns lateral movements whilst in possession of the ball in order to outwit an opponent. A reduction in stride length when dribbling is perhaps needed to effect controlled contact with the ball and propel it forward with the right amount of force by the swinging leg. The muscle activity required for kicking the ball and the action of synergistic and stabilizing muscles to facilitate balance while the kicking action is being executed are also likely to contribute to the added energy cost.

Perceived exertion rises while dribbling in parallel with the elevation in metabolism (Reilly and Ball, 1984). All-out efforts are likely to be limited by attaining the ceiling in perceived exertion and so top running speeds may not be attained in dribbling practices unless the frequency of ball contact is reduced. This in effect is done on occasions when a player kicks the ball ahead to allow himself space to accelerate past a stationary or more slowly moving opponent.

Blood lactate levels are also elevated as a consequence of dribbling the ball, the increased concentrations being disproportionate at the high speeds (see Figure 5.5). In the study of Reilly and Ball (1984), the lactate inflection threshold was estimated to occur at 10.7 km/h for dribbling but not until 11.7 km/h in normal running. This indicates that the metabolic strain of fast dribbling will be underestimated unless the additional anaerobic loading is considered.

Kawakami and co-workers (1992) measured $\dot{V}O_2$ using a telemetric system (K2, Roma) which players wore as a back-pack whilst executing game-related drills. The highest $\dot{V}O_2$ was 4.0 l/min whilst dribbling, though it varied between 2.0 and 4.0 l/min for drills such as 1 vs 1 and 3 vs 1 practices.

There are other game-related activities that require unorthodox movements. About 16% of the distance covered by players in a game is in moving backwards or sideways. The percentage is highest in defenders who may, for example, have to back up quickly under high kicks forward from the opposition's half or move sideways in jockeying for position prior to tackling. The added physiological costs of unorthodox directions of movements have been examined by getting nine soccer players to run on a treadmill at speeds of 5, 7 and 9 km/h, running normally, running backwards and running sideways (Reilly and Bowen, 1984). The extra energy cost of the unorthodox modes of running increased disproportionately with speed of movement.

Table 5.4 Mean (± s.d.) for energy expended (kJ/min) and ratings of exertion at three speeds and three directional modes of motion ($n = 9$)

Speed (km/h)	*Forwards*	*Backwards*	*Sideways*
		Energy expended	
5	37.0 ± 2.6	44.8 ± 6.1	46.6 ± 3.2
7	42.3 ± 1.7	53.4 ± 3.5	56.3 ± 6.1
9	50.6 ± 4.9	71.4 ± 7.0	71.0 ± 7.5
		Perceived exertion	
5	6.7 ± 0.1	8.6 ± 2.0	8.7 ± 2.0
7	8.0 ± 1.4	11.2 ± 2.9	11.3 ± 3.2
9	10.2 ± 2.1	14.0 ± 2.0	13.8 ± 2.5

Source: Reilly and Bowen, 1984.

Running backwards and running sideways did not differ in terms of energy expenditure or ratings of perceived exertion (Table 5.4). Clearly improving the muscular efficiency in these unorthodox modes of movement would benefit the player.

Summary

The physiological responses of soccer players to match-play denote that a combination of demands is imposed on soccer players during competition. The critical phases of play for an individual call for anaerobic efforts but these are superimposed on a background of largely aerobic sub-maximal activities. The intermittent and acyclical nature of activity during competition means that it is difficult to model game-related protocols in laboratory experiments. It is likely that field studies with a greater specificity to the game will be employed more in future investigations of the physiology of soccer. The work-rate and activity profiles can be used to design appropriate training protocols to optimize fitness and ensure that performance during play is enhanced. Whilst physiological considerations have a place in a systematic preparation for competition, performance ultimately depends on the quality with which individual skills and team tactics are executed.

REFERENCES

Ali, A. and Farrally, M. (1991) Recording soccer players' heart rates during matches. *Journal of Sports Sciences*, **9,** 183–9.

Bangsbo, J. (1994) The physiology of soccer: with special reference to intense intermittent exercise. *Acta Physiologica Scandinavica*, **150,** Suppl. 619.

Bangsbo, J. and Lindquist, F. (1992) Comparison of various exercise tests with endurance performance during soccer in professional players. *International Journal of Sports Medicine*, **13,** 125–32.

Bangsbo, J., Norregaard, L. and Thorso, F. (1991) Activity profile of professional soccer. *Canadian Journal of Sports Science*, **16,** 110–16.

Bangsbo, J., Norregaard, L. and Thorso, F. (1992) The effect of carbohydrate diet on intermittent exercise performance. *International Journal of Sports Medicine*, **14,** 207–13.

Catterall, C., Reilly, T., Atkinson, G. and Coldwells, A. (1993) Analysis of the work-rates and heart rates of association football referees. *British Journal of Sports Medicine*, **27,** 153–6.

Covell, B., El Din, N. and Passmore, R. (1965) Energy expenditure of young men during the weekend. *Lancet*, **1,** 727–8.

Ekblom, B. (1986) Applied physiology of soccer. *Sports Medicine*, **3,** 50–60.

Ekblom, B. and Williams, C. (1994) Nutrition and football. *Journal of Sports Sciences*, **12,** Suppl. 1.

Florida-James, G. and Reilly, T. (1995) The physiological demands of Gaelic football. *British Journal of Sports Medicine,* **29,** 41–5.

Gerisch, G., Rutemöller, E. and Weber, K. (1988) Sports medical measurements of performance in soccer, in *Science and Football* (eds T. Reilly, A. Lees, K. Davids and W. Murphy), E. & F.N. Spon, London, pp. 60–7.

Hughes, M. (1988) Computerised notation analysis in field games. *Ergonomics*, **31,** 1585–92.

Jacobs, I. (1981) Lactate, muscle glycogen and exercise performance in man. *Acta Physiologica Scandinavica*, Suppl. 495.

Kawakami, Y., Nozaki, D., Matsuo, A. and Fukunaga, T. (1992) Reliability of measurement of oxygen uptake by a portable telemetric system. *European Journal of Applied Physiology*, **65,** 409–14.

Miles, A., MacLaren, D. and Reilly, T. (1992) The efficacy of a new energy drink: a training study. Communication to the Olympic Scientific Congress, Benalmadena, Spain, 14–19 July.

Ogushi, T., Ohashi, J., Nagahama, H. *et al.* (1993) Work intensity during soccer match-play (a case study), in *Science and Football II* (eds T. Reilly, J. Clarys and A. Stibbe), E. & F.N. Spon, London, pp. 121–3.

Reilly, T. (1986) Fundamental studies in soccer, in *Sportspielforschung: Diagnose Prognose*, (eds H. Kasler and R. Andresen), Verlag Ingrid Czwalina, Hamburg, pp. 114–20.

Reilly, T. (1990) Football, in *Physiology of Sports*, (eds T. Reilly, N. Secher, P. Snell and C. Williams), E. & F.N. Spon, London, pp. 371–425.

Reilly, T. (1993) Science and football: an introduction, in *Science and Football II* (eds T. Reilly, J. Clarys and A. Stibbe), E. & F.N. Spon, London, pp. 3–11.

Reilly, T. (1994a) Motion characteristics, in *Football (Soccer)* (ed. B. Ekblom), Blackwell Scientific, London, pp. 31–42.

Reilly, T. (1994b) Physiological aspects of soccer. *Biology and Sport*, **11,** 3–20.

Reilly, T. and Ball, D. (1984) The net physiological cost of dribbling a soccer ball. *Research Quarterly for Exercise Sport*, **55,** 267–71.

Reilly, T. and Bowen, T. (1984) Exertional cost of changes in directional modes of running. *Perceptual and Motor Skills* **58,** 49–50.

Reilly, T. and Lewis, W. (1985) Effects of carbohydrate feeding on mental functions during sustained exercise, in *Ergonomics International '85* (eds I.D. Brown, R. Goldsmith, K. Coombes and M.A. Sinclair), Taylor and Francis, London, pp. 700–2.

Reilly, T. and Stirling, A. (1993) Flexibility, warm-up and injuries in mature games players, in *Kinanthropometry IV* (eds W. Duquet and J.A.P. Day), E. & F.N. Spon, London, pp. 119–23.

Reilly, T. and Thomas, V. (1976) A motion analysis of work-rate in different positional roles in professional football match-play. *Journal of Human Movement Studies*, **2,** 87–97.

Reilly, T., Hughes, M. and Yamanaka, K. (1991) Put them under pressure. *Science and Football*, **5,** 6–9.

Rohde, H.C. and Espersen, T. (1988) Work intensity during soccer training and match-play, in *Science and Football* (eds T. Reilly, A. Lees, K. Davids and W. Murphy), E. & F.N. Spon, London, pp. 68–75.

Saltin, B. (1973) Metabolic fundamentals in exercise. *Medicine and Science in Sport and Exercise*, **5,** 137–46.

Seliger, V. (1968a) Heart rate as an index of physical load in exercise. *Scripta Medica, Medical Faculty, Brno University*, **41,** 231–40.

Seliger, V. (1968b) Energy metabolism in selected physical exercises. *Internationale Zeitschrift für Angew Physiologie*, **25,** 104–20.

Smaros, G. (1980) Energy usage during football match, in Proceedings of the 1st International Congress on Sports Medicine Applied to Football, vol. 11 (ed. L. Vecchiet), D. Guanello, Rome, pp. 795–801.

Van Gool, D., Van Gerven, D. and Boutmans, J. (1983) Heart rate telemetry during a soccer game: a new methodology. *Journal of Sports Science*, **1,** 154.

Van Gool, D., Van Gerven, D. and Boutmans, J. (1988) The physiological load imposed on soccer players during real match-play, in *Science and Football* (eds T. Reilly, A. Lees, K. Davids and W. Murphy), E. & F.N. Spon, London, pp. 51–9.

Wagenmakers, A.J.M., Brookes, J.H., Conley, J.H. *et al.* (1989) Exercise-induced activation of the branched-chain 2-oxo acid dehydrogenase in human muscle. *European Journal of Applied Physiology*, **59,** 159–67.

Yamaoka, S. (1965) Studies on energy metabolism in athletic sports. *Research Journal of Physical Education*, **9,** 28–40.

Nutrition

6

Don MacLaren

Introduction

Soccer may be considered an endurance sport, incorporating periods of intense exercise interspersed with lower levels of activity over 90 minutes. The estimated energy requirements for a soccer game embracing both casual recreational play and top-class professional games are between 21 and 73 kJ (5–17 kcal) per minute (Reilly, 1990). For a 70 kg player the result could be the loss of approximately 100–200 g of carbohydrate. Since the body stores of carbohydrate are limited (approximately 300–400 g), this loss is significant. If muscle stores of carbohydrate are not adequately replenished, then subsequent performance will be impaired. The carbohydrate intake of elite soccer players is often inadequate and so the concentrations of carbohydrate in active muscle may become low.

The energy demands of soccer are such that there is likely to be a significant production of heat within the body. Even in cold conditions, considerable amounts of sweat are lost in an attempt to dissipate this heat, thus resulting in a degree of dehydration (Maughan, 1991). A mild degree of dehydration will impair skilled performance and affect strength, stamina and speed. An adequate fluid intake is necessary to offset the effects of dehydration.

Despite the fact that the major causes of fatigue for soccer players are the depletion of muscle glycogen stores and dehydration, players are forever looking for nutritional supplements to help improve their performance and aid recovery. Vitamins and minerals are the legal products

Science and Soccer. Edited by Thomas Reilly. Published in 1996 by E & FN Spon, London. ISBN 0 419 18880 0.

that players may consider, although substances such as creatine, sodium bicarbonate, caffeine, and alcohol will be briefly discussed.

In this chapter the nutritional requirements of soccer players are considered in terms of energy, carbohydrates, fluid intake and vitamins. A brief examination is also undertaken of some ergogenic substances. Where possible, references will be made to research in soccer, and recommendations stated for use before, during and after training or competition.

6.1 ENERGY

6.1.1 Sources of muscular energy

Human energy provides the basis for movement in all sports, and any successful performance depends on the ability of the athlete to produce the right amount at the right time. Sports differ in their energy requirements, and thus each sport imposes specific energy demands on the athlete. Soccer entails multiple sprints, yet is an endurance sport. Consequently the ability to generate energy rapidly as for a sprint is as necessary as the ability to generate sustained energy over the 90 minutes of a match.

Adenosine triphosphate (ATP) is the only usable form of energy in muscle contractions, where the ATP is broken down to ADP by the action of enzymes (ATPases) at the cross-bridge heads of muscle filaments:

$$\text{ATP} \rightarrow \text{ADP} + \text{Pi} + \text{Energy (30.5 kJ/mol)}$$

Adenosine triphosphate (ATP) is a high energy compound which is used as an immediate source of energy for muscle activity. During intense activity such as in a sprint, the ATP stores will be used first. The ATP stores are in very limited supply and are capable of providing energy for only a few seconds. Another related high-energy source, creatine phosphate (CP), is also found in skeletal muscle in small amounts. Although it cannot be used as an immediate source of energy, it can rapidly replenish ATP:

$$\text{CP} \rightarrow \text{Cr} + \text{Pi} + \text{Energy (43.1 kJ/mol)}$$
$$\text{CP} + \text{ADP} \rightarrow \text{ATP} + \text{Cr} + \text{Energy (12.6 kJ/mol)}$$

Creatine phosphate and ATP are known as the phosphagen stores, and are the rapid response energy systems for muscular activity. Together they are capable of supplying sufficient energy for about 8–10 s of high-intensity exercise. They are sometimes referred to as the ATP–PC system.

Because ATP and CP are found in very small amounts within skeletal muscle and can be used up in a matter of seconds, it is necessary to have alternative stores of energy. Carbohydrate, fat and protein comprise the other

Table 6.1 Energy stores for a 70 kg man (15% body fat)

Energy source	*Amount (g)*	*Energy (kJ)*
Fat (adipose tissue)	10 500	378 000
Muscle glycogen	350	5 600
Liver glycogen	100	1 600
Blood glucose	20	320
Protein	9 000	153 000

energy stores and can provide the body with enough ATP to last for many weeks of starvation. Table 6.1 summarizes the amount of energy stores in the body for a 70 kg person.

If the exercise is continued for more than a few seconds, a further energy system is brought into action. This system has been called the lactic acid system because lactic acid is produced as a consequence. Glycogen, from the muscle's store, is used rather than blood glucose because it is more readily available. The glycogen is broken down to release glucose-6-phosphate molecules which are then further broken down via glycolysis to produce pyruvic acid. Under normal steady-state conditions the pyruvic acid is oxidized to carbon dioxide and water, producing quite a lot of energy in the process. Under conditions of intense exercise, where the rate of pyruvic acid production is greater than its oxidation, or conditions where oxygen is inadequate, the pyruvic acid is converted to lactic acid:

Glycogen → G-6-P → → → → Pyruvic acid
Pyruvic acid → Lactic acid

If the exercise is prolonged and sustained at a steady state, then the complete oxidative breakdown of pyruvic acid is possible:

Pyruvic acid → Carbon dioxide + Water + Energy

The amount of energy produced from the breakdown of a glucose molecule to lactic acid results in the formation of 3 ATP molecules whereas complete oxidation results in the formation of 39 ATPs. The usefulness of the lactic acid system is that it is switched on very rapidly (within 1 second), so promoting ATP production in times where the phosphagens may be compromised and other energy stores are incapable of being used. Almost 50% of the energy during intense 6 s bouts of sprinting is derived from the breakdown of glycogen and the resultant build-up of lactic acid (Boobis *et al.*, 1987). Of course the end result is an accumulation of lactic acid which decreases the pH of the muscle cell and results in fatigue (MacLaren *et al.*, 1989).

Other sources of energy during prolonged steady-state exercise include glucose from the liver transported via blood, fatty acids from triglyceride stores in adipose tissue, and amino acids from a variety of tissues:

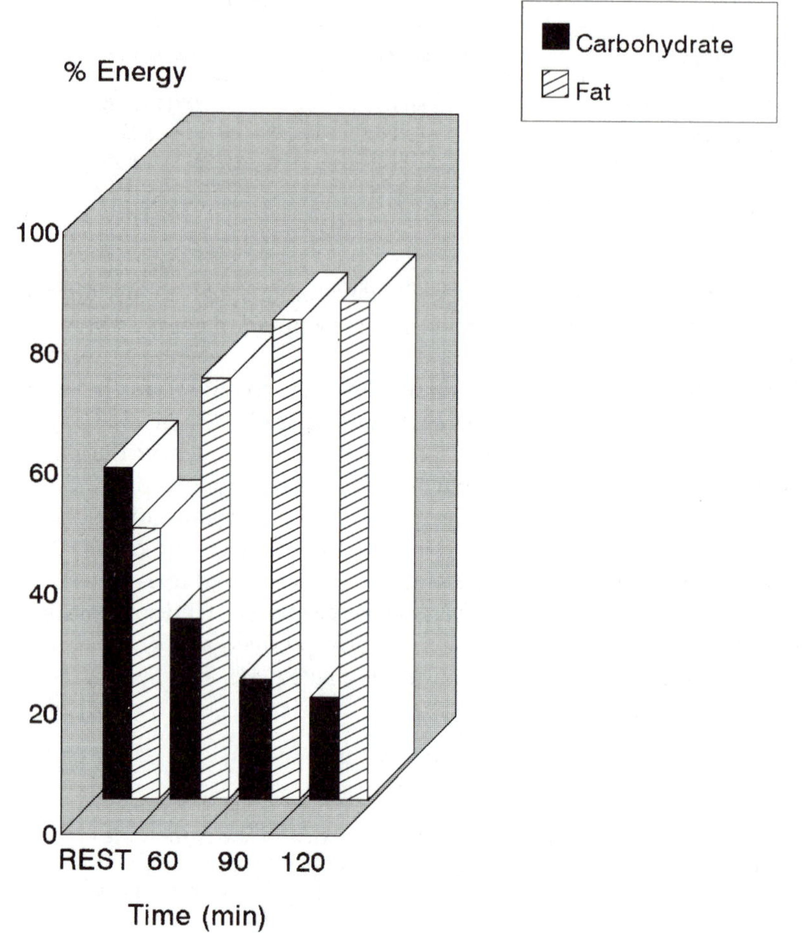

Figure 6.1 Relative contribution of carbohydrate and lipid as energy sources during prolonged moderately intense exercise.

Glucose → Carbon dioxide + Water + Energy
Fatty acid → Carbon dioxide + Water + Energy
Amino acid → Carbon dioxide + Water + Energy
Amino acid → Glucose → Carbon dioxide + Water + Energy

The relative contribution of these sources for energy supply to the muscle during prolonged exercise can be seen in Figure 6.1.

6.1.2 Nutritional sources

Carbohydrates and fats are the principal sources of energy for muscular activity in humans, although energy may also be obtained from protein and from alcohol. The amounts of energy contained within 1 gram of these nutrients are:

Carbohydrates	16 kJ (3.75 kcal)
Fats	37 kJ (9 kcal)
Protein	17 kJ (4 kcal)
Alcohol	29 kJ (7 kcal)

The data above clearly illustrate the advantages of fats as an energy store compared to carbohydrates. This advantage is reinforced on realizing that carbohydrates are stored bound with water whereas fats are not; three molecules of water are stored with every one molecule of glycogen.

Although alcohol appears to be a potentially useful source of energy, it can only be metabolized by the liver and not by muscle. It is therefore of no ergogenic benefit to the soccer player and is more likely to have adverse effects on performance.

Carbohydrates

Carbohydrates are composed of carbon, hydrogen and oxygen, of which the simplest form are monosaccharides such as glucose, fructose and galactose. They are also known as the simple sugars. Disaccharides are made up from two monosaccharide molecules, e.g. sucrose (table sugar) is made from glucose and fructose, maltose is made from glucose and glucose. The complex carbohydrates are polysaccharides of which starch is the plant storage form and glycogen the animal storage form. All the carbohydrates contain approximately the same amount of energy, i.e. 16 kJ/g. Table 6.2 provides examples of the types of food high in carbohydrates which are either complex carbohydrates or simple sugars, whilst Table 6.3 illustrates the carbohydrate foods based on their glycaemic index. The latter is an index of the ability of the food to raise blood glucose levels; a high glycaemic index is indicative of a food being digested rapidly and absorbed as glucose, and thus elevating blood

Table 6.2 Foods containing 100 g of either complex carbohydrates or simple sugars

Complex carbohydrates	*Simple sugars*
400 g white bread	100 g sugar
250 g wholemeal bread	150 g chocolate bar
400 g boiled pasta	130 g honey
400 g baked potato	150 g jam
300 g boiled rice	140 g shortcake biscuit
130 g crackers	170 g fruit cake
120 g Cornflakes	1 litre Coca-Cola
150 g Weetabix	2 litres lemonade
750 g baked beans	

Table 6.3 Classification of some foods based on their glycaemic index

High index	*Moderate index*	*Low index*
Bread	Pasta	Apples
Potato	Noodles	Beans
Rice	Crisps	Lentils
Sweetcorn	Grapes	Milk
Raisins	Oranges	Ice cream
Banana	Porridge	Yoghurt
Cereals	Sponge cake	Soup
Glucose		Fructose

glucose. It should be noted that simple sugars such as fructose do not have a high glycaemic index because they do not elevate blood glucose.

Fats

Fats are composed essentially of carbon, hydrogen and oxygen. Fat is stored in the form of triglycerides, which consist of a glycerol and three fatty acid molecules. They may be found in adipose tissue, between muscles, and within muscle cells. During exercise triglycerides in adipose tissue are broken down to glycerol and fatty acids, which are then transported to the muscles for oxidation in the mitochondria. Fatty acids have been classified as being saturated or unsaturated depending on whether the bonds between the carbon atoms are saturated or not. Saturated fatty acids are generally of animal origin and are solid at room temperature whereas unsaturated fatty acids are mainly from plant origin and are liquid at room temperature. From a health perspective it is advisable to consume fats which are unsaturated rather than fats which are saturated (suggestions are that 60% of fat intake should be in the form of unsaturated fatty acids).

Protein

Proteins are made up of 20 or so naturally occurring amino acids. These molecules consist of carbon, hydrogen, oxygen and nitrogen. It is the nitrogen that has to be broken and eliminated as urea if taken in excess. Proteins are essential in the body not only as structural components but also because all enzymes are proteins; many of the hormones that integrate metabolism are proteins, and so are large macromolecules such as haemoglobin. The human body has a minimum requirement of protein in quantity and also in quality. The requirement with regard to quality is concerned with the fact that out of the 20 amino acids which make up all the proteins in our body, eight of them are essential. The eight essential amino acids are needed in small amounts regularly since we are incapable of making them ourselves, and a deficiency

Table 6.4 Essential amino acids and foods high in protein

Essential amino acids	*Foods high in protein*
Histidine (children only)	Milk (whole or skimmed)
Isoleucine	Cheese and cottage cheese
Leucine	Yoghurt
Lysine	Meat
Methionine	Fish
Phenylalanine	Eggs
Threonine	Beans/peas/lentils
Tryptophan	Peanuts/soy products
Valine	Broccoli/sprouts

will ultimately lead to death. Table 6.4 shows the essential amino acids, and the foods rich in proteins. Essential amino acids can be found in meat, fish and some dairy products, as well as in plant products such as cereals, pulses and nuts.

6.1.3 Recommended energy intake

Among the most important questions asked of nutritionists are those concerned with how much food needs to be consumed per day, and what type of food should be eaten in order to remain healthy. The Department of Health and Social Security (DHSS) in the United Kingdom provides information regarding this in its Recommended Intakes of Nutrients. The information presented contains the approximate amounts of energy needed to be consumed in order to balance the energy expended, as well as the protein, mineral and vitamin requirements in order to lead a healthy existence (DHSS, 1979).

The energy requirements are based on the recommendations of a committee whose members examined the average energy expended by individuals of different sexes, varying ages and having different activity levels. For example, a 25-year-old man with a sedentary lifestyle is recommended to consume about 10 500 kJ (2500 kcal), whereas a man of similar age but who is considered very active should consume about 14 000 kJ (3333 kcal). Older people are recommended to have a lower energy intake than younger individuals, and males to have a higher energy intake compared with females of the same age and lifestyle.

The approximate energy intake of a 20–30-year-old elite soccer player is likely to be 8500–16 500 kJ (2033–4000 kcal) per day (Clarke, 1994). Lower values may reflect off-season periods whereas higher values are likely to be found during periods of training or match-play. Any significant variations from this resulting in either an excessive intake or reduced intake could lead to overweight or undernutrition, respectively. Clearly there are times when the soccer player may need to increase energy intake, such as during intense pre-season training, or to reduce energy intake, such as during periods of comparative inactivity in recovering from injury or surgery.

General recommendations regarding the contribution of carbohydrates, fats, and protein to the energy intake are as follows:

Carbohydrate	55–60%
Fat	25–30%
Protein	10–15%

This means that a soccer player whose energy intake is 16 MJ (16 000 kJ) should be consuming 550–600 g of carbohydrate (55 or 60% of 16 000 ÷ 16), 108–130 g of fat (25 or 30% of 16 000 ÷ 37), and 94–141 g of protein (10 or 15% of 16 000 ÷ 17). In what form should these nutrients be consumed? Recommendations based on health requirements suggest that about 60% of the carbohydrates should be in the form of complex carbohydrates, and 60% of the fats in the form of unsaturated fats. With respect to protein, the requirements for the essential amino acids should be met. This can be achieved by eating a variety of foods containing protein; vegans (who avoid meat, milk and dairy products) need to pay special attention to the correct mix of pulses and legumes in their diet.

Recommendations have been made regarding the minimum amounts of protein to be consumed per day, and these have stated a value amounting to 0.7 g/kg body weight. For a 70 kg person this would be 49 g of protein. Since the Western diet normally contains 15% energy intake from protein, an individual would have to eat less than 5470 kJ (1300 kcal) per day in order not to meet the recommendations; the value would be less than 8400 kJ (2000 kcal) if 10% of the energy intake was in the form of protein. A contribution of less than 10% energy intake from protein is unlikely except in cases of dieting, starvation or possibly carbohydrate loading.

The recommendations for protein intake of athletic populations is normally 1.0–2.0 g/kg per day. This higher value reflects the greater protein turnover as a result of increased activity and a loss of amino acids in sweat. The extremely high protein intake of some bodybuilders and weight trainers (in excess of 2.0 g/kg body weight per day) is unlikely to be of any significant benefit unless anabolic steroids are taken in combination. The recommendation for soccer players has been suggested to be in the range 1.4–1.7 g/kg per day (Lemon, 1994).

Dietary surveys of soccer players have highlighted the inadequacies of their dietary intakes with respect to nutrients (Table 6.5). Data collected in our laboratory on energy intakes of elite soccer players show that their carbohydrate intake tends to be too low whilst their fat intake is generally high. The importance of an adequate carbohydrate intake (i.e. 55–60% of total energy intake) cannot be overstressed.

The energy cost of playing First Division English League soccer has been estimated as being 69 kJ (16.4 kcal) per minute (Reilly and Thomas, 1979).

Table 6.5 Nutrient intake of soccer players

Reference	*No. of subjects*	*Energy (kJ)*	*Carbohydrates (%)*	*Fat (%)*	*Protein (%)*
Jacobs *et al.* (1982)	15	20 700	47	39	14
Short and Short (1983)	8	12 500	43	41	16

This represents an energy expenditure of 6210 kJ (1480 kcal) for a 90 minute match. If 60% of the energy were to come from carbohydrates, 227 g of carbohydrate would need to be consumed to replace it afterwards. How could this be consumed? Table 6.2 provides possible answers.

6.2 IMPORTANCE OF CARBOHYDRATES

6.2.1 Early studies

One of the earliest studies to show the importance of carbohydrates to athletic performance was carried out on marathon runners in the Boston marathons of 1924 and 1925 (Levine *et al.*, 1924; Gordon *et al.*, 1925). After the 1924 marathon the doctors undertaking the study found that the post-race blood glucose levels of six runners studied were decreased, and that there was a strong relation between their physical condition and their blood glucose concentration. As a result of these findings, the doctors encouraged a group participating in the 1925 marathon to go on a diet high in carbohydrate for 24 hours before the race and to eat some sweets after they had run about 24 km. The results proved very encouraging in so far as the runners improved their times, had a higher post-race blood glucose concentration than the previous year, and were in better condition.

The next studies reported on humans were performed in Scandinavia in 1939, where it was shown that when exhausted subjects were given 200 g of carbohydrate they were able to continue exercising for a further 40 min (Christensen and Hansen, 1939). The same researchers also showed that when subjects were on a high carbohydrate diet they could exercise for a longer period of time than when on a normal or a low carbohydrate/high fat diet.

6.2.2 Muscle glycogen

The introduction of the muscle biopsy technique to nutritional studies in the 1960s led to the discovery that a high carbohydrate diet undertaken for three days resulted in an elevated muscle glycogen content (Bergstrom *et al.*, 1967). The effect was an increase in time to exhaustion when cycling an ergometer at

Table 6.6 Relationship between carbohydrate intake, muscle glycogen concentration and time to exhaustion

Diet	*Muscle glycogen concentration*	*Time to exhaustion*
Mixed diet (50% carbohydrate)	1.75 g per 100 g wet muscle	115 min
Low carbohydrate (0% carbohydrate)	0.6 g per 100 g wet muscle	60 min
High carbohydrate (80% carbohydrate)	3.5 g per 100 g wet muscle	170 min

Source: Bergstrom *et al.*, 1967.

75% $\dot{V}O_2$max compared to a mixed diet, and to a diet high in fat (Table 6.6). The same research group also observed that a more pronounced effect could be obtained if the muscle glycogen levels were depleted by exercise before undertaking a high carbohydrate regime for three days. Under such circumstances, muscle glycogen concentrations of 4 g/100 g wet muscle were obtained.

This muscle biopsy technique was used to further demonstrate that muscle glycogen concentrations fell during prolonged exercise, and that the very low levels coincided with the development of fatigue. Thus it was concluded that depletion of muscle glycogen caused fatigue.

The exercise intensity is related to muscle glycogen depletion in specific muscle fibre types (Vollestad and Blom, 1985; Vollestad *et al.*, 1984). Figure 6.2 highlights the effects of exercise intensities of approximately 65% and 90%

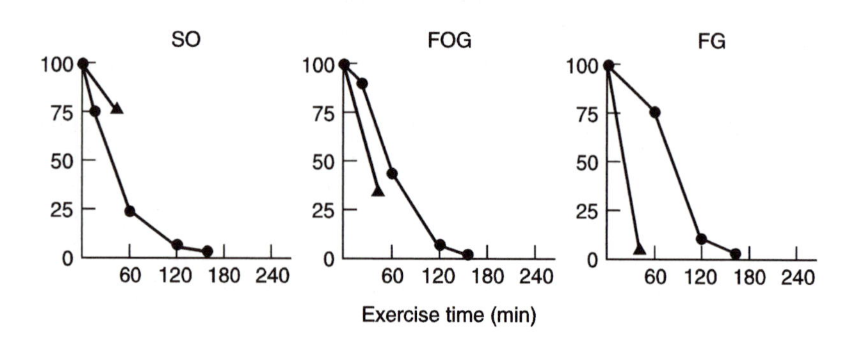

Figure 6.2 Rate of glycogen disappearance from three muscle fibre types during moderate, i.e. 65% $\dot{V}O_2$max (•), and intense, i.e. 90% $\dot{V}O_2$max (▲), exercise. (Data from Vollestad *et al.*, 1984.)

$\dot{V}O_2$max on the glycogen depletion of slow oxidative (SO), fast oxidative glycolytic (FOG) and fast glycolytic (FG) fibres. It can be seen clearly that at high exercise intensities glycogen stores were depleted in FG fibres, whilst all fibre types have depleted stores at the moderate exercise intensities. It is important to appreciate that repeated bouts of sprint activity (such as all-out sprints in soccer) will lead to glycogen depletion in the FG fibres, and thus impair sprint performance; lower exercise intensities will be possible, but not intense bouts.

Since there is much evidence relating muscle glycogen depletion to fatigue during prolonged and possibly to high-intensity exercise, there needs to be a consideration as to how much and when to consume carbohydrates in order to replenish the depleted stores. Costill *et al.* (1981) observed the effects of feeding meals containing 188, 325, 525 or 648 g of carbohydrate in a 24-hour period following a 16 km run at approximately 80% $\dot{V}O_2$max, which was in turn followed by five 1-minute sprints corresponding to an exercise intensity of approximately 130% $\dot{V}O_2$max. The exercise resulted in a 60% decrease in muscle glycogen content, which was only restored when 525–648 g of carbohydrate was consumed; the consumption of 188–325 g of carbohydrate failed to restore muscle glycogen content in a 24-hour period. Other studies have reported similar findings in so far as rates of glycogen resynthesis after depletion were maximal when 0.7–3.0 g/kg carbohydrate was consumed every 2 hours (Blom *et al.*, 1987; Ivy *et al.*, 1988).

The question regarding timing of the meal following exercise was addressed by Ivy *et al.* (1988). They showed that the highest rates of glycogen resynthesis occurred when carbohydrates were consumed immediately after prolonged exercise. If the carbohydrate is given two or more hours after the exercise, then restoration takes longer to achieve. This clearly has implications for the soccer player who trains or competes late in the evening and decides not to eat anything until breakfast, or the professional who trains in the morning, skips lunch and eats a hearty meal in the evening. It is unlikely that muscle glycogen stores will be sufficiently replenished. Drinking carbohydrate beverages immediately after a match or after training may be desirable under these circumstances, although the normal practice may be to delay this for some time. Coaches and soccer players would be well advised to give serious consideration to post-exercise nutrition.

6.2.3 Muscle glycogen and soccer

Is there any evidence that muscle glycogen depletion and/or carbohydrate feedings are pertinent to performance in soccer? Saltin (1973) examined the effects of pre-exercise muscle glycogen levels on work-rates of nine subjects during soccer play. Five of the players had normal muscle glycogen levels (96 mmol kg^{-1}), whereas four possessed low muscle glycogen levels (45 mmol kg^{-1}) before the start of the match. The reason for the low levels of muscle

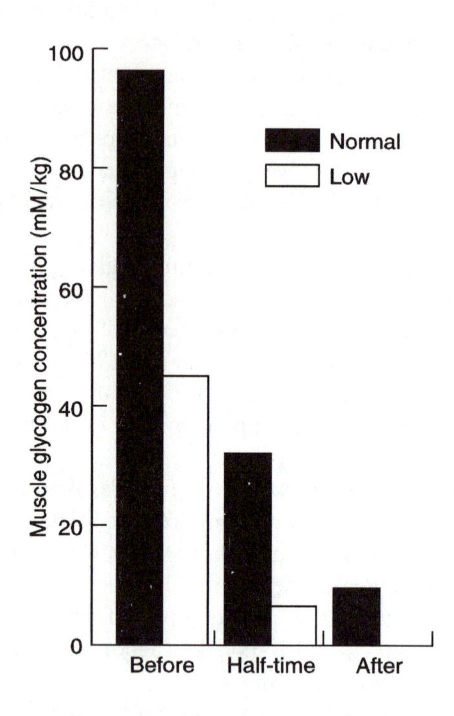

Figure 6.3 Muscle glycogen changes for nine players during a soccer match. (Data from Saltin, 1973.)

glycogen in the four players was due to the fact that they had a hard training session the previous day and had not consumed enough carbohydrates subsequently. Further muscle biopsy samples were taken at half-time and at the end of the match, and the players' movements were analysed using cine-film. A significant glycogen depletion occurred in the muscles during the match (Figure 6.3), and this probably contributed to the comparatively lower distances covered by the four players during the second half, i.e. 4100 m as opposed to 5900 m. The four players with low muscle glycogen spent 50% of the time walking compared to 27% for the other five players, and they only spent 15% of the time sprinting compared with 24% from the other group. Clearly, it is undesirable for soccer players to have reduced muscle glycogen stores prior to a match. Can this happen? Pre-season training camps, evening matches and so on could result in inadequate attention being given to nutritional considerations in aiding recovery of glycogen stores.

Muscle glycogen content has been reported to approach complete depletion after a Swedish First Division match (Karlsson, 1969, cited by Shephard and Leatt, 1987). A 63% reduction was found after a regular match undertaken by

Malmo FC players (Jacobs *et al.*, 1982), and a 50% decrease in elite Canadian soccer players during a simulated match (Leatt and Jacobs, 1988).

6.2.4 Hypoglycaemia

In spite of the considerable weight of evidence that muscle glycogen depletion is correlated with fatigue, there is some evidence that hypoglycaemia may be a causative factor in some individuals. Hypoglycaemia may be evident in prolonged exercise, and can contribute to fatigue. This normally occurs as a result of an inadequate restoration of the liver glycogen stores, as it appears that liver gluconeogenesis is unlikely to produce sufficient glucose for maintenance of blood concentrations once glycogen stores are low. The brain and the central nervous system are dependent on glucose for their metabolism. Blood glucose concentrations of about 3.0 mmol l^{-1} can cause nervousness and trembling, whilst levels below 3.0 mmol l^{-1} can lead to loss of consciousness.

For the skilled performer any slight malfunction of the brain will result in slow or inappropriate decision-making. In soccer, this may result in being caught in possession with the ball, incorrect weighting of a pass, mis-timed tackles or even missed goal scoring opportunities. One may speculate that lapses in concentration could be as a result of hypoglycaemia leading to poor decision-making.

Ekblom (1986) reported lower than normal blood glucose concentrations of 3.8 mmol l^{-1} in players at the end of a Swedish First Division match; indeed three players had values of between 3.0 and 3.2 mmol l^{-1}. These low levels could impair decision-making compared with those players' responses under normal resting blood glucose levels of approximately 5.0 mmol l^{-1}.

6.3 IMPORTANCE OF FLUIDS

Although reductions in muscle and liver glycogen are understood to be major causative factors in the onset of fatigue during prolonged exercise, the loss of body fluids leading to dehydration may be another important cause. Mild dehydration will impair performance and reduce the capacity for exercise; a decrease in body weight of 5% due to fluid loss (i.e. 3.5 kg for a 70 kg person) can result in a 30% decrease in physical work capacity (Saltin and Costill, 1988).

Fluid loss during exercise is associated with the need to maintain a relatively constant body temperature. Exercise results in an increased production of energy, both metabolic and heat, and it is as a consequence of this that body temperature becomes elevated. Maughan (1991) estimated that the rate of heat production of running a marathon in 2 h 30 min is approximately 80 kJ/min (20 kcal/min). Since the major biological mechanism for losing heat during exercise is by evaporation of sweat, this will result in the loss of 2 litres per hour if all the 4.8 MJ (1200 kcal) of heat is to be removed by this route. This rate is possible, and would lead to a total loss of 3 litres during a soccer match, as its intensity is comparable to the energy expenditure of marathon running (Reilly, 1990). A loss of 3 litres represents a 4% loss of body weight for a 70 kg person. It has been reported that a 2% body weight decrease due to dehydration results in impaired performance, and as previously stated a 5% decrease in body weight results in a 30% decrease in work capacity (Saltin and Costill, 1988). Clearly there is a need to ensure that dehydration is minimized.

Are dehydration and fluid balance likely to be of significance for soccer players? Despite the fact that the last four World Cups were held in countries where high temperatures were evident, i.e. Spain (1982), Mexico (1986), Italy (1990) and the USA (1994), comparatively little research has been reported on this topic. Mustafa and Mahmoud (1979) observed that the average evaporative fluid loss from the Sudanese team in World Cup qualifying matches was about 3% body weight. More recently Elias *et al.* (1991) reported that in the 1988 USA youth soccer tournament 34 players collapsed from heat exhaustion during six days. Training and playing soccer in hot and humid conditions accentuate heat stress-related problems if fluid ingestion is inadequate. Fluid replenishment is manifestly essential, and the most appropriate form must be considered.

6.3.1 Water

Prolonged exercise, such as playing soccer, in a warm environment results in a significant fluid loss. It would seem quite rational therefore to conclude that the best way of replenishing fluid loss would be to ingest water. Certainly evidence was available from the early 1970s that the rate of gastric emptying from the stomach became compromised if glucose was added to water (Costill and Saltin, 1974), and this led to the widespread favouring of water over carbohydrate drinks. Subsequently the gastric emptying characteristics of many different fluids have been studied, and the results confirm the inhibitory effects of adding calories to water (Murray, 1987).

The rate at which fluids are absorbed by the body is a combination of the rate of gastric emptying and the rate of fluid uptake by the small intestine. It is not advisable therefore to draw conclusions about fluid absorption based solely on gastric emptying rates despite the fact that gastric emptying is slower than fluid uptake by the small intestine. It is conceivable that whilst a dilute glucose drink may reduce the rate of gastric emptying in comparison to water, the glucose stimulation of fluid uptake by the small intestine results in an similar overall rate of fluid absorption.

Many studies have been performed to determine the advantages of water or carbohydrate-electrolyte solutions on performance either in the laboratory or in field conditions (for reviews see Murray, 1987; Maughan, 1991). Such studies have invariably concentrated on the enhancement of exercise time to exhaustion rather than on improvements in time to complete a set distance. The advantage of some form of fluid over a no-fluid treatment is well established. What is less clear is whether water has any advantages over a carbohydrate-based fluid or vice versa. When dehydration is believed to be the major factor impairing performance, such as in prolonged exercise in the heat, then ingesting as much water as is deemed necessary to offset dehydration is advisable. The consensus view in the UK is that a sports drink which contains an energy source in the form of carbohydrate together with electrolytes is more effective than plain water in maintaining performance (*British Journal of Sports Medicine*, 1993).

6.3.2 Carbohydrate drinks

Many recent studies on the beneficial effects of carbohydrate ingestion on performance have been reported (see Coyle, 1991). Factors such as type of carbohydrate, concentration of carbohydrate, and timing of ingestion have been considered. The primary purpose of carbohydrate ingestion during exercise is to maintain blood glucose concentrations, and if the exercise is prolonged, to maintain carbohydrate oxidation rates in the later stages. This would permit continuation of exercise at an adequate intensity. The purpose of carbohydrate ingestion after exercise is to restore the muscle and liver glycogen stores as quickly as possible.

When carbohydrate supplementation is provided during prolonged exercise, subjects can exercise for longer and produce greater power at the end of such performance than when given nothing or indeed when provided with water alone (Coggan and Coyle, 1991). So carbohydrate supplementation is recommended whenever the exercise is likely to be severe enough to significantly deplete glycogen stores and so impair performance. This certainly would apply in top-level soccer. But how much carbohydrate needs to be ingested?

In most of the studies that have been reported where carbohydrate ingestion improved performance subjects were given 30–60 g/h (Coggan and Coyle, 1991). These authors had previously reported that the glucose infusion rate necessary to restore and maintain blood glucose levels late in exercise and thereby delay fatigue by 45 min was 1 g/min or 60 g/h (Coggan and Coyle, 1987). It has been shown that glucose infusion can result in a glucose utilization rate of 1.8 g/min (MacLaren *et al.*, 1994). It appears that the maximum rate at which a carbohydrate can be utilized is approximately 120 g/h.

In order to consume 60 g/h of a carbohydrate, it is possible to do so in the following manner:

300 ml of a 20% solution
600 ml of a 10% solution
1200 ml of a 5% solution
2400 ml of a 2% solution

The first drink would appear to be too concentrated and may compromise fluid intake whilst the last drink is too large a volume to ingest in an hour. The two other drinks would appear to be reasonable; although 600 ml may not be sufficient in fluid intake on a hot and humid day, whilst a 10% concentration of glucose (but not maltodextrin) may inhibit gastric emptying. By drinking relatively large volumes of 5–10% carbohydrate solutions, most athletes can meet their carbohydrate needs and obtain 600–1200 ml/h of fluid. For the soccer player who uses 200 g of carbohydrate in a match and loses 2–2.5 l of fluid, a regular consumption of a carbohydrate-electrolyte drink is advisable. In what form is this best provided, and when should it be taken?

Ideally it would be prudent to drink 200 ml of 5–10% carbohydrate-electrolyte immediately before the match and 1200 ml/h of the same at intervals throughout the match. However, due to FIFA regulations this is not usually possible at international competitive matches and so a compromise is necessary. In the 1994 World Cup the regulations were amended to allow players to consume drinks at the side-line; the changes were made in the light of warnings that players might suffer acutely from dehydration. Drinking fluids at half-time is certainly advisable. The concentration and volume of fluid can be changed to suit the external environmental conditions, i.e. on a hot and humid day the emphasis should be on a greater volume of fluid with a dilute carbohydrate content (say no more than 5%), whilst on a cold and rainy day the emphasis should be on a smaller volume of fluid intake with a higher carbohydrate content (say 10%). New drinking strategies should never be tried out under match-play conditions; rather they should be used during training first.

No significant differences have been found in the ability of various forms of ingested carbohydrate to maintain blood glucose concentrations, change carbohydrate oxidation rates, or improve performance. These sources have included glucose, sucrose, fructose, maltose and maltodextrins (also known as glucose polymers). Fructose ingestion is problematical because it can cause diarrhoea due to its slower intestinal absorption. Maltodextrins have become popular because they are not as sweet tasting and they are osmotically less active than glucose and sucrose.

6.4 MEALS

6.4.1 Pre-competition

Soccer players should leave at least a 3-hour interval between a full meal and competition in order to minimize gastrointestinal problems such as nausea and a feeling of fullness. It is recommended that the stomach should be reasonably empty at the time of the match since the digestion and absorption of food will compete with the muscles for a good blood supply. Fatty foods are known to slow down the rate of gastric emptying and therefore should be avoided. The meal should be high in carbohydrates, preferably complex carbohydrates such as bread, cereals, pasta, rice, potatoes, fruits and vegetables. The actual amount of calories consumed will vary between individuals and how much they had eaten previously. Proteins are acceptable to eat as long as they are not fatty proteins, i.e. meat high in fats or containing fatty sauces, or fatty cheeses. This meal (and accompanying drinks) should contain about 5 g/kg body weight of carbohydrate; for a 70 kg soccer player this represents an intake of 350 g of carbohydrate. The accompanying drink could be high in carbohydrates, but not in the form of fructose.

The only time that carbohydrates should possibly be avoided is in the 30–60 min immediately before competition or training. For some individuals this timing of feeding may produce a rapid fall in blood glucose levels in the first 20 min or so of exercise and so impair performance (Costill *et al.*, 1977; Hargreaves *et al.*, 1984). Carbohydrate ingestion 5 min or so before exercise does not precipitate this problem.

The most important aspects of pre-competition meals are to elevate the body's carbohydrate stores, ensure hydration and yet provide satisfaction for the player. Trying out new foods or significantly altering eating patterns should be discouraged. Experimenting with new pre-competition meals should take place before training or before unimportant games.

6.4.2 Post-competition

The major considerations after competition are to replenish carbohydrate and fluid losses. As already mentioned, it is important to consume carbohydrates as soon as possible after exercise in order to achieve a quick and complete glycogen restoration. The first 2 hours post-exercise is the most crucial period for the ingestion of carbohydrates (Ivy *et al.*, 1988), since the glycogen-synthesizing enzymes are very active during this time. A recommendation would be to consume 1.5 g/kg body weight of carbohydrate within the first 30 min after competition or exercise; for a 70 kg player this represents 105 g of carbohydrate. Whether the carbohydrate is in solid or liquid form is immaterial and may be left to the preference of the player. Some athletes do not like to eat after strenuous exercise but are quite willing to drink. A concentrated carbohydrate beverage would prove invaluable to these players.

Post-exercise rehydration is best served by a predetermined schedule of fluid intake rather than for the player to rely on sensations of thirst; the thirst mechanism is unreliable in terms of adequate rehydration. Plain water will certainly promote significant rehydration, although increased urine production and a decreased thirst response may lead to failure to replace fully the intracellular water. It has been suggested that the addition of small amounts of sodium (in the form of salt) will increase the rate of recovery of fluid balance by helping to retain water and thereby promote a normalizing of plasma volume (Nadel, 1988). The use of commercial sports drinks can help to increase the voluntary consumption of fluid compared with water, and therefore may be beneficial in achieving rehydration post-exercise (Johnson *et al.*, 1988). The use of carbohydrate-electrolyte drinks may be of special advantage in recovery post-exercise, where the twin aims of rehydration and restoration of glycogen stores are needed.

6.5 ERGOGENIC AIDS

Ergogenic aids are substances, legal or illegal, which may lead to an improvement in athletic performance. Such substances include alcohol, caffeine, creatine, alkalinizers and vitamins. Anabolic steroids, amphetamines, erythropoietin and blood doping are illegal and are therefore not considered here. The agents discussed are not banned substances and may be taken in small doses.

6.5.1 Alcohol

Alcohol in small doses has been reported to improve hand steadiness and motor control in sports that require accuracy, i.e. archery, darts, shooting and snooker (Reilly and Halliday, 1985). Its use elsewhere is unlikely to enhance performance; indeed in moderate to large quantities it is likely to prove disadvantageous (American College of Sports Medicine, 1982). When ingested in large amounts, alcohol reduces exercise ability by impairing psychomotor coordination, increasing diuresis and so possibly leading to dehydration, and compromising the maintenance of body temperature, particularly in cold temperatures.

Two myths which are associated with alcohol concern its energy content and its carbohydrate content. Although alcohol does possess a higher amount of energy than carbohydrate (29 kJ compared with 16 kJ per gram), it is metabolized by the liver like fat, and is energetically not useful to muscle. Secondly, alcoholic drinks do not contain large amounts of carbohydrate, contrary to popular belief. A bottle of beer contains on average 10–15 g of carbohydrate (5–8 g in 'lite' beers). Alcohol drinks do not provide useful nutrients for enhancing sports performance. The facts that alcohol is a diuretic and results in increased urine water loss, and that alcohol may interfere with glucose metabolism by reducing gluconeogenesis, mean that it is an inappropriate drink for athletes.

There is no evidence that light social drinking will impair performance even on the following day. Moderate to large intakes will adversely affect endurance capacity, strength, speed and reaction times (Williams, 1991).

6.5.2 Caffeine

Caffeine is a drug that is found in several drinks (tea, coffee, cola) and foods (chocolate). It is a stimulant in high concentrations, and elicits a number of physiological and psychological responses linked with increasing endurance performance. The effects are mainly due to an elevation in plasma fatty acids and a consequent sparing of muscle glycogen (Giles and MacLaren, 1984). The IOC has banned caffeine and set a threshold level of 12 μg/ml for urine samples. This may be achieved by ingesting approximately 800 mg of caffeine, comparable to drinking six cups of coffee or 17.5 cans of Coca-Cola in a short time. A caffeine intake half this amount, i.e. approximately 350 mg, promotes fat utilization (Giles and MacLaren, 1984).

In spite of the evidence that caffeine stimulates the release of fatty acids and possibly spares muscle glycogen stores, the research findings are equivocal (Powers and Dodd, 1984). Consuming carbohydrate with caffeine negates the fatty acid stimulatory effect of caffeine, and so this combination would have no ergogenic influence. It should also be recognized that caffeine is a diuretic and would lead to dehydration.

6.5.3 Creatine

Short-duration high-intensity exercise requires the regeneration of ATP primarily from the breakdown of CP and from anaerobic glycolysis. A significant reduction of CP occurs after 6 s of cycle sprinting (Boobis *et al.*, 1987). A 16% decrease in CP after ten 6 s cycle sprints has been reported, and this coincided with a reduction in power output (Gaitanos *et al.*, 1993). Depletion is greater in FG than SO fibres (Greenhaff *et al.*, 1992). It seems that the availability of CP is one of the limiting factors for maintaining the high rates of energy necessary for this type of activity. Could supplementing the muscle creatine pool therefore result in enhanced all-out performance?

Harris *et al.* (1992) demonstrated that ingestion of 20–30 g of creatine monohydrate per day for more than two days increased the muscle creatine content by up to 50%. Subsequent studies have shown that creatine supplementation resulted in improved performance during repeated 30 s bouts of maximal isokinetic exercise (Greenhaff *et al.*, 1993b), during repeated bouts of 6 s sprint cycling (Balsom *et al.*, 1993), and during 300 m and 1000 m running time (Harris *et al.*, 1993). Furthermore, CP resynthesis during recovery from intense activity was enhanced following supplementation (Greenhaff *et al.*, 1993a). The general consensus is that creatine supplementation does enhance these types of activity when taken in doses in excess of 20–30 g/day. Whether such an intake is likely to promote soccer performance is uncertain, although a word of caution should be given in that the long-term effects of taking such large doses are also unknown.

6.5.4 Alkalinizers

Sodium bicarbonate and sodium citrate are alkaline salts that possess buffering properties in the human body when ingested. The theory behind their use is relatively simple. If a build-up of lactic acid occurs due to intense exercise, the resultant increase in acidity or reduction in pH is likely to be a contributory factor in the development of fatigue (MacLaren *et al.*, 1989). Alkalinizers increase the normal alkali reserve of the blood and help buffer the acid produced during exercise. Furthermore, the increase in bicarbonate ions in the blood may facilitate the efflux of hydrogen ions from the muscle and thereby maintain a higher pH in the muscle (MacLaren, 1986). Theoretically, alkaline salts should enhance performance in events maximizing the use of anaerobic glycolysis (i.e. intense exercise lasting between 30 and 120 s), and also intense intermittent exercise where removal of lactic acid from muscle during recovery is required. Maximal exercise tasks of less than 30 s and prolonged endurance tasks such as 5–10 km runs, which rely primarily on oxidative processes, generally do not benefit from the use of alkalinizers (Gledhill, 1984). Sodium bicarbonate ingestion may improve running time to exhaustion at an exercise intensity corresponding to a lactate concentration of 4 mmol l^{-1}; an increase of 17% in time to exhaustion from approximately 26 min to 31 min was observed in runners by George and MacLaren (1988). An alternative source of alkalinizer in the form of sodium citrate has been used with limited success (Parry-Billings and MacLaren, 1986).

Although soccer is a sport in which there are periodic bouts of intense exercise and the blood lactate levels have been reported to rise to 4.0–9.5 mmol l^{-1} (Bangsbo, 1994), it is unlikely that use of alkalinizers would be appropriate. This is due in particular to the prolonged nature of the game, and also the fact that gastrointestinal problems of nausea and diarrhoea can be associated with their ingestion.

6.5.5 Vitamins and minerals

Many of the vitamins play key roles in exercise metabolism although they do not possess any energy. The B vitamins are particularly noteworthy for their relevance to energy production in muscle and also in the recovery processes. Table 6.7 highlights the roles played by some of the vitamins. If vitamins are important in energy-producing reactions, then a deficiency of these vitamins will lead to an impaired performance, and an increase in these vitamins may lead to an enhanced performance. A deficiency in at least some of the vitamins will impair exercise performance although there is no evidence that vitamin supplementation improves performance in individuals eating a well-balanced diet (van der Beek, 1985). Physical activity increases the need for extra amounts of vitamins, but these can easily be obtained by consuming a balanced diet. The Department of Health and Social Security (DHSS) has provided

Table 6.7 Vitamins, their major functions, and their recommended daily intakes (RDI)

Vitamin	*RDI*	*Major role*
Vitamin A (retinol)	750 µg	Vision
Vitamin B_1 (thiamin)	1.4 mg	Carbohydrate metabolism
Vitamin B_2 (riboflavin)	1.7 mg	Electron transfer
Vitamin B_3 (nicotinic acid)	18 mg	ATP synthesis, fat synthesis
Vitamin B_6	No RDI	Amino acid, and glycogen synthesis
Folic acid	No RDI	Red blood cell synthesis
Pantothenic acid	No RDI	Fatty acid oxidation
Biotin	No RDI	Fatty acid, and glycogen synthesis
Vitamin B_{12}	No RDI	Red blood cell synthesis
Vitamin C	30 mg	Collagen synthesis
Vitamin D	2.5 µg	Calcium metabolism
Vitamin E	No RDI	Antioxidant
Vitamin K	No RDI	Blood clotting

guidelines (Table 6.7) for the minimum consumption of the vitamins A, thiamin, riboflavin, nicotinic acid, vitamin C and folic acid (DHSS, 1979).

Dietary surveys of athletic populations have shown that the estimated vitamin intakes usually exceed the recommended dietary intakes (Brotherhood, 1984; van Erp-Bart *et al.*, 1989). It appears that athletes do not usually suffer from an inadequate vitamin intake, and that vitamin supplementation may only be of value if there is likely to be a deficiency.

Similar findings have been reported with regard to mineral supplementation, in that no benefits have been determined. The only mineral which has presented as being problematical in athletes is iron. The so-called sports anaemia has been explained as being caused by a transitory dilutional effect of an expanded plasma volume due to training, and/or due to haemolysis of red cells as the foot strikes the floor in running. Other possible causes could be decreased iron absorption, increased iron losses in sweat, faeces and urine, or poor iron intake (Economos *et al.*, 1993). Female soccer players may be more susceptible to lower iron status as a result of menstrual blood loss. Anaemia will result in impaired aerobic performance and as a consequence may be treated by additional intakes of iron in combination with vitamin C. More in-depth reports can be found in the reviews by Newhouse and Clement (1988) and Haymes and Lamanaca (1989).

Summary

In order to maximize training it is essential for the soccer player and coach to give serious consideration to correct nutritional practices. It is not possible for a player to give 100% effort in training or during a match if inadequate attention has been devoted to nutrition. The following points may be considered in highlighting such nutritional practices.

- At least 55% of the energy intake should be in the form of carbohydrates – preferably complex carbohydrates. During intense training or game periods this value should be increased to 60–70% of energy intake.
- Carbohydrates should form the basis of the pre-competitive and post-competitive meals. The former should be consumed 2.5–3 h before playing, whilst the latter should be consumed within 2 h after playing. Carbohydrate beverages would be beneficial in these situations either to 'top-up' the meal or because the player does not feel like eating.
- Carbohydrate-electrolyte drinks are beneficial and should be taken before (say up to 60 min), during (every 15–20 min), and after (within 1–2 h) training or a match. Thirst is a poor indicator of the degree of dehydration.
- A protein intake of 1.4–1.7 g/kg body weight per day is desirable; this is likely if the player is consuming 15% of energy intake in the form of protein and is meeting the daily total energy intake requirements.
- A varied diet should ensure the players are likely to meet the recommended amino acid, vitamin and mineral requirements. Supplementation of these nutrients is unnecessary unless the diet is inadequate. There is no evidence that extra amounts of amino acids, vitamins or minerals will improve performance.
- Players should be discouraged from trying out new dietary practices prior to a match; these regimes should be tried out first before or during training.
- In spite of the potential benefits of some ergogenic aids, there is an uncertainty regarding the consequences of frequent use of these products.

REFERENCES

American College of Sports Medicine (1982) Position statement on the use of alcohol in sports. *Medicine and Science in Sports and Exercise*, **14,** ix–xi.

Balsom, P.D., Ekblom, B., Soderlund, K. *et al.* (1993) Creatine supplementation and dynamic high-intensity intermittent exercise. *Scandinavian Journal of Medicine and Science in Sports*, **3,** 143–9.

Bangsbo, J. (1994) The physiology of soccer – with special reference to intense intermittent exercise. *Acta Physiologica Scandinavica*, **151,** Suppl. 619.
Bergstrom, J., Hermansen, L., Hultman, E. and Saltin, B. (1967) Diet, muscle glycogen and physical performance. *Acta Physiologica Scandinavica*, **71,** 140–50.
Blom, P.C., Hostmark, A.T., Vaage, O. *et al.* (1987) Effect of different post-exercise sugar diets on the rate of muscle glycogen synthesis. *Medicine and Science in Sports and Exercise*, **19,** 491–6.
Boobis, L.H., Williams, C., Cheetham, M.E. and Wooton, S.A. (1987) Metabolic aspects of fatigue during sprinting, in *Exercise: Benefits, Limits and Adaptations* (eds D. Macleod, R. Maughan, M. Nimmo *et al.*), E. & F.N. Spon, London, pp. 116–40.
British Journal of Sports Medicine (1993) Consensus statement on fluid replacement in sport and exercise. *British Journal of Sports Medicine*, **27,** 34.
Brotherhood, J. (1984) Nutrition and sports performance. *Sports Medicine*, **1,** 350–89.
Christensen, E.H. and Hansen, O. (1939) Hypoglykamie arbeitsfahigkeit und ermuding. *Skandinavisches Archiv fur Physiologie*, **81,** 172–9.
Clarke, K. (1994) Nutritional guidance to soccer players for training and competition. *Journal of Sports Sciences*, **12,** S43–S50.
Coggan, A.R. and Coyle, E.F. (1987) Reversal of fatigue during prolonged exercise by carbohydrate infusion or ingestion. *Journal of Applied Physiology*, **63,** 2388–95.
Coggan, A.R. and Coyle, E.F. (1991) Carbohydrate ingestion during prolonged exercise: effects on metabolism and performance. *Exercise and Sports Science Reviews*, **19,** 1–40.
Costill, D.L. and Saltin, B. (1974) Factors limiting gastric emptying during rest and exercise. Journal of Applied Physiology, **37,** 679–83.
Costill, D.L., Coyle, E.F., Dalsky, G. *et al.* (1977) Effects of elevated plasma FFA and insulin on muscle glycogen usage during exercise. *Journal of Applied Physiology*, **43,** 695–9.
Costill, D.L., Sherman, W.M., Fink, W.J. *et al.* (1981) The role of dietary carbohydrates in muscle glycogen resynthesis after strenuous running. *American Journal of Clinical Nutrition*, **34,** 1831–6.
Coyle, E.F. (1991) Timing and method of increased carbohydrate intake to cope with heavy training, competition and recovery. *Journal of Sports Sciences*, **9,** S29–S52.
DHSS (1979) *Recommended Intakes of Nutrients for the United Kingdom*, HMSO, London.
Economos, C., Bortz, S.S. and Nelson, M.E. (1993) Nutritional practices of elite athletes: practical recommendations. *Sports Medicine*, **16,** 381–99.
Ekblom, B. (1986) Applied physiology of soccer. *Sports Medicine*, **3,** 50–60.
Elias, S.R., Roberts, W.O. and Thorson, D.C. (1991) Team sports in hot weather: guidelines for modifying youth soccer. *Physician and Sportsmedicine*, **19,** 67–80.
Gaitanos, G.C., Williams, C., Boobis, L.H. and Brooks, S. (1993) Human muscle metabolism during intermittent maximal exercise. *Journal of Applied Physiology*, **75,** 712–19.
George, K.P. and MacLaren, D.P. (1988) The effect of induced alkalosis and acidosis on endurance running at an intensity corresponding to 4 mM blood lactate. *Ergonomics*, **31,** 1639–45.
Giles, D. and MacLaren, D. (1984) Effects of caffeine and glucose ingestion on metabolic and respiratory functions during prolonged exercise. *Journal of Sports Sciences*, **2,** 35–46.

Gledhill, N. (1984) Bicarbonate ingestion and anaerobic performance. *Sports Medicine*, **1,** 177–80.

Gordon, B., Kohn, L.A., Levine, S.A. *et al.* (1925) Sugar content of the blood in runners following a marathon race. *Journal of the American Medical Association*, **185,** 508–9.

Greenhaff, P.L., Nevill, M.E., Soderlund, K. *et al.* (1992) Energy metabolism in single muscle fibres during maximal sprint exercise in man. *Journal of Physiology*, **446,** 528P.

Greenhaff, P.L., Bodin, K., Harris, R.C. *et al.* (1993a) The influence of oral creatine supplementation on muscle contraction in man. *Journal of Physiology*, **467,** 75P.

Greenhaff, P.L., Casey, A., Short, A.H. *et al.* (1993b) Influence of oral creatine supplementation on muscle torque during repeated bouts of maximal voluntary exercise in man. *Clinical Science*, **84,** 565–71.

Hargreaves, M., Costill, D.L., Coggan, A. and Nishibata, I. (1984) Effect of carbohydrate feeding on muscle glycogen utilization and exercise performance. *Medicine and Science in Sports and Exercise*, **16,** 219–22.

Harris, R.C., Soderlund, K. and Hultman, E. (1992) Elevation of creatine in resting and exercised muscle of normal subjects by creatine supplementation. *Clinical Science*, **83,** 367–74.

Harris, R.C., Viru, M., Greenhaff, P.L. and Hultman, E. (1993) The effect of oral creatine supplementation on running performance during short term exercise in man. *Journal of Physiology*, **467,** 74P.

Haymes, E.M. and Lamanaca, J.J. (1989) Iron loss in runners during exercise: implications and recommendations. *Sports Medicine*, **7,** 277–85.

Ivy, J.L., Katz, S.L., Cutler, C.L. *et al.* (1988) Muscle glycogen synthesis after exercise: effect of time of carbohydrate ingestion. *Journal of Applied Physiology*, **64,** 1480–5.

Jacobs, I., Westlin, N., Karlsson, J. *et al.* (1982) Muscle glycogen and elite soccer players. *European Journal of Applied Physiology*, **48,** 297–302.

Johnson, H.L., Nelson, R.A. and Consolazio, C.F. (1988) Effects of electrolyte and nutrient solutions on performance and metabolic balance. *Medicine and Science in Sports and Exercise*, **20,** 26–33.

Leatt, P.B. and Jacobs, I. (1988) Effect of a liquid glucose supplement on muscle glycogen resynthesis after a soccer match, in *Science and Football*, (eds T. Reilly, A. Lees, K. Davids and W. Murphy), E. & F.N. Spon, London, pp. 42–7.

Lemon, P.W.R. (1994) Protein requirements of soccer. *Journal of Sports Sciences*, **12,** S17–S22.

Levine, S.A., Gordon, B. and Derick, C.L. (1924) Some changes in the chemical constituents of the blood following a marathon race. *Journal of the American Medical Association*, **82,** 1778–9.

MacLaren, D.P. (1986) Alkalinizers, hydrogen ion accumulation and muscle fatigue: a brief review, in *Sports Science: Proceedings of the VIII Commonwealth and International Conference* (eds J. Watkins, T. Reilly and L. Burwitz), E. & F.N. Spon, London, pp. 104–9.

MacLaren, D.P., Parry-Billings, M., Gibson, H. and Edwards, R.H.T. (1989) A review of metabolic and physiological factors in fatigue, in *Exercise and Sport Science Reviews* (ed. K. Pandolf), vol. 17, Williams & Wilkins, Baltimore, Md, pp. 29–66.

MacLaren, D.P., Reilly, T., Campbell, I. *et al.* (1994) The hyperglycaemic glucose clamp technique, glucose utilization, and carbohydrate oxidation during exercise. *Journal of Sports Sciences*, **12,** 143.

Maughan, R.J. (1991) Fluid and electrolyte loss and replacement in exercise. *Journal of Sports Sciences*, **9,** 117–42.

Murray, R. (1987) The effects of consuming carbohydrate-electrolyte beverages on gastric emptying and fluid absorption during and following exercise. *Sports Medicine*, **4,** 322–51.

Mustafa, K.Y. and Mahmoud, N.E.A. (1979) Evaporative water loss in African soccer players. *Journal of Sports Medicine and Physical Fitness*, **19,** 181–3.

Nadel, E.R. (1988) New ideas for rehydration during and after exercise in hot weather. *Gatorade Sports Science Exchange*, **1**(3).

Newhouse, I.J. and Clement, D.B. (1988) Iron status in athletes: an update. *Sports Medicine*, **5,** 337–52.

Parry-Billings, M. and MacLaren, D.P. (1986) The effect of sodium bicarbonate and sodium citrate ingestion on anaerobic power during intermittent exercise. *European Journal of Applied Physiology*, **55,** 524–9.

Powers, S.K. and Dodd, S. (1984) Caffeine and endurance performance. *Sports Medicine*, **2,** 165–74.

Reilly, T. (1990) Football, in *Physiology of Sports* (eds T. Reilly, N. Secher, P. Snell and C. Williams), E. & F. N. Spon, London pp. 371–425.

Reilly, T. and Halliday, F. (1985) Influence of alcohol ingestion on tasks related to archery. *Journal of Human Ergology*, **14,** 99–104.

Reilly, T. and Thomas, V. (1979) Estimated daily energy expenditures of professional association footballers. *Ergonomics*, **22,** 541–8.

Saltin, B. (1973) Metabolic fundamentals in exercise. *Medicine and Science in Sports*, **5,** 137–46.

Saltin, B. and Costill, D.L. (1988) Fluid and electrolyte balance during prolonged exercise, in *Exercise, Nutrition, and Metabolism*, (eds E.S. Horton and R.L. Terjung), Macmillan, New York, pp. 150–8.

Shephard, R.J. and Leatt, P. (1987) Carbohydrate and fluid needs of the soccer player. *Sports Medicine*, **4,** 164–76.

Short, S.H. and Short, W.R. (1983) Four year study of university athletes' dietary intake. *Journal of the American Dietetic Association*, **82,** 632–45.

van der Beek, E.J. (1985) Vitamins and endurance training: food for running or faddish claims? *Sports Medicine*, **2,** 175–97.

van Erp-Bart, A.M.J., Saris, W.H.M., Binkhorst, R.A. *et al.* (1989) Nationwide survey on nutritional habits in elite athletes (Part II): mineral and vitamin intake. *International Journal of Sports Medicine*, **10,** (Suppl 1), S11–S16.

Vollestad, N.K. and Blom, P.C.S. (1985) Effect of varying exercise intensity on glycogen depletion in human muscle fibres. *Acta Physiologica Scandinavica*, **125,** 395–405.

Vollestad, N.K., Vaage, O. and Hermansen, L. (1984) Muscle glycogen depletion patterns in type I and subgroups of type II fibres during prolonged severe exercise in man. *Acta Physiologica Scandinavica*, **122,** 433–41.

Williams, M.H. (1991) Alcohol, marijuana and beta blockers, in *Ergogenics: the Enhancement of Sports Performance* (eds D.R. Lamb and M.H. Williams), Benchmark, Indianapolis.

Special populations

7

Thomas Reilly

Introduction

Soccer at the professional level of play is arguably the world's leading sport. The best players receive the adulation of fans and lucrative financial rewards from both employers and sponsors. Yet the popularity of soccer is reflected as equally in the number of active participants in the recreational game as in those who watch the elite play.

Participants in soccer include young boys and girls, those who play the game for purposes of 'keep fit' indoors or outdoors and a growing number of women and veteran players. Indeed the game has been modified for play by visually handicapped people with appropriate audio stimuli for locating the movement of the ball. The officials who regulate the game and control the players are usually neglected when aspects of play are considered.

This chapter is devoted to special populations within soccer. Attention is directed to specific groups including young and veteran players, women and participants in modified forms of the game. Finally, consideration is given to soccer referees, their roles and fitness requirements.

7.1 WOMEN'S SOCCER

Research into women's soccer has lagged considerably behind the men's game. This is because participation by women in the game has only recently been actively promoted. There was a step-wise increase in women's soccer clubs

Science and Soccer. Edited by Thomas Reilly. Published in 1996 by E & FN Spon, London. ISBN 0 419 18880 0.

throughout the majority of European countries throughout the 1980s; for example, the number of clubs registered with the Women's FA increased from 188 to 321 between 1980 and 1991. The women's game gained international credibility in 1991 when the First World Cup for women's teams took place in China. In the winning nation, the USA, the game is now widely played, as it is in Sweden, the host of the Second World Cup Finals in 1995. It still has to be developed in a majority of countries world-wide where women's participation in sport is restricted by cultural, domestic and economic circumstances.

It is probably only at the elite or professional level of play that female soccer players engage in strenuous sport-specific training programmes. In Danish soccer clubs, the training of women during the season consists of two to three sessions of 90 minutes each week. The national team players undergo additional training which consists of running (two to four sessions of 20–30 min per week) and general strength training (30 min once or twice a week). The training frequency is modified to suit individuals according to their domestic or occupational restraints (Jensen and Larsson, 1993). The intensity and quality of training are altered as the season progresses, starting with an emphasis on continuous running and culminating in short intermittent exercise later on. The programmes are comparable to those of elite British female soccer players, who train on five occasions each week during the season (Davis and Brewer, 1992).

The relative intensity of exercise at which women play soccer at an elite level approaches that of male counterparts and averages about 70% of $\dot{V}O_2$max (see Table 7.1). The overall energy expenditure has been estimated to be about 4600 kJ for a 60 kg player. This compares with an estimate of about 5700 kJ for a 75 kg male soccer player (Ekblom and Williams, 1994). Female players sustain

Table 7.1 Maximal oxygen uptake (ml kg^{-1} min^{-1}) mean values for women soccer players.

Team	*n*	*$\dot{V}O_2max$*	*Reference*
England national squad	14	52.2[a]	Davis and Brewer (1992)
Italian elite players	12	49.8	Evangelista *et al.* (1992)
Danish national team	10	57.6[b]	Jensen and Larsson (1993)
Australian elite players	10	47.9	Colquhoun and Chad (1986)
Canadian university team	12	47.1	Rhodes and Mosher (1992)
English university players	10	42.4	Miles *et al.* (1993)

[a] Estimated, after training.
[b] After training.

heart rates exceeding 85% of maximal values for two-thirds of the game but overall cover less distance than do the male players. Recreational players in small-sided games perform at about 70–75% VO_2max but play for shorter periods and can take time off for rests (Miles *et al.*, 1993).

The specific nutritional requirements of women soccer players were considered by Brewer (1994). Elite players were recommended to consume a diet high in carbohydrates. They were encouraged to consume foods high in calcium and iron in order to ensure general health. Those counselling female soccer players on nutrition should be aware of the possible existence of (or potential risk of inducing) eating disorders, particularly with regard to advice on weight loss.

The performance of female soccer players is not necessarily compromised by the phase of the menstrual cycle. In other sports, Olympic medals have been won by athletes during menses and throughout the follicular and luteal phases. Some women experience pre-menstrual discomfort, others dysmenorrhoea or painful menses. Participation in exercise programmes can attenuate menstrual discomfort and oral contraceptives are used to regulate the cycle by some sportswomen. There is evidence that women are more vulnerable to errors in the pre-menses phase and this has been reflected in the incidence of injury incurred by Swedish players (Moller-Nielsen and Hammar, 1989). Fewer traumatic injuries were noted in women players using oral contraceptive pills to reduce pre-menstrual and menstrual symptoms of discomfort.

Women on high training loads are prone to disruptions to the normal menstrual cycle. This may be indicated by a shortened luteal phase or by absence of menses, known as amenorrhoea. Competitive and personal stress are also implicated in amenorrhoea which is seen as a disturbance of the hypothalamic-pituitary-gonadal axis that regulates the menstrual cycle. A consequence of the low oestrogen levels associated with prolonged training-induced amenorrhoea is a loss of minerals (notably calcium) from bone. The osteoporosis observed in women distance runners is rarely found in women soccer players who tend to have less arduous training programmes and a higher energy intake.

The body composition of women soccer players tends to be closer to other team sports participants than to endurance athletes. Body adiposity tends to be only slightly lower than for women in the normal population of the same age. Percentage body fat values for Danish national players were 20.1 (17.5–25.0)% in mid-season and 22.3 (20.1–28.3)% at the start of systematic training (Jensen and Larsson, 1993). This compares with values of 24.7 (± 2.4)% for top players in the English League, 25.1 (± 5.4)% for university players and 26% for age-matched women in the general population (Reilly and Drust, 1994). England national squad members had values of 21.5 and 21.1% as an average at the start and end of a 12-month training programme (Davis and Brewer, 1993).

The muscle strength of women soccer players can be measured as peak torque during isokinetic movements. Top women soccer players tend to have greater muscle strength values than university players or non-games players (Reilly and Drust, 1994). The differences are pronounced at fast movement velocities, a characteristic also of male soccer players.

Despite the high aerobic demands imposed by competition on women soccer players, fitness test profiles suggest that the anaerobic system is more developed than the oxygen transport system. This may be due to an imbalance of emphasis in training or to a reluctance to engage in aerobic training on a systematic basis. Nevertheless, the maximal oxygen uptakes of Danish women soccer players have reached values of 57.6 (51.5–63.8) ml kg^{-1} min^{-1} in mid-season compared with 53.3 (48.0–60.8) ml kg^{-1} min^{-1} at the start of systematic training (Jensen and Larsson, 1993). Estimated $\dot{V}O_2max$ values of the England squad increased from a mean value of 48.4 ml kg^{-1} min^{-1} to 52.2 ml kg^{-1} min^{-1} during a 12-month period of training (Davis and Brewer, 1992). The average value of Australian national squad members was 47.9 ml kg^{-1} min^{-1} (Colquhoun and Chad, 1986) and elite Italian players had average values of 49.8 ml kg^{-1} min^{-1} (Evangelista *et al.*, 1992). These are appreciably higher than the mean of 42.4 (±4.3) ml kg^{-1} min^{-1} for university players (Miles *et al.*, 1993). A spread of values from various studies is shown in Table 7.1.

7.2 YOUTH SOCCER

The behavioural building blocks of soccer excellence are laid down during the growth process. For many sports the physiological determinants of performances are influenced by genetic factors: they must be carefully nurtured during childhood (approximately ages 4–12) and adolescence (ages 13–19 years) if exercise and sporting potential are to be realized. A huge problem for soccer scouts is the identification of soccer talent at an early age.

Developing talent poses questions about whether there is a 'golden age' for specialization, if there is an optimal timing for skills acquisition and so on. There are considerations for the negative consequences of high training loads on the growing skeleton, the possibilities of psychological 'burnout' through specializing too early and the concomitant parental and social pressures. There is some concern about the hours devoted to soccer by those excelling at an early age who are obliged to play for their school, representative teams and participate also in soccer schools for talented young players.

Chronological age is not a perfect marker of biological maturity. Consequently early developers may be at an advantage in sport because of their size. They may orientate towards particular roles (centre-back or centre-forward) due to their advantage in height. They may be demotivated and drop out of sport later on when their counterparts catch up. Late developers may emerge as potential champions only when growth is finished. Since a near 12 months difference in age can make an enormous difference in performance capability, some boys and girls may be disadvantaged in under-age competitions by virtue of having a birth date late in the year. In contact sports where strength is important this will lead to risk of injury. The alternative of soccer competitions where children are matched according to biological age is acknowledged to be unrealistic.

The most sensitive period for learning new movement patterns is probably between 9 and 12 years of age. The movement and muscle activity patterns of skilled kicking actions are evident in young soccer players by age 11. The practice in some sports is to start specialist training well before this time. Whilst the physiological systems that sustain prolonged training sessions can withstand the strenuous programmes prescribed, the growing skeletons of these elite children may not. The consequences of the stresses imposed on skeletal structures may show up later in the form of overuse injuries.

The $\dot{V}O_2$max increases with age but this is largely due to an increased body size. Anaerobic power and capacity are less well developed in children compared to aerobic power. This is reflected in the relative contribution of aerobic and anaerobic mechanisms in all-out effort. In children a 6-minute run is dependent almost entirely on aerobic metabolism. For a maximal effort of less than 60 seconds, children derive 60% of the total energy from anaerobic sources compared to 80% in adults. The relatively poor anaerobic capacity in children is confirmed by low levels of lactate production during intense exercise bouts and suggests a low glycolytic rate. There is especially limited potential in the prepubescent child for developing the anaerobic system. Evidence from magnetic resonance spectroscopy has confirmed that children are less able than adults to effect ATP rephosphorylation in anaerobic pathways during high-intensity exercise. Anaerobic capacities increase progressively during maturation until reaching adult levels after the teenage years.

The truism that children are not miniature adults is acknowledged by the national governing bodies of soccer; the rules for youth soccer are different in some respects from the game for adults, although the modifications vary from country to country. Competitive matches start at under-8, in the USA for example, where matches are divided into four quarters each of 12 minutes. At under-10 each half lasts 25 minutes, at under-12 this becomes 30 minutes which is progressively lengthened to 45 minutes at under-19. In tandem with these changes are modifications to the size of the pitch, the number of players and the allowances for substitutions. The regulations are intended to reduce the physiological strain on young players but the game still retains its intermittent high-intensity nature.

The physiological demands of German under-11 and under-12 matches were assessed by Klimt *et al.* (1992). Heart rates were in the range 160–180 beats/min, values comparable with elite adult players. Blood lactate levels remained in the 3–4 mmol l^{-1} range and reflect the completion of high-intensity efforts by children without major accumulation of lactate.

7.3 REFEREES

The referee is the key official in regulating behaviour of soccer participants by implementing the rules of play. This makes demands on mental faculties, visual perception, attention and decision-making. The referee has to be decisive and strict, yet employ discretion where appropriate. The other officials, the two linesmen, offer some assistance in cases of controversy but the ultimate decision is charged to the referee.

These decision-making stresses are superimposed on a relatively high level of physiological stress. The referee is expected to keep up with play whatever its tempo in addition to maintaining alertness throughout the game. These demands have implications for fitness required to officiate at a high level.

Referees in the Premier Division of the English League cover approximately 9.5 km during the course of the game. Of this total distance, on average 47% is covered at a jogging pace, 23% walking, 12% sprinting and 18% reverse running (Catterall *et al.*, 1993). Greater distances have been reported for top-class referees in Japan (Asami *et al.*, 1988). Mean values of 10.5 km were reported for seven foreign referees at international matches and 11.2 km for ten referees in the Japan National Soccer League. These figures compare favourably with distances covered by professional soccer players and exceed those reported for central defenders (Reilly, 1994).

The physiological strain incurred by soccer referees can be indicated by monitoring heart rate during a match. Short-range radio telemetry provides a convenient method of doing this as the equipment is lightweight and can be worn by the referee without interfering with him in any way. Measurements on 13 referees during top-class league matches indicated heart rates averaging 165 beats/min throughout the whole game (Figure 7.1). An estimate of the aerobic loading on referees would be 70–75% VO_2max.

A fatigue effect is evident in referees as indicated by a fall-off in work-rate towards the end of play. This happens despite the fact that the heart rate is maintained at about 165 beats/min (Catterall *et al.*, 1993). The fatigue is linked with diminishing energy stores within the active muscles. The urgency on the part of the losing team to move the ball quickly forward and press for a score means that the work-rate demands on the referee are unrelenting until the game is over.

The high level of exercise intensity associated with refereeing has consequences both for mental judgements and for fitness. Decrements in cognitive function are noted once the exercise intensity exceeds about 50% VO_2max (Figure 7.2). Decrements in less complex psychomotor functions are observed at only marginally greater exercise intensities (Reilly and Smith, 1986). It is likely, therefore, that decision-making is prone to error at the exercise intensities experienced by referees during professional and international matchplay.

The overall pattern of a referee's activity is acyclical and varies in parallel

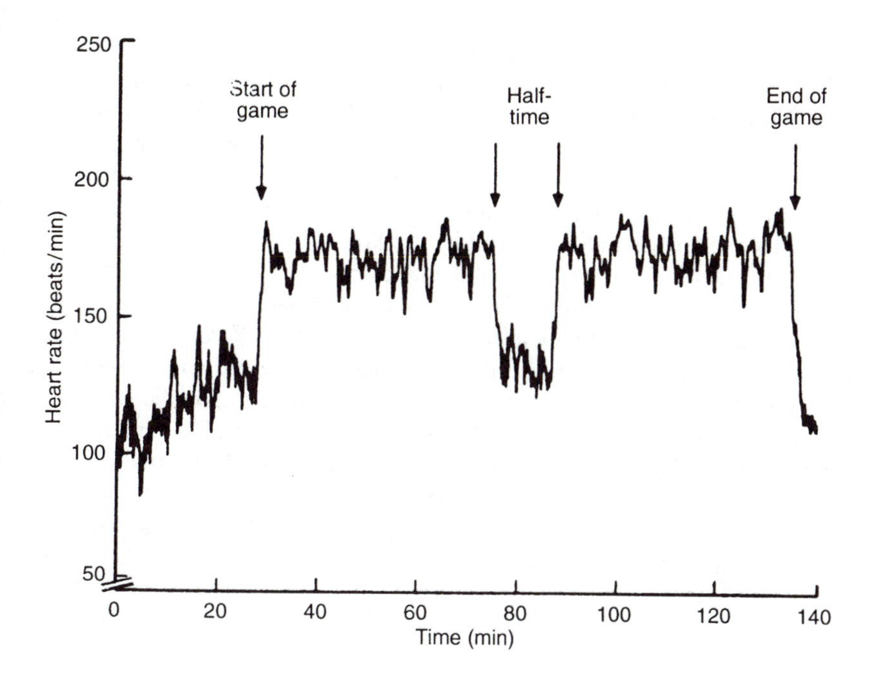

Figure 7.1 Heart rate of a referee throughout a Premier Division match in the English League. (Reproduced with permission from Catterall *et al.*, 1993 with kind permission of Butterworth-Heinemann Journals, Elsevier Science Ltd, Kidlington OX5 1GB, UK.)

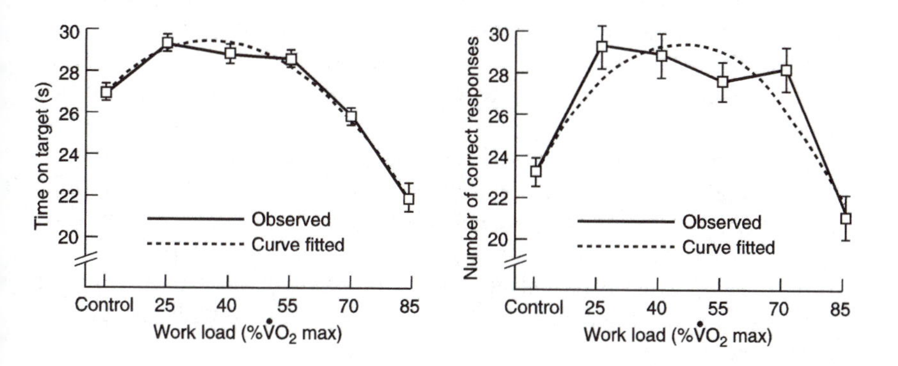

Figure 7.2 Effect of exercise intensity on psychomotor tasks (left) and on cognitive function (right) during exercise. (Data from Reilly and Smith, 1986.)

with the players' actions. Players incur additional energy expenditure when directly involved in executing match skills but they have a degree of choice in undertaking high-intensity efforts 'off-the-ball'. In contrast, the referee has to follow the play, irrespective of the intensity of previous movements and sometimes may not be able to keep up with it. The referee must, therefore, be able to anticipate the direction of play in advance. The referee does not display the acute-angled changes of direction which players make, but the overall work-rate profiles and the frequency of discrete events (a change in category of activity every 6 seconds) are broadly similar.

It seems the major physiological demands on referees are imposed on the oxygen transport system. Superimposed on these are short-term high-intensity runs. The work-rate profiles exhibited by referees can be interpreted to mean that aerobic fitness is an important requirement of soccer referees but they also need to be quick 'off-the-mark'. The national governing bodies recognize this and impose strict fitness standards on their officials. The need to pass these fitness tests is an ongoing concern of referees since some loss of fitness occurs due to the ageing process. The referees at professional soccer matches are often 15 years older than the average player and so the work-rate must be deemed appreciable. The fact that the referee usually has to hold down a professional job outside football is another source of stress. This brings them little sympathy from soccer spectators when refereeing decisions are not to their liking.

7.4 VETERAN PLAYERS

Professional soccer players in the 1990s have tended to maintain their careers for as long as possible. This contrasts with previous decades where it was the convention to contemplate retirement at about 30. Indeed many players now continue at the highest level well into their thirties, as evidenced by the ages of players at the 1994 World Cup Final. This is especially true of goalkeepers who have been traditionally able to stay in the game for longer than outfield players.

Another trend in recent years has been the tendency to continue playing at a recreational level following retirement from a professional career. Such participation may be for charitable or recreational reasons or for purposes of health. The emphasis on active sport as a means towards prevention of coronary heart disease has given rise to the growth of veterans' soccer, although this is largely a characteristic of developed nations with 'sport for all' promotional programmes.

Kohno *et al.* (1988) investigated the health status of senior Japanese citizens who had been playing soccer for about 2 hours a week for 40 years. These individuals had lower body fat levels than age-matched norms and had $\dot{V}O_2max$ values clearly superior to population values. These observations would support a life-long commitment to recreational soccer for purposes of maintaining sound cardiovascular health. Nevertheless the following qualifications were made. Matches should consist of 20 min halves; a small ball should be used; play should be on a grass or artificial grass surface; substitutes should be freely allowed and play should be against opponents of the same age.

The relatively high intensity of recreational soccer calls for attention to physical fitness if it is to be played safely. For this reason, it is important to get fit for the game before making a regular commitment to representing the team. This applies particularly in veterans' soccer in view of the greater health risks that are associated with the ageing process.

7.5 SMALL-SIDED GAMES

Modifications to the normal rules are made for young and for veteran players. Professionals may engage in 'conditioned games' during training sessions in order to emphasize particular aspects of skill or tactics. Games for recreational play also have their own local rules: this applies in particular to small-sized matches, with four to nine members on each side.

Small-sided games are especially suited to indoor leagues. Generally rules prohibit slide-tackling, playing the ball above waist (or shoulder) height or scoring from your own half of the pitch. Indoor matches tend to be for short periods, say 10–15 minutes each half, and are played at a high tempo. This is because the ball is out of play for only a small time, as it remains in play when hit off the side-wall or back-wall. The high tempo is a feature of the entertainment provided by professional indoor soccer in North America, where free substitution is allowed at all times. This form of play is common also in South America and Australia much more so than in Europe, where its use tends to be for recreational purposes. The extent to which errors and risk of injury are associated with a fast pace of play is unknown.

The physiological load imposed by 4-a-side soccer incorporates anaerobic and aerobic components. The mean relative exercise intensity exceeds 82% $\dot{V}O_2$max and mean blood lactate levels of 4–5 to 4.9 mmol l^{-1}. Time in possession of the ball is higher than in 11-a-side outdoor matches (MacLaren *et al.*, 1988). These observations lead to the conclusion that small-sided games provide an effective physiological stimulus for preparation for the normal 11-a-side match.

Male and female players may be used in different combinations to make up small-sided teams. Females have virtually the same physiological responses to recreational play in female-only and in mixed (2 males, 2 females) teams (Miles *et al.*, 1993). In the latter condition their direct participation in play tends to be reduced due to a male dominance in dictating play. This could be overcome by an appropriate matching of skill levels between the sexes.

Summary

The popularity of soccer is reflected in the range of participants at a recreational and competitive level. These include under-age and veteran players as well as women and male athletes. The game has been modified for play by visually impaired individuals. It has been modified for conditioned games, for indoor recreational purposes and for competitive tournaments. The match officials are rarely the subject of scientific investigations, despite the physiological and psychological stresses associated in particular with refereeing.

REFERENCES

Asami, T., Togari, H. and Ohashi, J. (1988) Analysis of movement patterns of referees during soccer matches, in *Science and Football* (eds T. Reilly, A. Lees, K. Davids and W.J. Murphy), E. & F.N. Spon, London, pp. 341–5.

Brewer, J. (1994) Nutritional aspects of women's soccer. *Journal of Sports Sciences*, **12,** 535–8.

Catterall, C., Reilly, T., Atkinson, G. and Goldwells, A. (1993) Analysis of the work rates and heart rates of association football referees. *British Journal of Sports Medicine*, **27,** 193–6.

Colquhoun, D. and Chad, K.E. (1986) Physiological characteristics of female soccer players after a competitive season. *Australian Journal of Science and Medicine in Sport*, **18,** 9–12.

Davis, J.A. and Brewer, J. (1992) Physiological characteristics of an international female soccer squad. *Journal of Sports Sciences*, **10,** 142–3.

Davis, J.A. and Brewer, J. (1993) Applied physiology of female soccer players. *Sports Medicine*, **16,** 180–9.

Ekblom, B. and Williams, C. (1994) *Foods, Nutrition and Soccer Performance*, E. & F.N. Spon, London.

Evangelista, M., Pandolfi, O., Fanton, F. and Faina, M. (1992) A functional model of female soccer players: analysis of functional characteristics. *Journal of Sports Sciences*, **10,** 165 (Abstr.).

Jensen, K. and Larsson, B. (1993) Variations in physical capacity in a period including supplemental training of the national Danish soccer team for women, in *Science and Football II* (eds T. Reilly, J. Clarys and A. Stibbe), E. & F.N. Spon, London, pp. 114–17.

Klimt, F., Betz, M. and Seitz, U. (1992) Metabolism and circulation of children playing soccer, in *Children and Exercise XVI: Paediatric Work Physiology* (eds J. Coudert and E. van Praagh), Masson, Paris, pp. 127–9.

Kohno, T., O'Hata, N., Morita, H. *et al.* (1988) Can senior citizens play soccer safely? in *Science and Football* (eds T. Reilly, A. Lees, K. Davids and W.J. Murphy), E. & F.N. Spon, London, pp. 230–6.

MacLaren, D., Davids, K., Isokawa, M. *et al.* (1988) Physiological strain in 4-a-side soccer, in *Science and Football* (eds T. Reilly, A. Lees, K. Davids and W.J. Murphy), E. & F.N. Spon, London, pp. 76–80.

Miles, A., MacLaren, D., Reilly, T. and Yamanaka, K. (1993) An analysis of physiological strain in four-a-side women's soccer, in *Science and Football II* (eds T. Reilly, J. Clarys and A. Stibbe), E. & F.N. Spon, London, pp. 140–5.

Moller-Nielsen, J. and Hammar, M. (1989) Women's soccer injuries in relation to the menstrual cycle and oral contraceptive use. *Medicine and Science in Sports and Exercise*, **21,** 126–9.

Reilly, T. (1994) Motion characteristics, in *Football (Soccer)* (ed. B. Ekblom), Blackwell Scientific, London, pp. 31–42.

Reilly, T. and Drust, B. (1994) The isokinetic muscle strength of women soccer players. Communication to the 10th Commonwealth and International Scientific Conference, Victoria, Canada.

Reilly, T. and Smith, D. (1986) Effect of work intensity on performance in a psychomotor task during exercise. *Ergonomics*, **29,** 601–6.

Rhodes, E.C. and Mosher, R.E. (1992) Aerobic and anaerobic characteristics of female university soccer players. *Journal of Sports Sciences*, **10,** 143–4.

PART TWO

Biomechanics and soccer medicine

Biomechanics applied to soccer skills

8

Adrian Lees

Introduction

Sports biomechanics offers methods by which the very fast actions which occur in sport can be recorded and analysed in detail. There are various reasons for doing this. One is to understand the general mechanical effectiveness of the movement, another is the detailed description of the skill, yet another is an analysis of the factors underlying successful performance. An important application of sports biomechanics within any sport, and soccer in particular, is the definition and understanding of skills. This can help in the coaching process and as a result enhance the learning and performance of those skills.

There are a wide range of skills which form the foundation of soccer performance. Those which have been the subject of biomechanical analysis are the more technical ones which are concerned directly with scoring. For example, in soccer, shooting at goal is an aspect of kicking and is the means by which goals are scored. Similarly, heading the ball and throwing-in can be important elements of attacking play, while goalkeeping skills are important in preventing scores.

Other skills are important in the game but have received much less attention in terms of biomechanical analysis. For example, kicking actions such as passing and trapping the ball, tackling, falling behaviour, jumping, running, sprinting, starting, stopping and changing direction

Science and Soccer. Edited by Thomas Reilly. Published in 1996 by E & FN Spon, London. ISBN 0 419 18880 0.

are all important skills in soccer but have received little detailed analysis. The skills in other codes of football have similarly received little attention in terms of biomechanical analysis.

This chapter looks at those skills in which biomechanics has been successfully applied in order to gain an insight into their mechanical characteristics.

8.1 KICKING

Kicking is without doubt the most widely studied skill in soccer. Although there are many variations of this skill due to ball type, ball speed and position, nature and intent of kick, the variant which has been most widely reported in the literature is the maximum velocity instep kick of a stationary ball. Essentially this corresponds to the penalty kick in soccer.

The mature form of the kicking skill has been described by Wickstrom (1975). It is characterized by placement of the supporting leg at the side and slightly behind the stationary ball. The kicking leg is first taken backwards and the leg flexes at the knee. The forward motion is initiated by rotating around the hip of the supporting leg and by bringing the upper leg forwards. The leg is still flexing at the knee at this stage. Once this initial action has taken place the upper leg begins to decelerate until it is essentially motionless at ball contact. During this deceleration, the lower leg vigorously extends about the knee to almost full extension at ball contact. The leg remains straight through ball contact and begins to flex during the long follow-through. The foot will often reach above the level of the hip during the follow-through. A kinetogram of the lower leg action of the kicking leg is given in Figure 8.1.

Kicking, like many of the skills in soccer, is developmental in nature and it

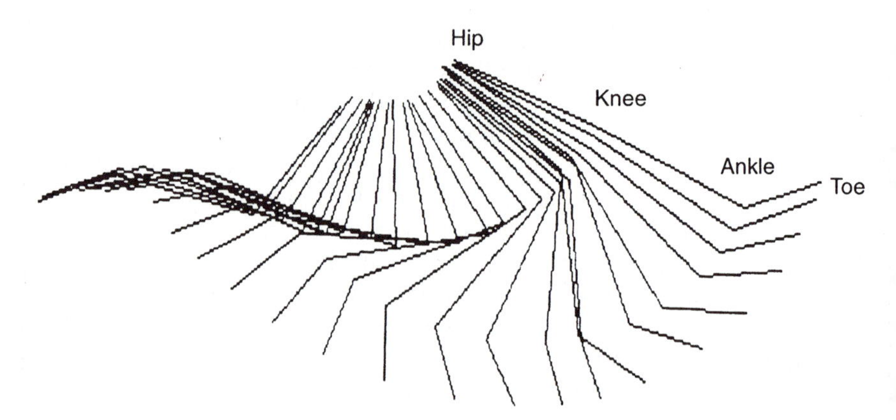

Figure 8.1 A kinetogram of the lower leg action during a soccer kick.

has been shown that it develops from an early age. Bloomfield *et al.* (1979) analysed the kicking action of young boys from the age of 2 to 12 years. They looked at various indicators of performance and were able to characterize six levels of development. These ranged from level 1 (average age 3.9 years) where the children often hit the ball with their knee or leg, to level 6 (average age 11.2 years) where the mature kicking pattern as described above had been achieved by 80% of the children. The intermediate ages for levels 2–5 were 4.11, 4.8, 6.11 and 8.2 years respectively. Although chronological age was not found to be a good predictor of level of skill development, the age ranges suggest that the skill develops rapidly between the age of 4 and 6 years. This has implications for skill development for children of a very young age and is an illustration of the role biomechanics can play in this area.

The definition of the mature skill above suggests that there are four stages to motion. The first is the priming of the leg during the backswing; the second is the rotation of both upper and lower leg forwards which occurs as a result of hip rotation and upper leg flexion; the third is when the upper leg decelerates and there is corresponding lower leg acceleration leading to impact with the ball; and the fourth is the follow-through. The two intermediate stages are the most important from a performance point of view. The interaction between the upper and lower leg can be seen in Figure 8.2 which shows the angular velocity of the upper and lower leg throughout the movement. On this graph each stage is marked, and it can be seen that during stage 2, both the lower and upper leg increase in angular velocity. The muscular energy for this must come from the muscles around the hip and upper leg. In stage 3, just before impact, there is an increase in lower leg angular velocity, while at the same time a decrease in the upper leg angular velocity. There appears to be a trade of energy between the two segments. A high angular velocity of the lower leg means a high foot velocity, and this is important in the production of a well-struck kick. It can be seen that in order to achieve high foot velocity energy must be built up in the early stage of the movement. About 50% of the lower leg angular velocity at impact is built up during stage 2, and the remaining 50% is transferred from the upper leg during stage 3. The energy of the upper leg is built up also in stage 2. Therefore the range of movement that the hips and leg move through, and the muscular strength applied during stage 2 will determine the maximal speed of the foot at impact.

It would be expected that there is a relationship between muscle strength and performance, due to the fact that the muscles are directly responsible for increasing foot velocity. Such a relationship has been found by several researchers. Cabri *et al.* (1988) found that there was a high correlation between knee flexor and extensor strength as measured by an isokinetic muscle function dynamometer and kick distance. There was also a significant relationship between hip flexor and extensor strength but this was lower than that for the knee. Similar results have been found by Poulmedis (1988) and Narici *et al.* (1988) who used ball velocity as measures of performance. If muscle strength

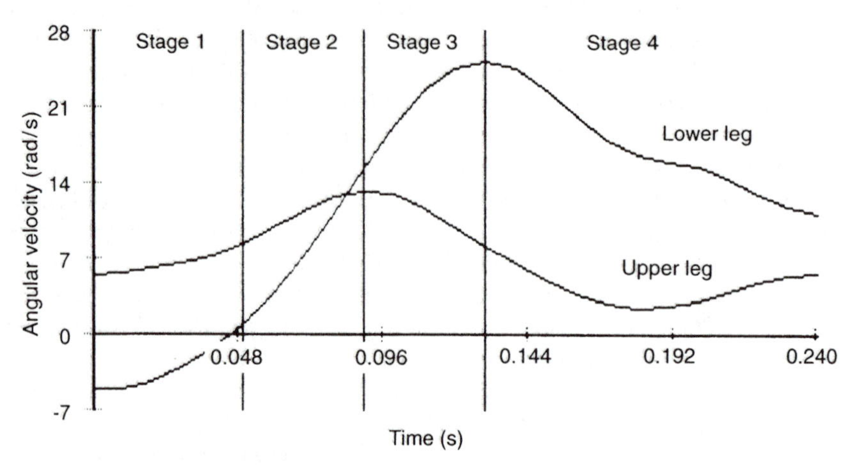

Figure 8.2 Angular velocity of the upper and lower leg during a soccer kick. The four stages of the kick are marked.

is related to performance then it would be expected that training should show positive effects on ball speed or distance. De Proft *et al.* (1988) found that over a season of specific leg muscle strength training, muscle strength increased and so too did kick performance as measured by kick distance. The correlations between leg strength and distance increased from the beginning to the end of the season.

Although the evidence reviewed above suggests that there is a good relationship between muscle strength and performance, there are other factors which contribute to successful kicks. These factors can be appreciated from a consideration of the relationship between foot and ball velocity before and after contact. By considering the mechanics of collision between the foot and ball (following the treatment of Plagenhoef, 1971), the velocity of the ball can be stated as:

$$V_{(\text{ball})} = V_{(\text{foot})} \frac{[M].[1+e]}{[M+m]} \qquad \ldots (1)$$

where V = velocity of ball and foot respectively, M = effective striking mass of the leg, m = mass of the ball and e = coefficient of restitution. The effective striking mass is the mass equivalent of the striking object (in this case the foot and leg) and relates to the rigidity of the limb.

The term $M/[M+m]$ gives an indication of the rigidity of impact and relates to the muscles involved in the kick and their strength at impact. Therefore one would expect that the best correlations with performance would be with eccentric muscle strength, and the data from Cabri *et al.* (1988) suggest that this is the case.

The term $[1+e]$ relates to the firmness of the foot at impact. Because the ball is on the ground, the foot contacts the ball on the dorsal aspect of the phalanges and lower metatarsals. The large force of impact serves to forcefully plantarflex the foot and it will do so until the bones at the ankle joint reach their extreme range of motion. At this stage the foot will deform at the metatarsal-phalangeal joint. There is little to prevent considerable deformation here and this will affect the firmness of impact and the value of e, the coefficient of restitution. Asami and Nolte (1983) measured the amount of deformation at both the ankle and the metatarsal-phalangeal joint and found that while the change in ankle joint angle did not correlate at all with ball velocity, the change in angle at the metatarsal-phalangeal joint correlated significantly with ball velocity. The conclusion from this is that the deformability of the foot should be reduced for powerful ball kicking and that this deformability is related to the deformation at the front of the foot. Contact between the foot and ball should be made as close as possible to the ankle joint and not on the toes of the foot.

The term $M/[M + m]$ would be expected to have a value of about 0.8 based on realistic data for the masses of the foot and the ball. The term $[1 + e]$ would be expected to have a value of about 1.5. Therefore the product of the two suggests that the ball should travel at about 1.2 times the velocity of the foot, in other words the ball leaves the foot faster than the foot is travelling. The relationship now becomes:

$$V_{(\text{ball})} = 1.2 * V_{(\text{foot})} \qquad \ldots (2)$$

This ratio of ball to foot velocity is an indicator of a successful kick. Any study which investigates kicking mechanics should be able to report this value. Foot velocities for competent soccer players are between about 16 and 22 m/s. Reported ball velocities are in the range of 24–30 m/s. For professional players, Asami and Nolte (1983) reported a mean foot velocity of 28.3 m/s and a mean ball velocity of 29.9 m/s. This gives a ball to foot velocity ratio of 1.06. They considered that a value greater than 1.0 is indicative of a good kick. Higher ratios are obtained for more rigid impacts where both M and e will be higher. Plagenhoef (1971) has reported values of up to 1.46 for this ratio.

For submaximal kicking there appears to be a relationship between foot and ball velocity. Zernicke and Roberts (1978) report a regression equation between the two variables of foot and ball velocity over a ball speed range of 16–27 m/s.

$$V_{(\text{ball})} = 1.23 * V_{(\text{foot})} + 2.72 \qquad \ldots (3)$$

This is reassuringly close to the relationship for maximal kicking (equation 2) suggested above on the basis of theoretical data.

A characteristic of all soccer place kicking and frequently seen in other codes is the angled approach to the ball. Isokawa and Lees (1988) investigated the effect of changing the angle of approach on foot and ball velocities. Six male subjects were required to take one step approach in order to kick a stationary ball from angles of 0, 15, 30, 45, 60 and 90 degrees. (The direction of the kick was 0°.) The foot and ball speed were measured from high-speed cine-film. Although there were no significant differences between approach directions, the trend in the data suggested that the maximum swing velocity of the leg was achieved with an approach angle of 30° and the maximum ball velocity with an approach angle of 45°. Therefore an approach angle between 30 and 45° would be considered optimum, and agrees with practical observations. The explanation for this finding is that with an angled approach, the leg also is angled to the ball in the lateral plane, and so can be placed more under the ball and make a better contact. It was noted above that a more solid impact position will produce higher ball velocities.

Other types of kick have also been studied. The side-foot kick is often used to make a pass. In order to make a side-foot kick the foot has to be angled outwards. This prevents the leg from flexing in the same way as it would for an instep kick. Therefore the foot velocity during a side-foot kick is lower than the instep kick. As contact with the ball is made on the firm bones of the lower leg and ankle, the foot provides a much better surface for the impact. The resultant ball velocity is much higher than it would be with the same foot velocity for an instep kick. A further advantage is that the flatter side of the foot allows a more accurate placement of the ball. Elliott *et al.* (1980) have investigated the developmental nature of the punt kick. They found similar results to those reported by Bloomfield *et al.* (1979) in terms of levels and ages of development of the punt kicking skill. The punt kick and drop kick have been compared by McCrudden and Reilly (1993). Using 20 adult males as subjects they found that the mean range of the best drop kick was 36.1 m while for the punt kick the range was 40.1 m. They did not measure velocities or angles of projection and so the reason for the superior range in the punt may well be due to angle of projection rather than higher ball velocity. They concluded that any recommendation to use the punt kick must be tempered by accuracy and desired angles of projection which mostly prevail in competitive play.

8.2 THE THROW-IN

The throw-in is both a method of restarting the game and a tactical skill. The long range throw-in can be performed from a stationary position or with a run-up. In a stationary throw-in the movement is performed with both feet together on the ground. The throw is initiated by bending the knees and taking the ball backwards with both hands behind the head. As the ball is travelling backwards with respect to the body there is an upward extension of the knee joint and a marked pushing of the hips both forward and upward. This serves to prime the upper body for the recoil which will propel the ball forwards. As the upper body starts to come forwards there is a sequential unfolding starting

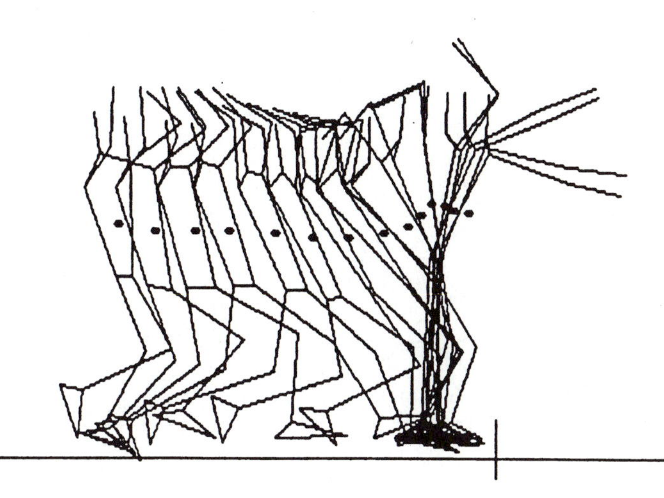

Figure 8.3 A kinetogram of a player performing a run-up throw-in (the ball is omitted from this illustration)

with the hips, then followed by the shoulders, elbows and finally the wrists and hands until ball release.

The sequencing of body segment motions is similar in the run-up throw-in and can be seen from the kinetogram in Figure 8.3 and the joint velocities in Figure 8.4. This sequential series of rotations about the medio-lateral axis serves to build up initial rotational velocity using the large muscles of the legs and trunk first, and then to transfer this energy out towards the distal segments in order to gain high end-speed velocity. This mechanism is identical in principle to that used to attain high foot velocity in the kick.

The advantage of the run-up in a running throw-in is that the ball has an initial forward speed. The running action means that one foot will be leading the other into the action. The movement goes from the rear to the front foot, but because both feet need to be on the ground, forward hip movement is restricted. The general segment motions of the upper body and their sequence are identical to those in the standing throw-in.

Kollath and Schwirtz (1988) investigated the long range throw-in action of skilled players with and without a run-up. They found that the mean distance achieved using the running throw-in was 24.1 m compared to 20.9 m for the standing throw-in. There were similar angles and heights of release in the two types of throw, and so the differences in range were attributed to differences in release speed which were 15.3 and 14.2 m/s respectively. The running throw-in is clearly superior to the standing throw-in. Within both types of throw, there were differences in the release parameters chosen to obtain similar ranges. Some players chose to use a low velocity and high trajectory while others used a high velocity and low trajectory. The low speed/high trajectory

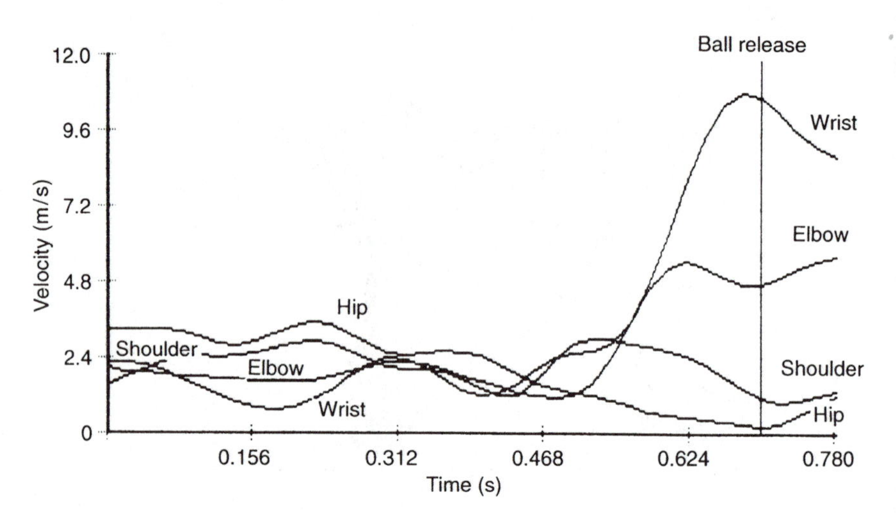

Figure 8.4 The joint velocities for a run-up throw-in as depicted in Figure 8.3. Figures 8.3 and 8.4 illustrate the rapid increase in velocity achieved by the hand, wrist and ball during the last phase of the motion.

throw might be used by players with poor muscular capability, or as a strategy to reach over a defensive wall or to ensure that the ball descends more vertically on to the awaiting players. The high speed/low trajectory throw will reduce the time the ball is in the air and might be used as a strategy to reduce the time for the opposition to regroup for the defence.

Other forms of the throw-in skill have appeared in order to take advantage of its attacking capability. One such variant is the 'handspring' throw-in which first appeared in US collegiate football during the 1980s (Messier and Brody, 1986). The player runs up with the ball in both hands, the ball is placed on the ground and the player rotates over it as in a handspring. After the feet hit the ground the body is rotating forwards with the arms and ball over the head. During the recovery from the handspring the ball is released. The rules state that for a throw-in to be legal the player must face the field of play, have both feet in contact with the ground, be on or outside the touchline at release and deliver the ball from behind and over the head using both hands equally. The handspring throw-in is a novel approach to the throw-in which does not contravene the rule. The advantage of this type of throw-in is that it is thought to have a greater velocity of release and hence a greater range.

Messier and Brody (1986) described the mechanics of the conventional running throw-in compared to the handspring throw-in. They studied 13 university level players performing the conventional throw-in and four performing the handspring throw-in. Data for the two groups at release are shown in Table 8.1.

These data show that the handspring throw-in achieves considerably

Table 8.1 Mechanical data for the conventional and handspring throw-ins at release

	Conventional	*Handspring*
Ball velocity (m/s)	18.1	23.0
Angle of release (deg.)	28.0	23.0
Distance (m)	29.3	44.0
Centre of mass velocities		
horizontal (m/s)	1.8	3.2
vertical (m/s)	0.9	–0.8
angular (rad/s)	4.5	5.5

greater range with a lower angle of projection than the conventional throw-in. A model for the conventional running throw-in is one which is characterized by the body as a whole moving forwards and upwards and rotating forwards. On to this is added the shoulder and elbow extension action. There is a sequencing of segmental actions going from the large to small segments distally, i.e. from trunk to upper arm to lower arm. The model for the handspring throw-in is one which is characterized by a body which is moving forwards at higher speed but dropping. It is also rotating forwards at a higher speed. The end-point velocity is therefore enhanced by the rotation of the whole body. There is sequential extension of the arms but this is less important. As a result the handspring throw-in is faster (23 compared to 18 m/s) with a lower angle of projection (23 compared to 28°). As a consequence of these release parameters it goes further. It would therefore be more suitable for playing strategies requiring a fast long-range ball.

8.3 GOALKEEPING

The goalkeeper has to anticipate attacks on goal and be positioned accordingly. There are a number of movement skills that the goalkeeper needs to master, but few of them have been subjected to biomechanical analysis. One exception is the diving motions made by goalkeepers in saving a set (penalty) shot reported by Suzuki *et al.* (1988). They analysed two skilled and two less skilled goalkeepers in terms of their ability to dive and save. They found that the more skilled keepers dived faster (4 m/s as opposed to 3 m/s) and more directly at the ball. In this case the skilled keeper was able to perform a counter-movement jump and launch himself into the air and then turn to meet the ball. The less skilled keeper failed to perform a counter-movement, thereby restricting his take-off velocity. He also failed to turn his body effectively to meet the ball. In this analysis both quantitative and qualitative methods were used to clarify the differences in performance between the two groups. No other goalkeeping actions have been studied in this way.

Summary

This overview has illustrated the way in which biomechanics can be applied to gain an insight into the performance of soccer skills. Many skills in the games of soccer are amenable to biomechanical analysis, but relatively few of them have been analysed in depth. There are still many opportunities for biomechanists to apply their analytical methods to soccer skills and to contribute to the development of soccer science.

REFERENCES

Asami, T. and Nolte, V. (1983) Analysis of powerful ball kicking, in *Biomechanics VIII-B* (eds H. Matsui and K. Kobayashi), Human Kinetics, Champaign, Ill., pp. 695–700.

Bloomfield, J., Elliott, B.C. and Davies, C.M. (1979) Development of the punt kick: a cinematographical analysis. *Journal of Human Movement Studies*, **6,** 142–50.

Cabri, J., De Proft, E., Dufour, W. and Clarys, J.P. (1988). The relation between muscular strength and kick performance, in *Science and Football* (eds T. Reilly, A. Lees, K. Davids and W.J. Murphy), E. & F.N. Spon, London, pp. 186–93.

De Proft, E., Cabri, J., Dufour, W. and Clarys, J.P. (1988) Strength training and kick performance in soccer, in *Science and Football* (eds T. Reilly, A. Lees, K. Davids and W.J. Murphy), E. & F.N. Spon, London, pp. 108–13.

Elliott, B.C., Bloomfield, J. and Davies, C.M. (1980) Development of the punt kick: a cinematographical analysis. *Journal of Human Movement Studies*, **6,** 142–50.

Isokawa, M. and Lees, A. (1988) A biomechanical analysis of the instep kick motion in soccer, in *Science and Football* (eds T. Reilly, A. Lees, K. Davids and W.J. Murphy), E. & F.N. Spon, London, pp. 449–55.

Kollath, E. and Schwirtz, A. (1988) Biomechanical of the soccer throw-in, in *Science and Football* (eds T. Reilly, A. Lees, K. Davids and W.J. Murphy), E. & F.N. Spon, London, pp. 460–7.

McCrudden, M. and Reilly, T. (1993) A comparison of the punt and the drop-kick, in *Science and Football II* (eds T. Reilly, J. Clarys and A. Stibbe) E. & F.N. Spon, London, pp. 362–8.

Messier, S.P. and Brody, M.A. (1986) Mechanics of translation and rotation during conventional and handspring soccer throw-ins. *International Journal of Sport Biomechanics*, **2,** 301–15.

Narici, M.V., Sirtori, M.D. and Morgan, P. (1988) Maximum ball velocity and peak torques of hip flexor and knee extensor muscles, in *Science and Football* (eds T. Reilly, A. Lees, K. Davids and W.J. Murphy), E. & F.N. Spon, London, pp. 429–33.

Plagenhoef, S. (1971) *The Patterns of Human Motion*, Prentice-Hall, Englewood Cliffs, NJ.

Poulmedis, P. (1988) Muscular imbalance and strains in soccer, in *Proceedings, Council of Europe Meeting on Sports Injuries and Their Prevention*, Papandal, The Netherlands, pp. 53–57.

Suzuki, S., Togari, H., Isokawa, M. *et al.* (1988) Analysis of the goalkeeper's diving motion, in *Science and Football* (eds T. Reilly, A. Lees, K. Davids and W.J. Murphy), E. & F.N. Spon, London, pp. 468–75.

Wickstrom, R.L. (1975) Developmental kinesiology, in *Exercise and Sports Science Reviews* (ed J. Wilmore), Academic Press, New York, pp. 163–92.

Zernicke, R. and Roberts, E.M. (1978) Lower extremity forces and torques during systematic variation of non-weight bearing motion. *Medicine and Science in Sports*, **10,** 21–6.

The biomechanics of soccer surfaces and equipment

9

Adrian Lees

Introduction

Sports biomechanics is concerned not only with the mechanical functioning of the human body in sport but also with the equipment and apparatus used. In all codes of football the equipment used has a major effect on the way the game is played. The ball itself is of a certain size, construction, weight and pressure, all of which affect the way it responds in play. The ground on which the game is played also affects the nature of the game. Surfaces for football, particularly soccer, have evolved in response to both performance and economic requirements. The controversy aroused by the introduction of the synthetic surface for soccer has led to biomechanical investigations into the performance and protection characteristics of all types of surfaces. The boot is an important piece of players' equipment and well-fitting boots can aid not only comfort but also provision of a positive interaction between player and surface to create traction for stopping, starting and turning. It can also aid player–ball interaction for passing and shooting. Boots must also protect the player and have a resistance to the stresses imposed on them during the game. Boots are not the only piece of equipment providing a protective function. Shin guards are essential in the modern game of soccer for protecting against kicks and blows, but have until recently been neglected in biomechanical investigations.

Science and Soccer. Edited by Thomas Reilly. Published in 1996 by E & FN Spon, London. ISBN 0 419 18880 0.

In this chapter an attempt is made to look at the major items of equipment used within football. The results of biomechanical research on such equipment are detailed.

9.1 FOOTBALL SURFACES

There has been much controversy in Britain concerning the use of artificial turf for playing soccer. It has been more readily accepted in North America, Scandinavia and the Middle East where there has not been the same long-standing traditions associated with the game, or where environmental considerations are important. In other parts of Europe particularly, there is little contemplation of anything other than a natural turf surface. Artificial turf is an issue in soccer not least because of the economic benefits that may accrue from its use, but also the pressure for its use in many parts of the world.

There is much conjecture concerning the merits of artificial surfaces, but only a little scientific evidence. Much of this was collected in England by a commission headed by Winterbottom (1985) and supported by the Football Association and the Sports Council. The first artificial pitch was installed in the UK in 1971, and the first Football League artificial pitch was installed at Queens Park Rangers (then in the Second Division of the Football League) in 1980. The opinions of players, managers and club chairmen were that soccer could be played to a high standard on artificial pitches but necessitated a modification to the playing of the game which often did not suit the 'British' game. Therefore a three-year moratorium was placed on the installation of artificial pitches for League soccer until Winterbottom's report had been considered fully. The report attempted to obtain scientific data on the comparative performance of both natural and artificial pitches. The author took several examples of each class of pitch from various levels of play and conducted a series of tests of performance characteristics. The tests were concerned with aspects of ball–surface interaction, player movement and player–surface interaction.

Two tests were used to establish the interaction between ball and surface. These were rebound resilience and rolling resistance. The former test established how a ball reacts after hitting the surface and relates to ball bounce, while the later test established how quickly a ball slows down when rolling over the surface. Rebound resilience was found to be 3–6% higher on artificial rather than real turf surfaces, although variation in ball type (all nominally at the same pressure) was between 3 and 7%. Variation in rebound resilience due to variations in ball pressure accounted for 4–5%. Other factors which affected the results were the spatial location on the pitch where there was less variation for artificial surfaces, and wetness which tended to reduce the rebound resilience. Various non-systematic effects were produced from the grass species used and the type of material used in the construction of artificial surfaces. In addition, as turf surfaces became older they became more compact and harder, thereby increasing the rebound resilience. The test for rolling resistance showed that this was about 20% less for artificial surfaces. This was reduced when the surface was wet. On turf surfaces spatial location could affect the results. Where the turf was more lush the rolling resistance was greater, and where worn it was lower. In both of these tests there would appear to be little difference between the two types of surface under optimum conditions. The turf surface would appear to be the more variable.

The tests used for assessing player movement were tests for torsional traction and sliding resistance between the boot and surface. The torsional traction was measured using a studded plate which was loaded on to the surface and rotated, simulating a player pressing the boot down on to the ground and twisting. The torque produced when slipping occurred was recorded. A torsional traction coefficient was calculated, and it was found that there was great variation in this for all types of surface. It ranged from 1.1 to 2.2 for turf surfaces, and 1.0 to 2.8 for artificial surfaces. It tended to be lower for sand-filled artificial surfaces compared to open weave surfaces. A major factor affecting these results was the type of stud and stud pattern, and the results presented on this topic are some of the few in any research which has considered the effect of these variables. Different stud types and patterns yielded a range of torsional coefficients from 1.7 to 2.5. It was concluded that the surfaces *per se* are not clearly different with respect to torsional traction. The test for sliding friction produced essentially the same results. There was found to be no difference between surface type, but factors such as surface pile, moisture, stud pattern and whether the front or rear of the shoe was used are all of some importance. Therefore other factors are more important with regard to sliding friction than the surfaces themselves.

Two tests were used to quantify player–surface interaction. These were based on impact characteristics of both the foot and the head. The former test used a specially constructed mechanical device known as the 'Stuttgart Artificial Athlete' which simulated the impact force produced during running. Again it was found that there was little difference between surface types, although artificial surfaces showed a little more deformation. Other important factors were whether the surface was sand-filled or open (the former absorbing energy better) and the degree of compaction of the turf. In both cases the effect of stud penetration had not been taken into account. The final test was of impact severity and was applicable to head and face impacts. A large spherical mass approximating to the head was dropped directly on to the surface from a small height. The impact deceleration was used to form an impact severity index. It was found that this was much higher on artificial surfaces, particularly the open weave type. It was thought that the reason for this was that the surface was 'bottoming out'. In other words the surface had compressed to its fullest extent and had begun to interact with the hard sub-surface layer. The addition of sand, moisture and a more shock absorbing sub-layer are construction techniques which can be used to reduce the impact severity of these surfaces.

The general conclusion of Winterbottom's committee was that in many respects there was little difference between the two types of surface, but in some important respects there were. In these respects, artificial surfaces could be designed to make their performance characteristics similar to those acceptable for natural surfaces, and the performance of pitches already laid could be controlled by the use of water to deaden a lively pitch and to provide a better energy-absorbing surface. Although artificial surfaces could be tailored to suit playing requirements, one feature of their performance does not readily match that of real turf, and that is its variability. Generally a natural surface is more varied both between surfaces and within an area of a pitch, and this is thought to be a crucial element in the game of soccer. In 1989 the Football League published its final report after the period of moratorium and concluded that artificial surfaces were not suitable for the playing of the game at a high level in the English League. They were deemed suitable for lower level play where economic advantages of the artificial surface were also of importance. This finding has been endorsed by the international authorities of the game (FIFA), and all competitive international matches are played on real turf. It should be noted that the conclusion reached by the Football League regarding the suitability of artificial surfaces was as much to do with subjective judgements of how the game should be played as with their performance, economic or injury characteristics.

9.2 THE SOCCER BALL

The full size ball is required to have an outer casing or cover which should be of leather or another approved material which does not prove dangerous to players. It should have a mass between 14 and 16 ounces (0.396 and 0.453 kg), a circumference between 27 and 28 inches (0.685 and 0.711 m), and an internal pressure of 0.6–1.1 atmospheres (60.6–111.1 kPa). The ball may be constructed in two main ways. The sewn ball is constructed of panels of leather or similar artificial material, while the moulded ball is made from rubber with cover panels bonded to the surface or from plastic with the cover panels painted on. The sewn ball may be treated to prevent the ingress of moisture, but some moisture will inevitably seep into it and increase its weight. All of these factors affect the way the ball will play. While its size only varies for junior play, its material and method of construction, and internal pressure may all vary. The effect of these variations on both performance and potential injury can be established.

The performance characteristics of the ball describe the way it flies through the air, bounces on the ground and reacts to being kicked or headed. The three main performance determinants of a soccer ball are its mass, its surface roughness and its internal pressure. The mass of the ball is restricted by the rules of the game. Small variations in mass can occur due to the ingress of water through the seams of the ball, or by absorption through the material. Both of these are less important nowadays due to the developments in ball materials and construction methods. Nevertheless, a heavier ball will have a lower velocity when kicked, although it will retain more of this velocity during flight. Of more importance are the aerodynamic forces acting on the ball. The aerodynamic forces are drag (air resistance) and lift. An important characteristic determining how air resistance will affect the ball's flight is Critical Reynold's number. This is a function of the diameter of the ball and its speed. If a ball has exceeded Critical Reynold's number then there is a reduced air resistance on the ball and it will fly further. For moderate to hard hit kicks this will be true and the ball will have a greater range than if Critical Reynold's number had not been exceeded. A lift force acts if spin is put on to the ball. The lift force may not always act vertically to lift the ball but may act horizontally, depending on the direction of spin, to cause the ball to swing away from its intended direction of flight. This spin swing effect is known as the 'Magnus effect', and is used tactically in corner kicks in soccer. The 'inswinger' and 'outswinger' are frequently used kicks from corner positions and the direction and amount of swing are determined by the direction and amount of spin put on the ball. In addition a curved ball flight is often used to get around the defensive wall from free kicks. Therefore a ball construction which allows good grip between foot and ball is important. The sewn method of construction is not only a practical method for manufacturing the ball but gives the opportunity for additional grip to be gained between the foot and ball to apply spin. In rugby football and American football the longitudinal shape of the ball enables it to be given a spin about the longitudinal axis by sweeping the foot across the ball during the punt kick. This spin causes the ball to drift towards the touch-line on its downward descent. This type of kick in rugby is often referred to as the 'torpedo' kick. Its more detailed mechanical description is given by Daish (1972).

The way in which the ball responds when bouncing from the ground depends on its internal pressure, ground characteristics and surface–ball frictional properties. The higher the internal pressure the better will be the bounce of the ball. However on soft turf surfaces, the condition of the surface is important. If the surface is too soft the surface dominates the behaviour of the ball. Only when the surface is reasonably hard will the true effects of the ball pressure be seen. When a ball bounces on the ground, the ball tends to skid, slightly reducing its forward velocity. How much is lost depends on the frictional interaction between the ball and surface. If the surface is well lubricated by water the ball will skid more easily and lose less forward velocity. As the ball also loses vertical velocity when bouncing, the effect of a ball bounce is to lose considerable energy. It will slow down and bounce less high. The interaction of the ball surface and surface condition serves to alter this effect in any one case. Therefore the predictability of a ball's bouncing behaviour will affect the playing of the game. This has been noted when discussing surfaces above.

The force imparted to the player during contact with the ball can lead to injury. Although repeated kicking of the ball will lead to overuse injuries of the toes and ankle (Masson and Hess, 1989), the main concern is usually in heading the ball. The possible injurious effect of heading the ball has been the subject of recent biomechanical investigations as a result of potential legal cases over the misuse of equipment for young players. This has furnished some useful data on ball characteristics.

Levendusky *et al.* (1988) investigated the impact characteristics of a stitched and moulded soccer ball and measured the force of impact using a force platform. They found that for velocities of impact of about 18 m/s the force of impact was about 6% higher in the stitched rather than the moulded ball. This finding has implications for the risk of injury of the players when heading a ball. Armstrong *et al.* (1988) continued this investigation by considering the effect of ball pressure and wetness on the impact force. They found that if a ball was wet it could increase the impact force by about 5% due to the extra weight as a result of water retention, and if a ball had a pressure increase from 6 to 12 psi (1 psi = 6975 Pa) there would be an increase in impact force of about 8%. These results clearly show the effect that poor combinations of conditions could have for the impact load sustained by the head during heading.

Levendusky *et al.* (1988) gave examples from the literature of where heading the ball can cause damage due to (i) surface deformation of the head leading to a broken nose, eye damage and lacerations, (ii) damage due to direct impact causing compression waves travelling through the brain creating high internal pressures, and (iii) rotational accelerations causing shearing between the brain and the skull. The levels of impact which are likely to cause injury are about 80 g for loss of consciousness and 200 g for fatalities. For rotational accelerations values greater than 5500 rad/s are likely to lead to a loss of consciousness.

Burslem and Lees (1988) investigated the acceleration on the head during a moderate speed header (ball velocity about 7 m/s) and found that accelerations were about 60 g, and rotational accelerations about 200 rad/s. Clearly there is more danger from the direct impact. In a mathematical simulation of impact Townend (1988) estimated that the average acceleration of impact was about 25 g, but increased with the reduction of mass of the player and the increase in mass of the ball, supporting the results of Armstrong *et al.* (1988). The conclusion that can be drawn here is that although heading is below the injury threshold, it is sufficiently close to it for care to be shown, particularly in dealing with young children in the development of the skill. The skill of heading can lead to greater head and neck rigidity thereby reducing the effect of the impact. This skill must be taught properly and carefully, and a reduced ball mass should be used for children.

9.3 THE BOOT

The football boot has evolved along traditional lines with features being added gradually to take account of the requirements of players and the trends within the game. The typical football boot is one which is still based on a leather construction, generally cut below the ankles, and with a hard outsole to which studs are attached. The thinness of the outsole provides the boot with its flexibility, while its hardness provides a firm surface for the attachment of studs. The studs may be either moulded as a part of the boot or detachable, and great variety is seen in sole stud patterns. Boots have a firm heel cup but do not usually include a heel counter as found in running shoes. Some boots have a raised heel to provide both heel lift and a midsole for shock absorption. Most boots will have a foam insock to aid in the provision of comfort and fit.

Although manufacturers take a systematic approach to boot design, there have been virtually no reported scientific investigations of football boot performance which have then been fed back into design. Even in soccer there has been little attempt to apply systematic investigations in order to improve boot performance. Notable exceptions are the work reported by Valiant (1988) and Rodano *et al.* (1988). Both of these studies have presented data on the vertical and horizontal forces acting on the boot. Essentially the vertical force serves to press the studs into the ground and to compress the sole of the boot, whereas the horizontal forces serve to provide traction and to deform the boot by the action of the foot on the boot leading to deformation of the heel cup, stretching or even splitting of the boot material.

There are some general principles governing the function of footwear which can be applied to the football boot. The boot in football, as indeed any form of footwear, provides an ergonomic function. It must be comfortable to wear and not be an encumbrance to the player or the play required of an individual. It must (1) relate to the demands of the game, (2) provide protection for the foot and (3) enable the foot to perform the functions demanded of it. These aspects can be considered in turn.

9.3.1 The demands of the game

The demands of the game on the boot can be established by notation analysis techniques. While these have been conducted in soccer and rugby in order to investigate the physiological and strategic demands and strategic development (Reilly and Thomas, 1976; Treadwell, 1988), few of these data can be used for an ergonomic assessment of the requirements of the boot. Therefore the functions that the boot is required to perform are based on anecdotal evidence and the experience of players.

Soccer studies such as those by Lees and Kewley (1993) provide a suitable model for assessing the demands of the game from the perspective of footwear. They looked at the physical demand which is placed on the boot during soccer

Table 9.1 Distances (m) of major movement categories

Activity	*Reilly and Thomas (1976)* m	%	*Withers et al. (1982)* m	%
Walking	2150	24.7	3026	27.0
Jogging	3187	36.8	5139	45.8
Cruising	1810	20.8	1506	13.4
Sprinting	974	11.2	666	5.9
Backing	559	6.5	874	7.9
Total distance	8680		11 211	

playing and training. They did this by identifying the major categories of playing movements made during a game of soccer, and recording their frequency of occurrence during both training and playing. The actions made by a player in soccer are many and varied and each is likely to put a unique demand on the strength of the boot. In order to obtain an indication of the role the boot has in the game it is necessary to investigate the types and numbers of actions which are made during the game. In an early study, Reilly and Thomas (1976) performed a motion analysis of the different positional roles in professional soccer. Although differences were found between positions of play the authors averaged the distances covered among the different positions to give the data shown in Table 9.1. Also in this table are further comparative data from Withers *et al.* (1982). Both sets of data show differences, but the general trends in distances covered, and their respective percentages for each activity, can be clearly seen.

In addition to this perspective both studies reported on the frequency of occurrence of some other more specific actions. Reilly and Thomas (1976) reported that the average number of jumps per individual per game was 15.5 and shots was 1.4. Withers *et al.* (1982) reported that the average number of tackles per individual per game was 13.1, jumps 9.4, turns 49.9 and contacts with the foot 26.1. Despite the differences between the two studies, the general trends are evident and clearly indicate that the general locomotor use of the boot is by far the most frequent. However, it might be expected that the more forceful actions (i.e. jumping, turning, sprinting) would be the actions putting the greater strain on the boot. Although small in number, the intensity of these actions could well be a critical factor which determines the life of the boot.

An estimate of the demand put on the boot was obtained by measuring the horizontal force on the boot during each of the categories of movements. The data from these two approaches were then combined to give an overall estimate of the demand on the boot, and related to the problems experienced by the players. The horizontal data were presented as a vector plot together with a 'stress clock' (Figure 9.1). The stress clock was produced by adding up the magnitudes of the force during foot contact which appeared in each of

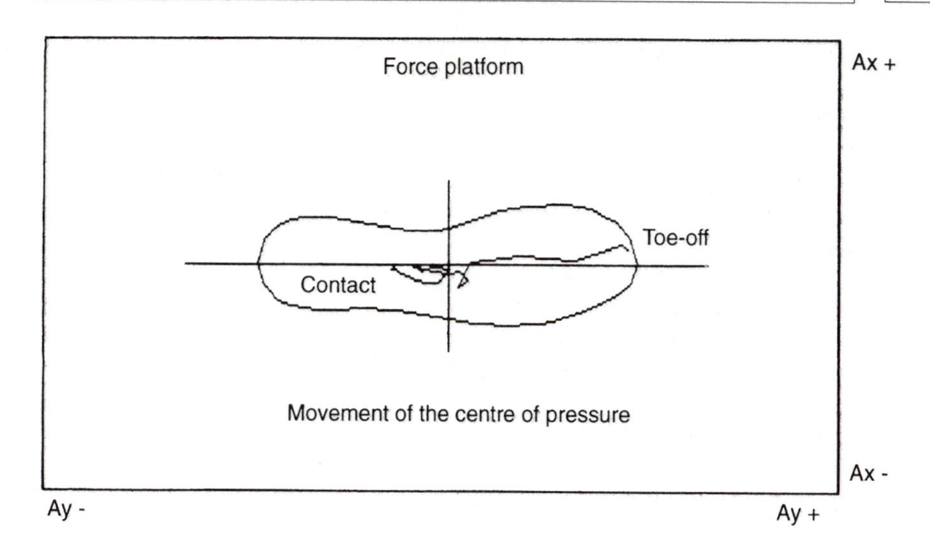

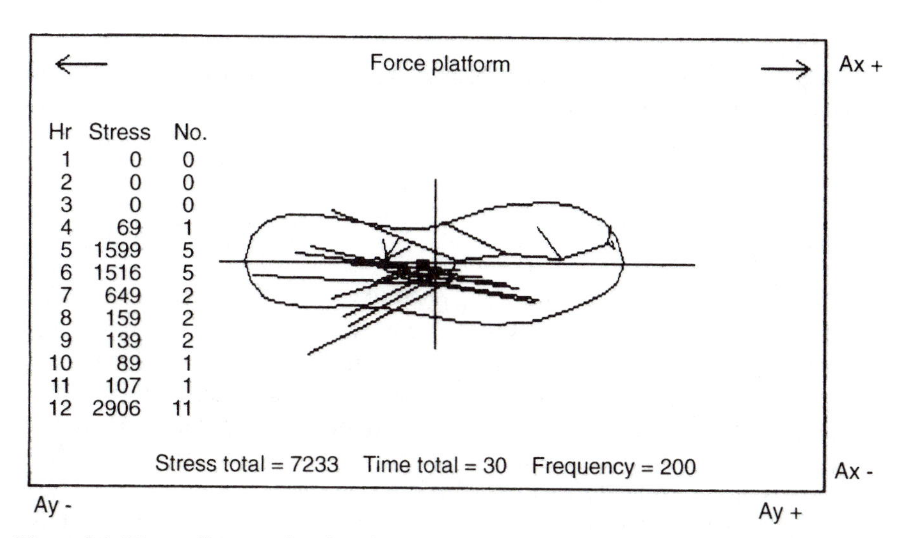

Figure 9.1 Upper diagram showing the movement of the centre of pressure across the boot during running from midsole contact to toe-off. Lower diagram showing the horizontal friction forces acting on the boot and a stress clock indicating the 'hourly' directions where most stress is located.

12 30° or 'hourly' segments. The 'hour', total stress and number of counts were given on the graphical output, together with their total and the sample rate. The accumulated force was converted to a 'severity index' for each of the playing actions. The direction of force was related to the occurrence of splits in the forefront and outside regions, corresponding very well with the main directions of the stress on the boot. It was estimated that over a period of 90

minutes playing or training the stress on the boot was three times greater in training than in playing. This has consequences for the type of boot that is used for both types of play. It was concluded that this approach to the assessment of the demand on the boot is useful, being one which had not previously been described in the literature. A subjective evaluation of the direction of stress had also provided a useful insight into the nature of the directions of stress on the boot.

The occurrence of splits in the forefront and outside regions corresponds very well with the main directions of the stress on the boot. Some of the playing actions identified were not amenable to the measurement of force (e.g. dribble, trap and tackles) and so these were omitted from the analysis. Others were more relevant and were included (e.g. locomotor actions). The average stress level on the boot which serves as a severity index can be integrated with the number of actions of each type occurring during both playing and training. If this is done the accumulated severity for 90 minutes of activity is:

professional training	161 kN
professional match-play	58 kN
amateur match-play	50 kN

While these data must be interpreted with care as they cover only a selection and not all of the actions occurring in the game, results illustrate that the demand put on the boot is likely to be considerably (i.e. three times) more severe in training than in match-play.

9.3.2 Protection

The boot must be comfortable, be a good fit to the foot, but at the same time protect from external forces, spread the pressures over the sole of the boot, and control foot movement, particularly rear foot movement. Comfort is a difficult term to define objectively. It often relies on subjective experiences of players. It is something that can change with time, or with conditions of play (e.g. wetness, foot microclimate, properties of the boot material). The fit of a boot is related to the type of last (the foot shape used for boot construction) and the materials used for its construction. There are substantial ethnic differences in foot shape, and it is unlikely that a boot designed for an American, Italian or oriental foot will fit a British foot well. Manufacturers either use standard lasts, or lasts which have been developed for other types of footwear (e.g. running shoes). Certain types of leather (e.g. kangaroo leather) have the properties of yielding to accommodate different foot shapes, while still providing a strong material, resistant to splitting. This helps to improve fit and enhance comfort.

The boot should be constructed so as to protect the foot from external forces which may arise from the ground, other players or by contact with the ball. When the foot contacts the ground the typical ground reaction force

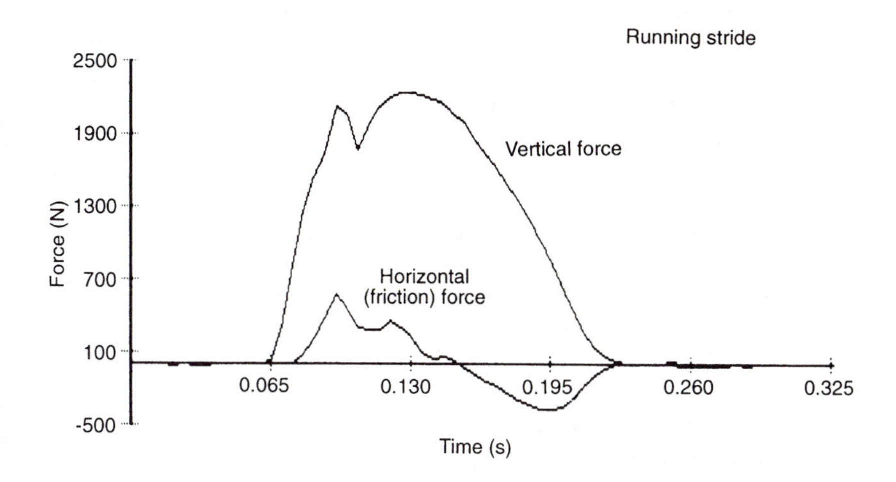

Figure 9.2 A typical vertical ground reaction force in running.

exceeds 2.5 times body weight (Figure 9.2). This force can increase as a result of running speed or type of landing action used. The force will also be higher on hard as opposed to soft grounds. The boot should have built into it materials designed to reduce the effect of these forces, but they often do not. In addition the boot should be able to distribute the force so that it is not concentrated in certain areas, such as for example under the heel, or more particularly under the head of the first metatarsal. The positioning of studs is particularly critical in this regard, as well as the method of attachment of stud to the boot. The foot is susceptible to knocking and treading by the feet of other players, and so the material of the boot should be able to provide protection to the foot from this. The use of sound or padded leather is necessary. When the ball is kicked, there is a contact force in excess of 1000 N (1.5 times body weight). This force will deform the foot but could also lead to bruising on the dorsal aspect of the foot. The force can also be a function of ball wetness, ball inflation pressure and ball construction as noted above (Levendusky *et al.*, 1988).

In running shoe design, great attention is paid to the reduction of the shock force associated with heel strike. This force is characterized by a sharp force peak whose magnitude can reach about three times body weight in sprinting (see Figure 9.2). This force can also be assessed by the use of acceleration measures on the lower tibia (Lafortune, 1991). Peak tibial deceleration in running can be up to 10 'g' (100 m/s^2). While these techniques have not been applied to study field games, the types of boot construction with thin soles suggest that this is an unimportant factor for players. While it may be less important due to the generally softer surfaces that soccer competitors play upon, nevertheless it is still a feature of any heel contact locomotor action.

This becomes progressively more important as the ground becomes harder. It is also likely to become more important as players become more used to the softness of everyday shoes and gradually lose the ability to withstand repeated hard heel impacts. The use of shock-absorbent materials placed in the heel of the boot is a standard method for reducing the severity of impact, but this is rarely provided in the boot. In addition, such protective materials would serve to raise the heel, which would put less stress on the calf muscle–tendon complex which experiences functional shortening from habitual wearing of raised heel footwear in everyday life.

During the game the most frequently used action of the boot is a normal running stride. This can be seen from the data in Table 9.1. In such a stride the foot typically contacts the ground on the lateral border of the heel. The foot then rolls over, and goes into a position of pronation. The amount of movement of the rear foot can be affected by the boot. In running shoes the shoe may actually increase this range of pronation, and special construction methods are required to control it. The boot generally does not have these anti-pronation devices, and players with excessive rear foot movement may benefit from some rearfoot control.

Of major interest in this context is the role of the boot for protecting against ankle inversion or eversion sprains. The boot was traditionally made with a high ankle support. The advent of a faster running game has led to a preference for the low-cut soccer-type boot. This boot allows greater movement of the subtalar joint, and may as a consequence lead to more frequent and more severe ankle injuries.

The ankle joint is one of the most vulnerable joints for a soccer player, and the boot is often relied upon to protect this joint from an inversion/eversion sprain or more serious damage. The role of the boot in protecting the ankle joint was investigated by Johnson *et al.* (1976). They investigated the torsional stiffness of different designs of boot uppers. They modelled the lower leg by a mass-spring-dashpot system which gave the joint its load response characteristics. The boot added another resistive layer to the outside of the ankle allowing the natural stiffness of the joint to be supplemented by the properties of the boot. The low-cut boot protected the subtalar joint, while the higher-cut boot protected both this and the ankle joint. In a simulation of the effect of using materials with differing stiffness, they found that if a low-cut boot was used it should be made of low-stiffness material. This was because the subtalar joint had a certain amount of mobility, and if the ankle was turned a low-cut boot would allow the subtalar joint to accommodate most of the movement. If the low-cut boot was of stiff construction, then the boot would transfer some of the load away from the subtalar joint to the ankle joint. As this does not have any degree of flexibility in the inversion/eversion direction the additional load would be taken up by the collateral ligaments, leading to a greater likelihood of ligamentous damage. On the other hand the high-cut boot should be made with stiff material because it already has a protective function with regard to the ankle joint and collateral ligaments. The stiffer the material, the more the load is taken by the boot material rather than the ligaments themselves. It should be noted, however, that the high-cut boot with stiff material is only about twice the stiffness of the low-cut, low-stiffness material boot, and that for a severe inversion movement even the high-cut boot would be insufficient to prevent damage occurring.

9.3.3 Performance

During performance, the boot must allow the player to perform without encumbrance and if possible to enhance the playing of the game. The boot must allow the player to run easily, and so lightness is a major consideration. However, there is some incompatibility between the lightness of materials and their ability to protect the foot. The boot must not inhibit the normal joint movement in many phases of the game, particularly running. The low-cut boot has shown itself to allow normal joint function both in terms of plantar- and dorsiflexion, and in supination and pronation. The lack of protection against excessive ranges of motion has been indicated above.

The studs are important for providing traction on a variety of surfaces. The grip provided is a function of the depth of penetration of the stud and the firmness of the turf. Very wet turf, possibly having a high surface water content, will mean that short studs fail to penetrate into the firmer ground underneath. On the other hand very hard turf will not allow good penetration, and lead to pressure areas on the foot at the heel or forefoot of the boot. Studs of varying length help to overcome some of these problems, but studs are also a source of injury to other players. In games such as rugby the traction requirements of the game differ according to playing position. The traction needed to push in a scrum is over five time that required in running. As all players must be reasonably able to apply force in a scrummage, the traction provided by the boot is overspecified for many playing actions.

Surprisingly the role of the boot in providing traction with the ground is one area which has received little attention. The amount of grip provided by a surface is an important component of playing quality. If there is too little grip the players will slip and fall, while if there is too much there is a danger that players will suffer knee and ankle injuries as their feet become locked during turns and manoeuvres. In a report for the Football Association, Winterbottom (1985) initiated an investigation into the effects of stud configuration which was later extended by the Football League (Football League, 1989). He found that the relationship of traction between boot and surface was a very complicated one, and identified two categories of movement important to players. These were sliding and turning movements. He found that the sliding resistance was affected by turf wetness as well as stud configuration. Differences between extreme conditions were as much as 300%. The torsional traction coefficients for different boot sole types ranged from 2.5 at the highest to 1.0 at the lowest. Therefore the type of sole and the stud configuration can lead to a 250% change in the degree of traction offered. This clearly should be matched with the pitch conditions, but is only ever done subjectively, and is an area which deserves greater research attention.

A point which is little considered is the ability of the boot to provide a sound dorsal surface for kicking. The foot makes contact with the ball on its dorsal surface (Asami and Nolte, 1983) and is very deformable under the large force applied during the kick. This deformation leads to reduced kicking effectiveness. The boot could strengthen the foot during this action, but would require a high flexion stiffness of the sole. However, this high stiffness would inhibit the normal flexion of the foot during the more common locomotor actions. The sole of the boot then could be designed with a hinge locking mechanism, and behave in a similar fashion to the elbow joint, for example, providing flexion in one direction but strength in the other.

9.4 SHIN GUARDS

The shin guard is used to protect the lower leg from impact injuries. These injuries can range from the severe (such as direct contact between the opponent's boot and the leg as in a poorly executed tackle) to the minor (such as bruises and scratches from glancing blows). The shin guard offers protection from some of these injuries.

The need for shin protection was evident in the early days of soccer when players used to put rolled up newspapers down the front of their socks. The shin pad evolved which was also inserted between the sock and the leg. It consisted of a foam backed plastic shield which was reinforced by wooden strips placed longitudinally down the guard. Recent trends in equipment design have spawned a wide range of different shin guard types, all following similar principles but more sophisticated in construction. A typical contemporary shin guard is integrated within an oversock which also incorporates shells to cover the ankle joint. The protective guard is constructed with a hard outer casing and a softer inner layer. The material used for the outer casing is usually thermoplastic moulded to the curvature of the leg, with a shock-absorbing inner material made of EVA (ethylene vinyl acetate) or other foam type material.

There are no performance standards for shin guard protection, although a European Standard is in the process of being introduced. Therefore there are no accepted means whereby the protective quality of the shin guard can be quantified. Manufacturers base their design on the logic of material behaviour, the efficacy of construction methods and on the opinions of players.

An attempt to quantify shin guard performance was performed by Cooper (1992). He tested the ability of five types of contemporary shin guards to reduce the impact from a direct blow. The methodology used followed the methods used for the testing of cricket pads (for which there is a British Standard: BS 6183, part 1 1981, British Standards Institution, London). This involved dropping a weighted mass directly on to the pad from a set height and monitoring its deceleration value. The pad was placed over an aluminium leg form which was rigidly held and hit by a 5 kg mass which was dropped from a height of 40 cm. The face of the striking mass was hemispherical of diameter 7.3 cm. The deceleration was measured by an accelerometer attached to the striking mass. These conditions were designed to simulate the energy of impact delivered by a cricket ball. A similar series of conditions were used in Cooper's study except a wooden leg form was used for convenience, and the diameter of the striking mass was smaller to represent the characteristics of the striking boot. The shin guards tested showed a reduction in deceleration ranging from 28 to 56% relative to the impact deceleration obtained from impacts on the wooden leg form without a shin guard in place. There is clearly a large difference between shin guard types. The poorer guard was constructed of a thermoplastic outer casing with a foam inner layer, while the better guard was of a similar thermoplastic outer shell but with an EVA inner layer. The outer layer serves to spread the load reducing the local pressure, while the shock-absorbent inner layer serves to reduce the effect of the impact load.

The shin guard also absorbs energy, but its capacity for doing this is restricted by the quantity of material within the shin guard construction, which is generally small. Therefore the shin guard can reduce the effect of bruising, glancing blows and scraping by the ground or an opponent's studs. It is unlikely to be effective against high energy direct blows which may lead to fracture. Nevertheless the shin guard provides an important protective function and its design and materials used in construction make it an important piece of equipment for the player.

Summary

This overview of the biomechanics of football equipment has shown that there are many factors which interact to affect the role of the equipment within the game. The equipment itself has mechanical characteristics which are subject to variation, but which can be reasonably well quantified. The interaction between the player and the equipment is also a source of variation. This is more difficult to quantify and to predict its effect on both performance and the protection afforded. Nevertheless, a good understanding of the general principles can be gleaned from the above examples. This should help with the application of sports biomechanics techniques to a wider range of equipment and football codes in the future.

REFERENCES

Armstrong, C.W., Levendusky, T.A., Spryropoulous, P. and Kugler, L. (1988) Influence of inflation pressure and ball wetness on the impact characteristics of two types of soccer balls, in *Science and Football* (eds T. Reilly, A. Lees, K. Davids and W.J. Murphy), E. & F.N. Spon, London, pp. 394–8.

Asami, T. and Nolte, V. (1983) Analysis of powerful ball kicking, in *Biomechanics VIII-B* (eds H. Matsui and K. Kobayashi), Human Kinetics, Champaign, IL, pp. 695–700.

Burslem, I. and Lees, A. (1988) Quantification of impact accelerations of the head during the heading of a football, in *Science and Football* (eds T. Reilly, A. Lees, K. Davids and W.J. Murphy), E. & F.N. Spon, London, pp. 243–8.

Cooper, S. (1992) A preliminary investigation into the shock attenuating characteristics of soccer shin guards. Unpublished dissertation, School of Human Sciences, Liverpool John Moores University

Daish C.B. (1972) *The Physics of Ball Games*, EUP, London.

Football League (1989) *Commission of Enquiry into Playing Surfaces: Final Report*, The Football League, Lytham St Annes.

Johnson, G., Dowson, D. and Wright, V. (1976) A biomechanical approach to the design of football boots. *Journal of Biomechanics*, **9,** 581–5.

Lafortune, M.A. (1991) Three dimensional acceleration of the tibia during walking and running. *Journal of Biomechanics*, **24,** 877–86.

Lees, A. and Kewley, P. (1993) The demands on the boot, in *Science and Football II* (eds T. Reilly, J. Clarys and A. Stibbe), E. & F.N. Spon, London, pp. 335–40.

Levendusky, T.A., Armstrong, C.W., Eck, J.S. *et al.* (1988) Impact characteristics of two types of soccer balls, in *Science and Football* (eds T. Reilly, A. Lees, K. Davids and W.J. Murphy), E. & F.N. Spon, London, pp. 385–93.

Masson, M. and Hess, H. (1989) Typical soccer injuries – their effects on the design of the athletics shoe, in *The Shoe in Sport* (eds B. Segesser and W. Pforringer), Wolfe Publishing, London.

Reilly, T. and Thomas, V. (1976) A motion analysis of work-rate in differential roles in professional football match-play. *Journal of Human Movement Studies*, **2,** 87–97.

Rodano, R., Cova, P. and Vigano, R. (1988) Design of a football boot: a theoretical and experimental approach, in *Science and Football* (eds T. Reilly, A. Lees, K. Davids, and W.J. Murphy), E. & F.N. Spon, London, pp. 416–25.

Townend, M.S. (1988) Is heading the ball a dangerous activity? in *Science and Football* (eds T. Reilly, A. Lees, K. Davids and W.J. Murphy), E. & F.N. Spon, London, pp. 237–42.

Treadwell P.J. (1988) Computer aided match analysis of selected ball games (soccer and Rugby union), in *Science and Football* (eds T. Reilly, A. Lees, K. Davids, and W.J. Murphy), E & F.N. Spon, London, pp. 282–7.

Valiant, G.A. (1988) Ground reaction forces developed on artificial turf, in *Science and Football* (eds T. Reilly, A. Lees, K. Davids, and W.J. Murphy), E & F.N. Spon, London, pp. 406–15.

Winterbottom, Sir W. (1985) *Artificial Grass Surfaces for Association Football: Report and Recommendations*, The Sports Council, London.

Withers, R.T., Maricic, Z., Wasilewski, S. and Kelly, L. (1982) Match analyses of Australian professional soccer players. *Journal of Human Movement Studies*, **8,** 159–76.

Injury prevention and rehabilitation

10

Thomas Reilly and Tracey Howe

Introduction

Soccer entails physical contact in the course of tackling or contesting possession of the ball with opponents and this inevitably leads to injury of varying severity. A majority of injuries are unintentional, resulting from an error on the part of the player concerned or by another player. The error may lead to an accident (or unplanned event) and some of these accidents lead to injury. Inflicting injury intentionally on another player is severely punished both by the laws of the game and, where the evidence is clear-cut, by civil law also.

There are many extrinsic factors which may cause injury, besides the behaviour of players. These include the state of the pitch, the weather conditions, inappropriate choice of footwear and inattention to warm-up. There are also intrinsic factors which embrace the mental state of the player, the level of fitness and the existence of predisposing factors such as muscle weakness or a previous injury.

Detailed considerations of injuries in soccer are provided in texts such as Ekblom (1994), Kulund (1982) and Lillegard and Rucker (1993). The aim of this chapter is to outline the most common injuries that occur to players, consider some predisposing factors and preventive measures. The main methods of treating injuries are delineated. First it is important to define an injury and examine the incidence of injury in soccer.

Science and Soccer. Edited by Thomas Reilly. Published in 1996 by E & FN Spon, London. ISBN 0 419 18880 0.

10.1 FACTORS AFFECTING INJURY OCCURRENCE

There is little in the way of standardization in the presentation of injury statistics. There is no common definition that has been generally accepted by those studying the epidemiology of sports injury. In consequence, it is difficult to make comparisons between analyses carried out in earlier decades with statistics from the game as currently played. Unless methodologies are similar, it is also impossible to make inferences about differences between countries.

Generally, analyses of injuries tend to be retrospective studies of the records held at the professional clubs or compiled by the medical team. Records usually detail the type of injury and the timing of the occurrence along with concise descriptive detail. The period of treatment is recorded and the diary is maintained until the player is fully recovered to play in competition. A player may feign injury as an excuse for poor performance but persistence in doing so may compromise that player's selection for the next game. Consequently, the operational definition of an injury might be one that prevents the player from training for two consecutive sessions. An injury that prevents the player from competing for a sustained period would constitute a severe injury.

For comparison with other sports in terms of risk, information in addition to the frequency of occurrence is required. The incidence of injury refers to the occurrence of new injuries in a particular time frame; the prevalence of injuries refers to the overall number of sufferers at a particular time. Exposure to injury risk may be expressed from statistics referring to injuries per 1000 hours of play or injuries per player exposure. Each game provides a reference in that it can be considered as exposing 22 players to injury over 90 minutes.

Soccer provides a different profile from running in that a majority of injuries occur during match-play or competition (Figure 10.1); in contrast, three out of four injuries to runners are attributable to training error. Training practices in soccer can be devised whereby collisions and full tacking are discouraged. Besides players may reduce the intensity of their efforts in practices if they know their place in the team is not in jeopardy. Runners may be reluctant to do this, feeling their own performances in racing may be adversely affected by a disturbed training programme.

It seems that the level of competition is influential in the incidence of injury. Player exposure risks (Table 10.1) show that the risk is higher in the first team of an English League Premier Division side, next highest in the A and youth team, and lowest in the reserves matches. In these cases injury was defined as being unable to train for two sessions on successive days.

The same trends are evident in the statistics compiled by Ekblom (1994). Injuries in games were about three times more frequent than in practice conditions, for both male and female players. The injury rate was two to three times higher in male players than in females and was considerably greater in adult professionals than found amongst youth players.

Each sport has its own characteristic profile of injuries. The majority of

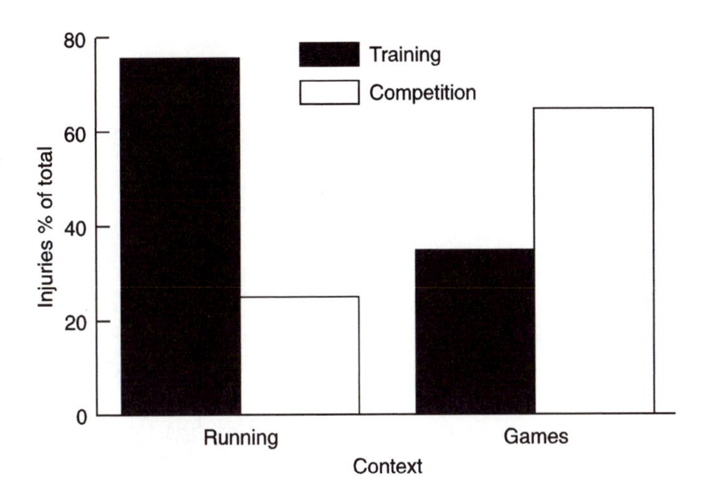

Figure 10.1 Injuries attributable to training and competition in runners and in games players (soccer, rugby, hockey and handball). (Data based on survey of Reilly and Stirling, 1993.)

injuries to soccer players are soft tissue (tendon and muscle) and joint trauma. These occur predominantly in the lower limbs; the joints most frequently affected are the knee and ankle. The muscle injuries are a combination of locomotor functions of running off the ball during play or bruising due to physical contact with other players. Soccer is not immune to trunk, head and upper limb damage. Back injury may be more disabling than lower limb muscle injury (Figure 10.2) and facial damage can occur due to opponent's elbows or clashes of heads in contesting possession of the ball in the air. Concussion is a risk in the latter case. Its seriousness is acknowledged explicitly in rugby football as players cannot resume play if concussion is suspected. Some injuries promote changes in the rules of play or in their implementation by the referee. Examples were the use of elbows in the English

Table 10.1 Injuries and player exposure risks for three categories of representative competition in a professional Premier Division English club over one full season

	Number of games	*Player exposures*	*Injuries*	*Injury per player exposure*
First team	48	528	45	1/11.7
Reserves	45	495	19	1/26.1
A and youth team	44	116	7	1/16.6
Aggregate	137	1139	71	1/16

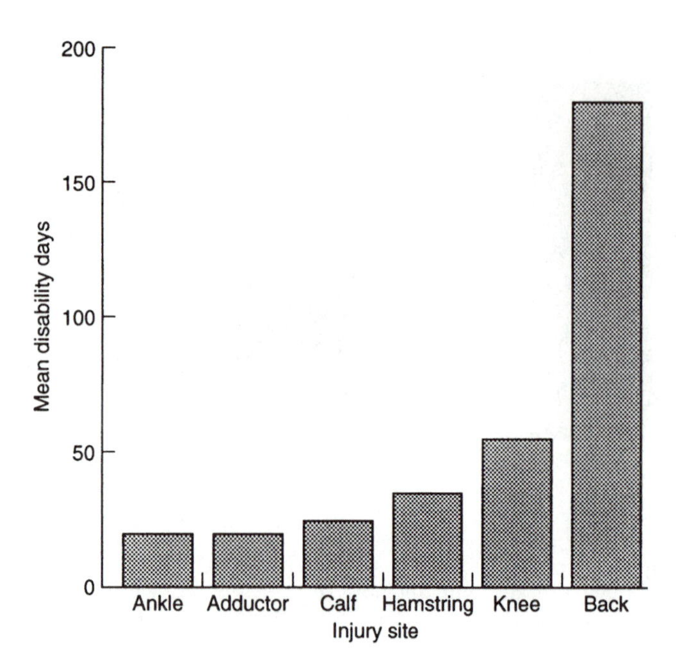

Figure 10.2 Mean disability time for severe injuries to professional soccer players. The back and knee joints have the longest period off training.

League during the 1994 season and the tackle from behind, outlawed in the 1994 World Cup.

10.2 AETIOLOGY OF INJURIES

As football is a contact sport players frequently sustain direct blows to the body, e.g. during a block tackle or from a collision with another player. These blows result in contusion injuries, a disruption of the blood vessels within the soft tissues leading to haematoma formation, or bone fractures.

Indirect injuries result from forces generated within the musculoskeletal structures during an activity. Damage may be sustained to muscles, tendons, ligaments, joint structures and bone. These types of injury often occur during the early or late stages of a game due to inadequate warm-up, poor flexibility or fatigue, as discussed previously.

Overuse injuries are caused by continued or repetitive actions or as a result of exposure of a structure to high loads. These types of injuries occur as a result of training errors, biomechanical abnormalities, inadequate or inappropriate footwear and terrain. Training errors include inadequate warm-up, excessive training regimes, sudden increases in the duration, frequency or intensity of training, and inadequate rehabilitation from injury. Biomechanical abnormalities include leg length discrepancies, soft tissue inflexibility, incorrect biomechanical alignment and joint stiffness. Problems with footwear include poor shock absorption qualities, poor grip and poor fit. Terrain may be responsible for injuries, e.g. if too much running is done on cambered road surfaces or hills or too lengthy a session on sandhills. Stress fractures, microfractures in cortical bone, resulting from excessive tensile loading, may be classified in this category of injury (Corrigan and Maitland, 1994). Such injuries are common in the metatarsals, fibula and tibia.

10.3 SOFT TISSUE INJURIES

Overstretching of a ligament results in a sprain and overstretching of a muscle results in a strain. There are three categories of ligament sprain and four categories of muscle strain. In Grade I sprains only a few fibres of the ligament are damaged. In Grade II sprains more fibres are damaged and the ligament is partially ruptured. In Grade III sprains the ligament is totally ruptured, the integrity of the joint is compromised and surgical repair is required. Grade I strains are minor injuries where only a few muscle fibres are damaged and the muscle sheath is intact. In Grade II strains the muscle sheath is still intact but there is considerable damage to blood vessels. In Grade III strains the muscle sheath is partially torn, bleeding is diffuse and a large area of muscle is involved. A Grade IV strain is a complete rupture of the muscle belly, a palpable gap is present and surgical repair is necessary. Sprains and strains may be chronic or acute injuries.

Whilst muscle damage may entail rupture of muscle fibres which is severe enough to prevent training for some days or weeks, microtrauma to muscle may not cause immediate pain. Soreness may be delayed to peak 48–72 hours after completing training regimes that employ eccentric contractions of active muscle groups (Newham *et al.*, 1983). Examples of such regimes include repetitive bounding drills or routines referred to as plyometrics. This describes repeated stretch–shortening cycles of active muscle groups. Damage to muscle is indicated by increased release of creatine kinase which leaks through the muscle membrane into the bloodstream and by disruption of the myofibrils within the muscle (Newham *et al.*, 1983). It seems that no permanent damage is done by such drills and players habituate to it with repeated sessions.

Muscle cramp is a condition where the muscle goes into spasm and fails to relax. It is seen mostly towards the end of a game or during extra time. The muscles mostly affected are the calf muscle (gastrocnemius) and the quadriceps. The condition is relieved by stretching the affected muscle. It is associated with reduced energy stores and hydration states within the muscle (Edwards, 1988).

Tendinitis is an inflammation of a tendon within its sheath. Tenosynovitis is inflammation of the synovial lining of the tendon sheath. Both conditions are accompanied by tenderness, swelling and pain on contraction of the muscle that causes movement of the tendon within its sheath. A common site is the Achilles tendon, often caused by ill-fitting boots or poor flexibility. Achilles tendinitis and plantar fasciitis (insertion tendinitis of the plantar aponeurosis at the calcaneous) are injuries which the soccer player shares in common with runners. These are associated with the high force levels through these soft tissues.

Shin splints is an overuse injury, often affecting soccer players who train on hard surfaces or with inappropriate shoe studs. Whilst 'shin splints' is a blanket term describing pain in the anterior lower leg, it mainly refers to a musculotendinous inflammation of the medial margin of the tibia (Lennox, 1993).

Bursae are sacs of synovial fluid which occur at points of friction, e.g. where the patellar tendon rubs over the tibia. Bursitis is an inflammation of such a structure. A bursa may become inflamed as a result of a direct blow, e.g. the prepatellar bursa during a fall on the knee. This is often a painful condition which is slow to resolve.

10.3.1 Injuries of the thigh

Hamstring strains are commonly reported in soccer players. Generally, damage is at the musculotendinous junction although injury may occur within the belly of the muscle. Mostly these injuries result from a quick forceful stretch of the muscle in an attempt to accelerate, decelerate or stretch for the ball. Injuries to the quadriceps may be a result of imperfect kicking or a blocked tackle. The muscle group may suffer contusions from direct blows. Occasionally a large haematoma within the quadriceps will require surgical drainage. Myositis ossificans, in which part of the muscle tissue becomes ossified, is a complication of a thigh muscle injury, particularly following a direct blow to the muscle (Lennox, 1993).

Injuries to the adductors of the thigh are also a feature of soccer. These muscles are stretched when the kicking leg crosses the non-kicking leg. Injuries may occur on either side and are associated with tightness in this muscle group. Rectus femoris, iliopsoas and rectus abdominis as well as adductor longus may be associated with groin pain. The condition referred to as 'Gilmour's groin' implicates the inguinal lining in groin pain.

The iliotibial band refers to a thick sheet of fascia that runs down the side of the leg between the iliac crest of the hip crossing the knee joint to insert on the lateral condyle of the tibia. Running may cause an overuse injury to this structure which is known as iliotibial band friction syndrome (Corrigan and Maitland, 1994). The soreness is due to friction on the iliotibial band when it slides backwards and forwards over the lateral femoral condyle when flexing and extending the knee joint.

10.3.2 Injuries of the leg

Jumper's knee is another overuse injury sometimes incurred by soccer players. The syndrome gets its name due to its frequency in jumpers and basketball players. It includes tendinitis, degeneration and sometimes partial rupture of the patellar tendon. It manifests in anterior knee pain, tenderness over the patella and is aggravated by contracting the knee extensors.

10.4 JOINT INJURIES

Injuries to the joints may result from direct blows but more commonly occur as a result of forces generated within the musculoskeletal structures during an activity.

10.4.1 Knee joint

The knee joint is particularly vulnerable in a sport such as soccer, being a hinge joint with long levers on either side of the joint. Besides the articulation between femur and tibia, the knee also includes the articulation between the patella (knee cap) and the anterior surface of the distal end of the femur. There are ligaments on the medial and lateral sides of the knee and a pair of ligaments within the joint but due to a fold in the joint capsule they are not enveloped by it. The cruciate ligaments lie within the knee joint and cross each other in the shape of an X. The anterior cruciate ligament runs upwards posteriorly and laterally from the anterior surface of the intercondylar area of the tibia to the medial surface of the posterior aspect of the lateral femoral condyle. Its function is to withstand forces displacing the femur backwards with respect to the tibia. The posterior cruciate ligament runs downwards anteriorly and medially from the posterior surface of the intercondylar area of the tibia to the lateral surface of the anterior aspect of the medial femoral condyle. Its function is to withstand forces displacing the femur forwards with respect to the tibia. Thus the integrity of the cruciate ligaments is important in securing the stability of the joint.

Injury to the anterior cruciate ligament (ACL) is a major threat to the career of the professional soccer player. Contemporary methods of ACL reconstruction after rupture, using the middle portion of the patella tendon or carbon fibre grafts, have meant that players may return to match-play despite a rupture of the ligament (Vierhout, 1993). A prolonged period of rehabilitation for restoring the strength of the muscles around the joint and proprioception is needed after surgery (Shelbourne and Nitz, 1990; Anderson *et al.*, 1991). Particular emphasis should be placed on the hamstrings as these muscles have a similar function to the ACL, preventing forward displacement of the tibia with respect to the femur. Unfortunately, the performance of repaired or reconstructed ACLs is not as good as that of intact ACLs due to the diminished proprioceptive feedback which may impair function and does not protect the joint sufficiently from re-injury.

The medial and lateral collateral ligaments prevent valgus and varus subluxation of the knee joint respectively. A blow to the outside of the knee, forcing the joint to open medially (valgus subluxation), will damage the medial collateral ligament. Conversely, a blow to the medial side of the knee, forcing the joint to open laterally (varus subluxation) will damage the lateral collateral ligament. Such injuries commonly occur during block or high tackles and result in instability of the knee joint.

Damage to the medial meniscus, historically referred to as 'cartilage' damage in soccer players, is another hazard of match-play. It is usually associated with weight-bearing when the foot is fixed to the ground and the knee joint is rotated medially in relation to the foot. This can happen when a player is hit from the side in a tackle and the knee joint is rotated whilst the foot is still on the ground. A swivel-boot was designed to allow the foot also to rotate in such an event in American Football but the design was never taken seriously for soccer. Recent advances in the surgical management of such injuries, employing arthroscopic techniques, together with aggressive rehabilitation programmes have allowed players an early return to competition (Vander Schilden, 1990).

A most damaging injury to soccer players is where the medial meniscus, medial ligament and anterior cruciate are injured together, a combination known as O'Donoghue's triad. The stability of the joint will be severely affected as a result. A long period of rehabilitation in which the strength of the quadriceps and hamstrings is built up usually follows surgery (Kannus and Jarvinen, 1990).

It is uncommon for footballers to fracture their patella; however, a direct blow to the medial side of the patella, e.g. in a high tackle, can cause it to dislocate laterally.

10.4.2 Ankle joint

Ankle sprains constitute about 20% of all injuries to soccer players (Ekblom, 1994). Most injuries affect the lateral ligaments of the joint, due to inversion and plantarflexion of the ankle. Injuries to the deltoid ligament on the medial side also occur but these are less common, being associated with pronation and outward rotation of the foot.

'Footballer's ankle' is a condition where bony growths, exostoses, develop on the anterior and posterior margins of the tibia and talus. These are thought to be due to repeated trauma to the joint (O'Neill, 1981). During powerful kicking the ankle joint is in a position of full plantarflexion. This results in the apposition of the posterior aspect of the lower border of the tibia and the posterior aspect of the talus. Conversely, during the push-off phase of running the anterior borders of the tibia and talus hit each other. The force of this is increased during acceleration (Corrigan and Maitland, 1994). These exostoses can be dislodged within the joint leading to persistent pain in the ankle. Surgical removal is often necessary (Biedert, 1993).

10.4.3 Foot injuries

Injuries to the feet are inevitable in soccer players due to direct blows from opponents when shots are blocked or to physical contact in tackling or contesting possession. They may also be due to faulty footwear or interactions with the playing surface. 'Turf toe syndrome', for example, refers to a sprain of the plantar capsule of the metatarsophalangeal joint of the big toe. Its cause is forceful dorsiflexion of this toe because of increased friction between the shoe and a hard or artificial playing pitch. The joint is especially painful during the push-off in a fast run.

Dorsiflexion to 90° at the metatarsophalangeal joint of the big toe is needed at push-off. Dorsiflexion is limited at the big toe of soccer players by a chronic condition known as hallux rigidus (Ekblom, 1994). This is a chronic injury due to repeated minor injury to this joint. Use of a stiffer sole may reduce pain but surgery is sometimes carried out to treat the problem.

The feet are subject to a host of minor niggling injuries following match-play. Sub-ungual haematoma underneath the toenail can produce acute pain. Blisters on the soles of the feet and on the toes can be extremely discomforting. They are especially seen when players start pre-season training or use new boots. Tendons and ligaments on the feet are also subject to strain. One danger of carrying these conditions into strenuous training or match-play is that they may cause other musculotendinous injuries due to asymmetry as the player favours the use of the most comfortable limb.

10.4.4 The shoulder joint

Dislocation of the shoulder joint results from a fall on the outstretched arm. Often this involves tearing of the rotator cuff muscles which give the shoulder joint its stability and control of movement. This injury may delay or prevent a player from returning to training or competition as the upper limbs play an important role in the maintenance of balance during running, kicking and tackling.

10.5 BONY INJURIES

Direct and indirect blows to the body may result in fracture. Fractures of the ribs may occur when one player lands on another following a collision or a tackle. This injury can be debilitating and may require a long period out of the game until the fracture has healed. Fractures of the clavicle and bones of the upper extremity occur from direct blows or falls on the outstretched arm. Blows to the head during contesting the ball in the air often result in fractures of the facial bones and skull. These injuries may have serious consequences and often necessitate surgery.

Fractures of the lower limb, especially the tibia and fibula, occur as a result of a block tackle. Such fractures require long periods of rehabilitation. Players frequently 'fall over the ball'. In this instance a player's time to react is dramatically reduced due to their body being close to the ground and fractures of the lower portion of the tibia and fibula may result.

10.6 TREATMENT AND REHABILITATION

The recovery from injury depends on accurate diagnosis in the first instance, proper first-aid and secondary treatment, a planned period of rehabilitation and a graded progress towards return to competition. These are subjects on which a great deal has been written (e.g. Reilly, 1981; Kulund, 1982; Harries *et al.*, 1994) and are outside the scope of this text.

Knowledge of first-aid is essential for both training and paramedical staff in attendance at practice sessions and matches. It is important to have basic first-aid facilities available on-site. There must also be a quick route of access to specialist medical facilities through the club's network of consultants or local hospitals. Modern medical imaging techniques, notably nuclear magnetic resonance imaging (MRI), have enhanced diagnostic facilities for soft tissue and joint injuries. Arthroscopy has enabled exploratory surgery and visualization of intact and damaged structures within joints prior to decisions about the wisdom of open surgical interventions.

The body has its own mechanisms of repairing damaged tissue, whether this is bone, tendon, ligament or muscle. For bone to reunite in correct biomechanical alignment, the separated portions must be repositioned correctly. This frequently involves surgical insertion of plates, screws and wires to hold bone ends and fragments together. This procedure accelerates the repair process and rehabilitation. Ligaments and tendons have much poorer blood supply than skeletal muscle and so their recovery takes longer. Mature skeletal muscle has a great capacity for regeneration and this process starts very soon after the damage to its cells occurs. The repair processes include formation of non-contractile collagenous fibres as well as the regeneration of new muscle cells.

The primary aim of the immediate treatment of soft tissue injuries is to control haematoma formation and avoid further damage to the soft tissues. The body part should be rested (**R**) for 24–48 hours and ice (**I**) applied intermittently with compression (**C**) and elevation (**E**) of the body part to reduce blood flow and subsequent oedema (**RICE**).

Although immobilization is essential immediately post-injury even short periods of immobilization (48 hours) may have devastating effects on the body. Muscle tissue begins to atrophy, the biomechanical properties of ligaments decrease and changes in the histological properties of articular cartilage occur. To prevent these changes it is essential that an active exercise regime is commenced as soon as possible. This should be prescribed and monitored by a chartered physiotherapist and should be within the limits of pain tolerance of the injured player. Such a regime will improve local circulation, facilitate the reabsorption of the haematoma and tissue exudate, maintain or restore muscle strength, fatigue resistance and flexibility of the injured muscle, proprioception of the injured joint and coordination of the limb. Progress may be monitored regularly using standard performance tests or more complex performance measures such as peak muscular torques during isokinetic movements (Perrin, 1993).

Electrotherapeutic modalities may also be employed by the chartered physiotherapist: these include ultrasound, transcutaneous electrical nerve stimulation (TENS), interferential therapy (IFT), pulsed short-wave diathermy (PSWD), laser and percutaneous neuromuscular stimulation. These modalities may be used for their analgesic effects, to reduce inflammation, to accelerate healing and maintain or improve muscle strength and size. The chartered physiotherapist may also use manual techniques that include mobilization and manipulation techniques, massage and proprioceptive neuromuscular facilitation (PNF) (Corrigan and Maitland, 1994).

Drugs such as non-steroidal anti-inflammatories (NSAIDs) have analgesic, antipyretic and anti-inflammatory properties and thus have a role in the treatment of soft tissue injuries (Stankus, 1993). Topical creams may be applied to facilitate absorption of superficial haematomas.

Orthotic devices or supports may be required to enable the player to bear weight on an injured lower limb or return to training and protect the injured body part against re-injury (Kannus and Jarvinen, 1990). Such devices may include crutches, walking sticks, knee braces, ankle supports, strapping and orthotic inserts inside shoes.

10.7 PREVENTIVE MEASURES

It is axiomatic to state that prevention is easier than cure. Identification of injury predisposition is a first step towards prevention, although this is often neglected even at the highest level of soccer play. Besides, games players are reluctant to recognize intrinsic factors that are responsible for incurring injuries and attribute about 50% of their injuries to chance (Figure 10.3).

Players who carry muscle weaknesses into competition are likely to experience situations where the muscle fails. Such weaknesses can be identified if players have a regular profiling of their muscle strength capabilities (Kannus

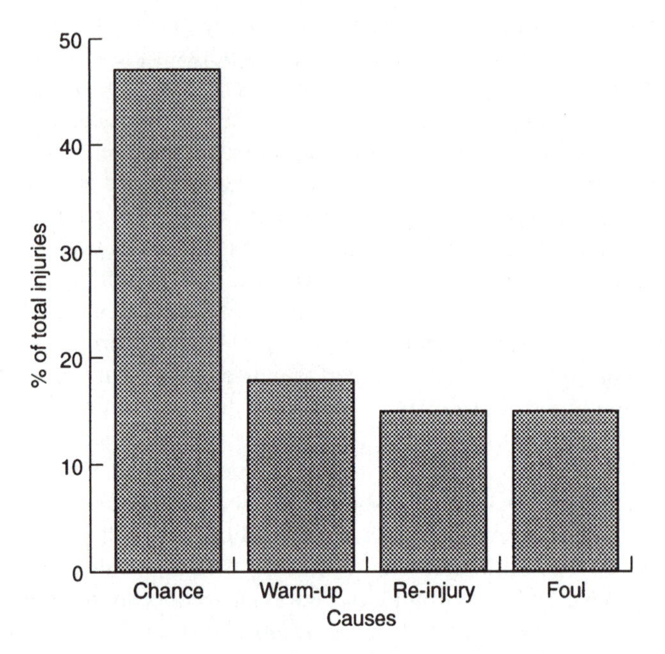

Figure 10.3 Causes to which games players attribute their injury. (Data based on survey of Reilly and Stirling, 1993, including rugby, hockey and handball as well as soccer players.)

and Jarvinen, 1990). This facility is available for teams with a systematized 'sports science support' programme. Muscle strength profiling should also show up asymmetries between left and right limb, the weaker of which is the side most likely to be affected in locomotor sports. Asymmetry may also be reflected in improper hamstrings to quadriceps ratios. Soccer players may acquire strong quadriceps but must also balance this by training the hamstrings. It seems that attention should be paid to eccentric as well as concentric contractions in training, in view of the eccentric role of the hamstrings in actions such as kicking a ball.

Strength angle profiles can be determined using isokinetic force data at a selection of angles throughout the range of movement at a particular joint (Perrin, 1993). This is especially relevant in avoiding re-injury, since reduction of strength may be evident only in a restricted range of motion. This could be corrected by recommending isometric exercises for the range of motion where muscle strength had been reduced.

In a study of Swedish soccer players, personal factors such as joint stability, muscle tightness, inadequate rehabilitation and lack of training were deemed responsible for 42% of all injuries observed (Ekstrand and Gillqvist, 1982). In an extension of this research, Ekstrand (1982) reported that 67% of soccer players had tight muscles and such players were vulnerable to injury. Tightness was pronounced in the hamstring and hip adductor muscle groups. As a programme of flexibility training among Swedish professional soccer players over a complete season was found to reduce the incidence of injury, there is little doubt that flexibility is an important protective factor against injury.

Stretching muscles prior to training and match-play affects flexibility in the short term. Flexibility routines can be incorporated into the warm-up. As flexibility is particular to each joint rather than representing a whole-body characteristic, it is important that the stretching routine is appropriate for footballers (Reilly and Stirling, 1993). The incidence of injuries over a season was less in games players who paid attention to jogging, technique work (to rehearse game skills) and lower body flexibility exercises than those who warmed up for the same duration but who had a more general type of warm-up regime. Warm-up is especially important in cold conditions in order to raise body temperature for the more strenuous training drills to follow.

Summary

The emphasis placed in this chapter has been on the occurrence of the major injuries in soccer, the main methods of treatment and the importance of preventive strategies. Whilst some injuries can be prevented and the risks of re-injury reduced, damage due to reckless play (especially by opponents) cannot be anticipated. Nevertheless, training practices can include collision-avoidance drills and routines that improve the ability to ride tackles safely. This demonstrates that training and coaching staff can contribute towards injury prevention and that the medical care of the player is a team effort. This team includes coaching personnel and management as well as chartered physiotherapist, paramedical and medical personnel.

REFERENCES

Anderson, C., Odensten, M. and Gillqvist, J. (1991) Knee function after surgical or nonsurgical treatment of acute rupture of the anterior cruciate ligament. A randomised study with a long-term follow-up period. *Clinical Orthopaedics and Related Research*, **264**, 255–63.

Biedert, R. (1993) Anterior ankle pain in football, in *Science and Football II* (eds T. Reilly, J. Clarys, and A. Stibbe), E & F.N. Spon, London.

Corrigan, B. and Maitland, G.D. (1994) *Musculoskeletal and Sports Injuries, Butterworth-Heinemann, Oxford.*

Edwards, R.H.T. (1988) Hypotheses of peripheral and central mechanisms underlying occupational muscle pain and injury. *European Journal of Applied Physiology*, **57**, 275–81.

Ekblom, B. (1994) *Football (Soccer)*, Blackwell Scientific, Oxford.

Ekstrand, J. (1982) Soccer injuries and their prevention. Medical Dissertation No. 130, Linköping University.

Ekstrand, J. and Gillqvist, J. (1982) The frequency of muscle tightness and injuries in soccer players. *American Journal of Sports Medicine*, **10,** 75–8.

Harries, M., Williams, C., Stanish, W.D. and Micheli, L.J. (1994) *Oxford Textbook of Sports Medicine*, Oxford University Press, New York.

Kannus, P. and Jarvinen, M. (1990) Nonoperative treatment of acute knee ligament injuries. *Sports Medicine*, **9**(4), 244–60.

Kulund, D. (1982) *The Injured Athlete*, Lippincott, Philadelphia.

Lillegard, W.A. and Rucker, K.S. (1993) *Handbook of Sports Medicine*, Andover Medical Publishers, London.

Lennox, C.M.E. (1993) Muscle injuries, in *The Soft Tissues. Trauma and Sports Injuries* (eds G.R. McLatchie and C.M.E. Lennox), Butterworth-Heinemann, Oxford, pp. 98–100.

Newham, D.J., Mills, K.R., Quigley, B.M. and Edwards, R.H.T. (1983) Pain and fatigue after concentric and eccentric contractions. *Clinical Science*, **64,** 55–62.

O'Neill, T. (1981) Soccer injuries, in *Sports Fitness and Sports Injuries* (ed. T. Reilly), Faber and Faber, London, pp. 127–32.

Perrin, D.H. (1993) *Isokinetic Exercise and Assessment*. Human Kinetics, Champaign, IL.

Reilly, T. (1981) *Sports Fitness and Sports Injuries*, Faber and Faber, London.

Reilly, T. and Stirling, A. (1993) Flexibility, warm-up and injuries in mature games players, in *Kinanthropometry IV* (eds W. Duquet and J.A.P. Day), E. & F.N. Spon, London, pp. 119–23.

Shelbourne, K.D. and Nitz, P. (1990) Accelerated rehabilitation after anterior cruciate ligament reconstruction. *American Journal of Sports Medicine*, **18**(3), 292–9.

Stankus, S.J. (1993) Inflammation and the role of anti-inflammatory medications, in *Handbook of Sports Medicine* (eds W.A. Lillegard and K.S. Rucker), Butterworth-Heinemann, Oxford.

Vander Schilden, J.L. (1990) Improvements in rehabilitation of the postmenisectomized or meniscal-repaired patient. *Clinical Orthopaedics and Related Research*, **252,** 73–9.

Vierhout, P.A.M. (1993) Arthroscopic reconstruction of the anterior cruciate ligament in professional soccer players using the Leeds–Keio ligament, in *Science and Football II* (eds T. Reilly, J. Clarys and A. Stibbe), E & F.N. Spon, London, pp. 425–8.

Psychology and injury

11

Frank Sanderson

Introduction

Injury blights the lives of many soccer players and affects the fortunes of many teams. That the personal and economic costs of injury are considerable is evidenced, for example, by the injury sustained by Paul Gascoigne in the 1991 FA Cup Final. The injury placed his future in the balance and has had major financial consequences for both the Tottenham and Lazio clubs. His subsequent injuries compounded the problems for Gascoigne, his clubs and his country.

As the costs of injury have risen, there has been increasing investment in the prevention and treatment of injuries. Not only are sports equipment and playing surfaces better than ever, but also there are more specialists emerging in sports medicine. But how much attention is actually being paid to psychological aspects of injury? Are there psychological dispositions which make injuries more likely to occur? What about the psychological implications of injury? And are there lessons we can learn from psychology which will help in the prevention and treatment of injury?

The aim of this chapter is to explore answers to these questions, and in the process, promote an awareness that there is an important psychological dimension to injury.

Science and Soccer. Edited by Thomas Reilly. Published in 1996 by E & FN Spon, London. ISBN 0 419 18880 0.

11.1 INJURY PRONENESS

It is quite common, of course, to be 'injury prone' because of physical or physiological factors, e.g. recurring injury due to an anatomical weakness, perhaps caused by previous injury. In this section, we are concerned with psychological factors which might cause a player to be injury prone.

11.1.1 Life stress and injury

Some researchers have argued that life stress is cumulative in its effects, enhancing the likelihood of injury by affecting the player's concentration on the task in hand. Bramwell and co-workers' (1975) finding that those players with the greatest accumulated life change (as caused by stressful life events, such as 'death of a spouse', 'getting divorced' or even 'getting married') were more likely to experience injury can be explained in terms of stress-induced attentional narrowing. More recently, it has been found that the frequency of injury is related to negative rather than total or positive life stress (Passer and Seese, 1983; R.E. Smith *et al.*, 1990) The latter also reported that only those athletes low in coping skills and in social support demonstrated a significant relationship between life stress and injury. Hanson *et al.* (1992) found that coping resources, negative life stress, social support and competitive anxiety were predictive of injury severity. May *et al.* (1985) suggested that athletes' experience of psychological stress has negative effects on their self-esteem, concentration and general emotional balance.

11.1.2 Personality and injury proneness

There are obvious connections between an individual's personality and the susceptibility not only to illness but also injury. For example, neurotic individuals tend to be cautious, timid, indecisive and easily stressed. These characteristics are likely to predispose the individual to injury in that the nervous player might, for example, be less than fully committed in a 50–50 tackle and sustains an injury as a consequence. Research evidence in support of this hypothesis is limited. Reilly (1975) found a relationship between apprehensiveness amongst professional soccer players and the number of joint injuries sustained in a season. Jackson *et al.* (1978) found that 'tender-minded' American Football players were more likely to be injured and that the more 'reserved' players tended to have the most severe injury. Others have failed to find a relationship between personality factors and injury (e.g. Kraus and Gullen, 1969; Brown, 1971).

A player's personality influences the way he or she tends to perceive things. **Perceptual style** is claimed to be important in that some people are more sophisticated perceivers than others. So-called field independent individuals have excellent 'vision' (in the sense that a good midfield soccer player is said to have good vision). They are the ones who tend to 'ride' the tackles and stay out of trouble. Field dependent types cannot easily focus on relevant visual information and may sustain injury as a consequence.

Nideffer (1989) has identified the related concept of 'attentional style'. He argued that individuals have a preferred attentional style, e.g. a broad external focus or a narrow internal focus, and that, under stress, they become more heavily dependent on this style, irrespective of its appropriateness in the situation. The notion of the individual with a preferred narrow focus displaying a stress-induced 'tunnel-vision' and thereby being more injury prone has intuitive appeal.

Research suggests that the intervening variable between personality and injury is stress in that an individual's personality may predispose him or her to experience stress in a wide variety of situations. Support for this link between stress and injury was provided by Davis (1991) who found that swimmers and American Football players experienced a reduction in the incidence of injury after they had commenced a regular progressive relaxation programme.

11.1.3 Anxiety reactions and injury

Sanderson (1981a) has outlined possible links between injury proneness and the individual's subconscious attempts to cope with anxiety reactions. Anxiety can be caused, for example, by the player's ambivalence about competition and aggression. Positive feelings about participation are encouraged because there are rewards, such as victory and prestige, to be had. However, the game also offers discomforting aggression and possible injury. The anxiety and tension which are associated with such conflict can seriously affect performance and increase the likelihood of real or imagined injury, as an unconscious means of reducing tension.

Injury resulting from counterphobia

A player who finds the game very anxiety-inducing may attempt to counteract the anxiety by meeting it head-on, by being overtly aggressive and fearless – the physical version of 'whistling in the dark'. Soccer will have some individuals of this kind, although Moore (1967) maintained that they tend to be attracted to high-risk sports such as downhill skiing, boxing, rugby or motor racing.

Closely related is the situation where an anxious player uses injury as a badge of courage, an overt sign of masculinity. He lacks real confidence, needing the visible scars of battle to confirm his manhood.

Injury as a weapon

Conversely, there is another kind of athlete who uses injury as a means of punishing another or others in an indirect way. Ogilvie and Tutko (1971) gave the example of the reluctant player forced to play because of an athletically frustrated father. By being injured, he can accomplish several objectives: he can make his father feel guilty for pressurizing him; he can frustrate his father's displaced aspirations; and he can avoid the undesired competition. A soccer player in dispute with the club could well use injury in this kind of way to frustrate the management. The player may wish to cause difficulties for the team and/or the coach because of real or imagined grievances – a particularly effective technique when the player is valuable. The player may lack courage for a confrontation, and so reacts in this indirect but effective way.

Injury as an escape

There are players who fear competition so much that injury provides an ideal way of reducing anxiety. With injury, whether real or imagined, the feared competition can be avoided, perhaps even without squad membership being jeopardized. The ego can also be kept intact: had there not been so much injury, the player can believe that they would have been (even more) outstanding. A player's disability can be used by the team-mates as a rationalization for any shortcomings in the team's performance.

Psychosomatic injury

Unconscious and powerful psychological forces can sometimes precipitate psychosomatic injury. The player frequently complains of injury and yet no organic reasons can be obtained to substantiate the claim. The player does not respond to conventional treatment. If that player is a key member of the squad, it is extremely frustrating to all concerned and may lead to a build-up of resentment in the team, thereby exacerbating the individual's psychological problems. Once the underlying emotional difficulties are resolved, then the physical problems disappear.

In each of the cases discussed above the fundamental cause of the injury is psychological, with a recurring theme being that the individual, often unconsciously, is attempting to cope with anxiety reactions.

It is advantageous that all those concerned with ensuring the player's complete recovery should have as much information about the individual as possible. Communication amongst players, coaches, trainers, physiotherapists and physicians is essential. It is possible for vulnerability profiles to be established on the basis of comprehensive information, thus allowing positive injury prevention.

Table 11.1 Symptoms associated with the psychologically vulnerable player

Symptoms	*Comments*
Discrepancy between ability and aggressiveness	A player with modest ability who is overly aggressive is vulnerable
Success phobia	Fear of failure is a common phenomenon and well understood, but the incidence of fear of success should not be underestimated
Uninhibited aggressiveness	The player presents a danger not only to him/herself, but to others
Feelings of invulnerability	Associated with reckless behaviour
Excessive fear of injury	The apprehension causes over-cautious play, paradoxically making injury more likely in, say, 50–50 tackles
Extensive history of injuries	Repeated injury may indicate physical and/or psychological vulnerability
Concealment or exaggeration of injuries	This indicates probable underlying psychological problems
Marked anxiety proneness	The over-nervous player's performance is detrimentally affected and injury is more likely

11.1.4 Symptoms of injury proneness

Lysens *et al.* (1989) examined the development of accident prone and overuse prone profiles of young players. Whilst acknowledging the importance of physical traits in predisposing a player to injury, they stressed that psychological factors need to be considered, even in relation to profiling the overuse prone player.

Various symptoms associated with the psychologically vulnerable player have been suggested, and they are listed in Table 11.1.

11.1.5 Exercise addiction

Some players become obsessed with exercise to the extent that they risk a variety of health problems, including multiple overuse injuries. Wichman and Martin (1992) pointed out that such players are difficult to treat, often requiring a catastrophe (e.g. consequential marriage break-up) for them to appreciate that they have a problem.

11.2 PSYCHOLOGICAL IMPLICATIONS OF INJURY

> The player's life becomes uncertain; the sense of control over oneself, so necessary for outstanding performance, is gone. He/she questions the reasons for the injury and how to prevent a recurrence. Being injured threatens not only a player's physical well-being, but acts as a threat to the player's self-concept, belief system, social and occupational functioning, values, commitments, and emotional equilibrium.
>
> *Danish, 1986, p. 346*

We are concerned here with the psychological problems that can ensue from injury and that can prevent or delay recovery. Many players sustain injuries that trigger significant negative psychological reactions. Weiss and Troxel (1986), in a survey of injured collegiate/elite players, found disbelief, fear, rage, depression, tension and fatigue to be common emotional responses. Common somatic complaints were upset stomach, insomnia and loss of appetite. Many found great difficulty in coping with the enforced inactivity, the powerlessness and the long rehabilitation. Such negative reactions can lead to recurrent injury problems, an inadequate recovery period and extended rehabilitation.

The implications of injury were profound for Graham Tutt, the Charlton Athletic goalkeeper who was forced to retire after a kick in the face at the age of 20:

> It's impossible to erase from my memory the moment of impact and pain when the boot of . . . made contact with my face at full force. The physical and mental scars of that accident will be with me for the rest of my life . . . I reached the ball and at the split-second that . . . was poised to strike it. His boot whacked me in the face . . . I couldn't hear anything. Everything was hazy and strange.
>
> 'He's kicked my eye out' was the immediate thought that ran through my mind. There was also a great deal of blood spurting from my nose and more blood coming from my cheek. My eyelid was split as well. But the most frightening thought was that I had lost an eye. (Harris and Varney, 1977, 78 ff.)

In a long and detailed account of the post-trauma events, Tutt did not indicate that his psychological 'scars' were appreciated by those involved in the treatment. In fact, when Tutt reached the dressing room, the manager's reaction was hardly sensitive:

> By this time both my eyes were closed and I was coughing blood, as he said to me, 'Can you go on?' (Harris and Varney, 1977)

His severe long-term problems with his 'nerves' seem to have gone totally untreated.

11.2.1 Factors affecting psychological reaction to injury

The individual's history of injury

If the injury background is extensive then psychologically negative reactions are likely to be more intense. Frustration, anger, resignation and despair may be intensely felt, creating an apathetic attitude during the recovery phase. In this kind of psychological state, the recovery phase is likely to be seriously extended and the chances of re-injury enhanced (Sanderson, 1981b).

The nature of the injury

Other things being equal, the psychological trauma will tend to increase as a function of the severity of the injury, but only to the extent that the individual is aware of the severity. As Stein (1962) has noted, it is often the case that the graver the injury, the fewer are the emotional complaints. This is partly because of immediate post-trauma shock which can leave a player amnesic and anaesthetized against feelings of pain. When full awareness returns, the process of rationalization has already begun. Additionally, there is a finality about a severe injury which, in a sense, can lessen or eliminate the feelings of anger and irritation which normally accompany the injury. Reactions are influenced by the degree of disability and how apparent it is to others. The soccer player, like other athletes, tends to have a sophisticated body image based on an awareness of the beauty and integrity of his body. If he is physically damaged, his body image, an integral part of his ego, may be threatened to an intolerable degree. The psychological implications can be extensive.

The nature of the sport

Few participants in sport fully expect to be injured but it is clear that the likelihood of injury varies markedly across sports, with soccer being one of the more dangerous sports. Injury in golf is relatively rare, whereas the risk of disabling injury in football is thousands of times higher than in underground mining (Mongillo, 1968). It seems reasonable to hypothesize that the psychological trauma associated with a particular injury will be a decreasing function of the general level of risk entailed in the sport. All else being equal, a particular sports injury will generate more emotional trauma in a low-injury risk golfer than in the high-injury risk soccer player.

The nature of the injury interacting with the nature of the sport

The interaction is important in that, for example, the psychological effect of a cut eye would be greater for a boxer than, say, for a free-style wrestler. Although both sports here are contact sports in which injury risk is relatively high, the nature of boxing ensures that a cut eye is a particularly devastating injury, encouraging greater psychological repercussions.

The level of competition in which the injury has occurred

The casual soccer player who plays infrequently and who sprains his wrist is unlikely to find the experience as traumatic as the player who earns his living from the game. The implication of this is that the psychological effects of injury are only worthy of serious study in relation to what might be termed 'serious' sport. The recreational skier who fractures his leg may suffer considerable personal trauma but it is of no general significance. Physical fitness is also important; the player who is physically fit, having become adapted to fairly severe physical stresses before injury, can more easily adapt to the demands of a physical rehabilitation programme – he is less likely to 'acquiesce' to the disability (Bender *et al.*, 1971).

Where the injury is so severe as to end the career of the player, there may be long-term psychological implications, particularly for those at the professional level. Kleiber and Brock (1992) found lower self-esteem and life satisfaction amongst former professional athletes 5–10 years after they had sustained career-ending injury.

The player's personality

The key importance of personality in reaction to injury can be demonstrated by reference to the theories of Hans Eysenck who has studied the structure of personality. As detailed in Chapter 17, Eysenck (1967) identified the two major independent dimensions of Extroversion and Neuroticism, along which the personality of individuals can vary. From Eysenck's theory, we can deduce that personality might affect the player's reaction to injury as shown in Table 11.2. Crossman and Jamieson (1985) found that players who overestimated the seriousness of an injury displayed more anxiety and greater feelings of inadequacy, anger, loneliness and apathy. Overestimation was more common amongst those players competing at lower levels. They suggested that such players might benefit from counselling concerning coping strategies.

Neurotic extroverts have been found to be particularly vulnerable to neurotic breakdown when sustaining injury towards the end of their careers as their over-valued physical abilities go into decline (Little, 1969). Little concluded:

Table 11.2 Effect of personality on reaction to injury

Personality type	*Characteristics*	*Reaction to injury*
Stable Extrovert	Sociable, carefree	Straightforward
Neurotic Extrovert	Impulsive, assertive, reckless, changeable, pain tolerant	Tendency to under-react to injury-inadequate rehabilitation and resumption of activity too early
Stable Introvert	Reflective, cautious	Straightforward
Neurotic Introvert	Nervous, unconfident, pessimistic	Tendency to over-react to injury, leading to lengthy rehabilitation

> Like exclusive and extensive emotional dependence on work, on key family relationship bonds, intellectual pursuits, physical beauty, sexual prowess or any other over-valued attribute or activity, athleticism can place the subject in a vulnerable pre-neurotic state leading to manifest neurotic illness in the event of an appropriate threat, or actual enforced deprivation, especially if abrupt and unexpected. (1969, p. 195)

11.2.2 Psychological hardiness

Kobasa (1979) has described a kind of individual who has abundant 'coping resources' in response to stressful events. Psychological hardiness is characterized as 'a constellation of characteristics such as curiosity, willingness to commit, seeing change as a challenge and stimulus to development, and having a sense of control over one's life'. It could be argued that Eysenck's self-assured stable extrovert matches this definition most closely.

Tunks and Bellissimo (1988) stated that, 'some individuals seem able to transform calamities into opportunities for growth while others transform everyday hassles into overwhelming adversities'. Whilst it is undoubtedly true that some players will be more psychologically hardy than others, it is important to recognize that coping skills can be learned even by those with a tendency to 'catastrophize'. This latter group represent the greatest challenge to the caregivers in that they will need active and sympathetic support during rehabilitation, whereas the psychologically hardy players will tend to take injury in their stride.

11.2.3 Attribution and injury

Studies of accident victims and those recovering from major illness suggest that causal attributions can play a significant role in psychological and physical recovery. By the same token, it is reasonable to expect that attributions would have a potentially important mediating role in recovery from athletic injuries. For example, if a player focuses on blaming an opponent for the injury, the resultant frustration, anger and sense of injustice could well hinder the recovery process.

Grove *et al.* (1990) investigated attributions for rapid or slow recovery from sports injuries. The attributions elicited from players were related to hypothetical events. The authors acknowledged that there is a need for research which examines the attributions for actual progress of injured players during rehabilitation.

The related construct of 'locus of control', the extent to which individuals perceive that they have power over what happens to them, is relevant here (see Chapter 17, p. 283). Those with an internal locus of control perceive themselves to be responsible for their own lives, whereas those with an external locus perceive themselves to be at the mercy of outside forces such as fate, luck and powerful others.

The practical implications of this idea in relation to injury are immediately apparent. An internal locus of control would generally seem to be more functional and mature.

Achievement motivation is higher in those who believe they are in control of their own destiny. Those acting in support of athletes would generally wish to encourage the internal disposition, thereby enhancing the likelihood that the injured player will be optimally committed to rehabilitation. Those players with an internal locus of control who are undergoing treatment would tend to be more aware of the importance of their own efforts in achieving full recovery.

Dweck (1986) described internal locus individuals as 'mastery oriented', as opposed to those exhibiting 'learned helplessness'. Learned helplessness exists when the individual, having experienced repeated failure (or repeated injury) together with inappropriate attributions, concludes that 'nothing I do matters' and that failure is unavoidable.

Relevant research has concentrated on *health* locus of control and health behaviour, health status and exercise adherence (O'Connell and Price, 1982; Dishman, 1986). Results have generally been equivocal although Slenker *et al.* (1985) found that joggers were significantly more internal than non-exercisers.

11.2.4 Individual differences in reaction to pain

Historically, a stimulus–response (S–R) model of pain was accepted, i.e. that the intensity of the pain is a reliable guide to the severity of the injury. It is now more widely recognized that many variables other than the severity of injury determine the reaction to pain. Pain perception can be influenced by: health, age, personality, suggestion (the doctor who warns the patient that 'this is going to be painful' encourages perception of pain), expectancy, experience, attentional focus (in a competitive soccer match a gashed shin may go unnoticed), and cultural factors.

Having considered psychological factors which will influence the individual's reactions to injury, the next question is what are the potential psychological reactions?

11.2.5 The player's psychological reactions to injury

Following injury, the player is faced with a range of threats which have the potential to delay or prevent the process of healing.

- Threats to body image: players typically have a positive body image and attach great importance to their physical appearance/integrity. An injury, particularly a disfiguring injury, threatens the sense of being a whole and undamaged person (what Eldridge, 1983, has called narcissistic disfigurement).
- Threats to self-esteem: closely linked with body image, self-esteem is also affected by, for example, the loss of autonomy and control, the need to depend on others and the loss of feelings of personal invulnerability.
- Threats to lifestyle: loss of important roles, separation from friends, thwarting of plans, uncertainty about the future and, perhaps, loss of income.

There is no doubt that some players suffer extreme mental torment as a consequence of disabling injury. Kubler-Ross (1969) noted a five-stage grief process which Rotella and Heyman (1986) argued has relevance to the player who experiences traumatic injury (Table 11.3). Recognizing that the reaction to injury can be traumatic, Rotella and Heyman (1986) emphasized the need for positive 'crisis intervention' to help the athlete to be focused on immediate practical and manageable concerns rather than seeing the injury as an 'overwhelming, engulfing catastrophe'.

The experience of serious injury need not be overwhelmingly traumatic if the individual is psychologically hardy like Alan Shearer of Blackburn Rovers and England. Half way through the 1993–4 season he ruptured his cruciate ligament. He was initially 'bewildered' by the diagnosis: 'It was unbelievable. I was expecting it to be something. . . . but nothing that serious. It's one of the worst injuries of the game'. But he quickly adjusted: 'It took me a few days to come to terms with what was happening, but it wasn't too bad after that'. He had the benefit of strong support from family and the club, as well as a phone call from Paul Gascoigne to say that he had recovered from a similar injury and that there was no reason why he (Shearer) should not. He was clearly well briefed about the rehabilitation process. 'There were dark times sure. Times when the knee never seemed to be getting any better. But I expected them and I was ready for them. I kept in touch with the lads. I went into training every day . . .' (quoted in Hulme, 1994).

Table 11.3 Five-stage grief response applied to sport

Stage	*Description*
1. Denial	A defence mechanism whereby the player downplays or ignores the reality of injury. 'There's no problem . . . it'll be okay'
2. Anger	Sometimes indiscriminate but often directed at the perceived cause, e.g. a particular opponent, the coach, inwardly to self
3. Bargaining	The player attempts to rationalize away the injury, indicating that the reality has not been fully accepted
4. Depression	Reality begins to emerge and the player finds it difficult to imagine making a full recovery
5. Acceptance	A necessary stage to be reached for the rehabilitation process to be effective. 'I'm injured but life goes on'

Based on Kubler-Ross, 1969.

11.3 THE TREATMENT PROCESS

11.3.1 Intervention strategies

It needs to be stressed that the physician and the physiotherapist are not treating an injury, but are treating an injured player. There may be complex psychological antecedents to the injury and there are certainly psychological consequences. The specialist in sports medicine should be sensitive to such factors in order that the treatment process is most effective.

A.M. Smith *et al.* (1990) detailed various psychological intervention strategies in relation to the most frequently reported emotional responses.

- Coping with frustration: clear explanation of treatment is necessary, allowing the player to make informed decisions. Well-defined and achievable goals should be provided and revised in the light of progress.
- Coping with depression: within the constraints imposed by the injury, there should be a resumption of activity, which serves to encourage the perception of progress and reduce helplessness. In this context, it is important that relationships are maintained with the team and the coach.
- Coping with anger: discussion with the player about the source of the anger

may reduce misunderstandings and irrational beliefs. The latter should be replaced with positive, realistic and rational thoughts.
- Coping with tension: first, there should be identification of the causes of the tension (e.g. fear of re-injury), followed by the use of mental imagery or relaxation procedures.

A.M. Smith *et al.* (1990), on the basis of their research into the emotional responses of athletes to injury, stressed the importance of the prompt recognition of emotional disturbance in facilitating the rehabilitation process and the safe return to competition. Davis (1991) provided evidence of the beneficial effects of progressive relaxation during team work-outs on reducing the incidence of injury.

11.3.2 Guidelines for support staff

There follows a list of broad guidelines to help the medical team and other support staff to be most effective in treating the injured player.

- Provide the player with 'quality time': make him/her feel that you really want to help by being supportive and reassuring and committed (Silva and Hardy, 1991).
- Take every opportunity to educate the patient. Be good at communication: reduce anxiety and uncertainty by providing accurate and clear information about the diagnosis and the prescribed treatment, including a recovery timetable (Ford and Gordon, 1993). Kahanov and Fairchild (1994) found evidence of miscommunication between athletes and trainers in that 52% of a sample of injured athletes did not understand how the rehabilitation process related to their injury.
- Listen carefully to the player: you are likely to gain information which will improve the treatment. The player may even suggest rehabilitation techniques which are more effective than the official ones.
- Negotiate a treatment plan: help the player set challenging, attainable and measurable goals. A simple approach to goal-setting involves answering three questions (Wiese and Weiss, 1987): Who? Will do what? By when?
- Show commitment.
- Liaise with other support staff.
- Be patient about the player's return to competition and about his behaviour. The rehabilitation process must take its proper course, with no premature return to competition. Coaches or team members can sometimes put undue pressure on a key player who is undergoing treatment for an injury – the question 'When are you going to be back?' can make the injured player feel that he/she is letting the team down. The highly competitive extrovert player, impatient to return to competition and used to being the centre of attention, can be exasperating to treat (Silva and Hardy, 1991).
- Recognize the motivational role of relaxation, positive self-talk, imagery

and social support (Rotella and Heyman, 1986; Grove and Gordon, 1991; Green, 1992).
- Know the sport: this is not a problem at professional clubs where the club doctor and paramedical staff will be very familiar with soccer and the particular club culture.
- Use peer modelling – the pairing of an injured player with someone who has successfully recovered from a similar injury – and injury support groups whereby players find comfort from sharing their concerns with other injured players (Wiese and Weiss, 1987).
- Get patients actively involved early with the aim of making the player self-directed. The more the player is actively involved, the more he/she will feel ownership of the treatment process and attain the appropriate 'mind-set' for recovery (Green, 1992).
- Follow-up early and frequently.

11.3.3 Exercise compliance

There has been a large amount of research in recent years into compliance or adherence to rehabilitation programmes – so-called secondary prevention. Detailed examination of adherence research is not possible here. However, those with a particular interest in the area should read Dishman (1986) for a useful review of research on exercise compliance.

Examination of the available research suggests the following conclusions.

- Blue-collar workers, smokers and overweight individuals are less likely to adopt and maintain a fitness regimen (Dishman, 1986).
- High self-motivation is associated with compliance/adherence (Fisher and Hoisington, 1993).
- For some individuals, perceived inconvenience and lack of time are associated with dropping out (Fisher and Hoisington, 1993).
- Support from significant others, e.g. health professionals, coaches and family members, is particularly important for adherence to rehabilitation (Rotella and Heyman, 1986; Fisher and Hoisington, 1993).
- Achievement goals and feelings of well-being are relatively important for adherence to the programme (Dishman, 1986).

In the context of adherence in the rehabilitation of sports injuries the following research findings may be useful. Eichenhofer *et al.* (1986) found that players high on somatic (physical) anxiety have difficulty adhering to prescribed treatment. Fisher *et al.* (1993) reported that adherers tended to perceive greater social support, be higher in self-motivation (intrinsic motivation) and believe they work harder. Duda *et al.* (1989) noted that adherers amongst a sample of intercollegiate players believed in the effectiveness of the treatment, perceived greater social support, were more self-motivated and were more focused on personal mastery and improvement in sport. They suggested that injured players should be assessed for degree of self-motivation and that special (goal-setting) support be given to those who are low in this variable. They also argued that the presence of social support networks such as family, friends and fellow players is likely to facilitate the treatment process.

Weiss and Troxel (1986) maintained that players high in self-esteem are more likely to persist in rehabilitation programmes. Lampton *et al.* (1993) found that injured athletes low in 'self-esteem certainty' and with high ego involvement in tasks tended to miss the most rehabilitation appointments. Given that injury itself can negatively affect self-esteem, it is imperative that those concerned with the rehabilitation process seek to promote the player's functional self-esteem and confidence.

Worrel (1992) explained how behavioural techniques involving the setting of short-term functional goals and cognitive techniques aimed at promoting positive thoughts and actions can be used to enhance the likelihood of compliance to rehabilitation. Ievleva and Orlick (1991) examined slow and fast healers from knee and ankle injuries and found that those athletes using cognitive techniques, i.e. goal-setting, positive self-talk and the use of healing imagery, tended to be the fastest healers.

Rapport with the injured player is most important: as Danish (1986) observed, 'when patients do not feel understood, do not find the health care professional warm and friendly, and are intimidated by the technical terminology, non-compliance is likely' (p. 347).

11.3.4 Goal-setting during rehabilitation

The sports medicine specialist may be involved in goal-setting in connection with a player's rehabilitation programme or, perhaps, in the context of stress management. As long as it is remembered that goals should be specific, measurable (implying quantifiable feedback), relevant, achievable yet challenging (implying frequent evaluation and re-setting) and accepted by the player, then there is a sound basis for helping the player (see also Chapter 12, p. 197). Particularly when the player is recovering from serious injury, there needs to be careful and frequent monitoring, and perhaps adjustment of goals. The following extract from Locke and Latham (1985) gives an indication of the kinds of goals which might be set.

> Strength goals are typically achieved by working with weights. . . . specific goals could be set for the amount of weight lifted with each muscle group and/or the number of repetitions, based on each individual's size, weight and capacity. Stamina goals may be set in terms of various types of repetitive activities that increase cardiovascular capacity, such as jogging, stroking, kicking, sprinting and rowing. These goals will involve maintenance of a specific pace (e.g. 7 min/mile) and/or exerting a specific amount of force (as in rowing) for a specific amount

of time (5 min, 1 hour etc.) Goals may also involve attaining a specific pulse rate (at rest, during exercise, etc.) . . . Skill goals will typically involve a specific number of correct repetitions of a specific task component.

In the rehabilitation context, goals could relate to, for example, strength, endurance, cardiovascular fitness, range of joint movement and psychological factors such as confidence, state anxiety and ability to concentrate.

Summary

The successful treatment of an injured player requires that support staff are sensitive not only to the physical, but also to the psychological antecedents and consequences of injury. Personality, exposure to life stress and psychodynamic factors have been linked with injury proneness. The psychological implications of injury can be profound, and are affected, for example, by the severity of the injury, the player's history of injury and the player's personality.

It is important that the support staff acknowledge the psychological dimension in the treatment process. Giving the player full attention, emphasizing good communication, and using motivation techniques to maximize the likelihood of compliance with the treatment process are necessary.

REFERENCES

Bender, J.A., Renzaglia, G.A. and Kaplan, H.M. (1971) Reaction to injury. in *Encyclopedia of Sports Sciences and Medicine* (ed. L.A. Larson), Macmillan, New York, pp. 1001–2.

Bramwell, S.T., Matsuda, M., Wagner, N.D. and Holmes, T.H. (1975) Psychosomatic factors in athletic injuries: development and application of the Social and Athletic Readjustment Rating Scale (SARRS). *Journal of Human Stress*, **1**, 6–20.

Brown, R.B. (1971) Personality characteristics related to injuries in football. *Research Quarterly*, **42**, 133–8.

Crossman, J. and Jamieson, J. (1985) Differences in perceptions of seriousness and disrupting effects of athletic injury as viewed by players and their trainer. *Perceptual and Motor Skills*, **61**, 1131–4.

Danish, S.J. (1986) Psychological aspects in the care and treatment of athletic injuries, in *Sports Injuries: the Unthwarted Epidemic* (eds P.F. Vinger and E.F. Hoener), PSG Publishing, Boston, Mass., pp. 345.

Davis, J.O. (1991) Sports injuries and stress management: an opportunity for research. *The Sport Psychologist*, **5**, 175–82.

Dishman, R.K. (1986) Exercise compliance: a new view for public health. *Physician and Sportsmedicine*, **14(5)**, 127–45.

Duda, J.L., Smart, A.E. and Tappe, M.K. (1989) Predictions of adherence in the rehabilitation of athletic injuries: an application of personal investment theory. *Journal of Sport and Exercise Psychology*, **II,** 367–81.

Dweck, C.S. (1986) Motivational processes affecting learning. *American Psychologist*, **41,** 1040–8.

Eichenhofer, R. Wittig, A.F., Balogh, D.W. and Pisano, M.D. (1986) Personality indicants of adherence to rehabilitation treatment by injured athletes. Paper presented to the Midwestern Psychological Association Conference, Chicago, May.

Eldridge, W.E. (1983) The importance of psychotherapy for athletic-related orthopedic injuries among adults. *International Journal of Sports Psychology*, **14,** 203–11.

Eysenck, H.J. (1967) *The Dynamics of Anxiety and Hysteria*, Routledge & Kegan Paul, London.

Fisher, A.C. and Hoisington, L.L. (1993) Injured athletes' attitudes and judgements toward rehabilitation adherence. *Journal of Athletic Training*, **28,** 48–50; 52–4.

Fisher, A.C., Scriber, K.C., Matheny, M.L. *et al.* (1993) Enhancing athletic injury rehabilitation adherence. *Journal of Athletic Training*, **28,** 312–18.

Ford, I. and Gordon, S. (1993) Social support and athletic injury: the perspective of sport physiotherapists. *Australian Journal of Science and Medicine in Sport*, **25(1),** 17–25.

Green, L.B. (1992) The use of imagery in the rehabilitation of injured athletes. *The Sports Psychologist*, **6,** 416–28.

Grove, J.R. and Gordon, A.M.D. (1991) The psychological aspects of injury in sport, in *Textbook of Science and Medicine in Sport* (eds J. Bloomfield, P.A. Fricker and K.D. Fitch), Blackwell Scientific, London, pp. 176–86.

Grove, J.R., Hanrahan, S.J. and Stewart, R.M.L. (1990) Attributions for rapid or slow recovery from sports injuries. *Canadian Journal of Applied Sports Sciences*, **15,** 107–14.

Hanson, S.J., McCullagh, P. and Tonymon, P. (1992) The relationship of personality characteristics, life stress and coping resources to athlete injury. *Journal of Sport and Exercise Psychology*, **14,** 262–72.

Harris, H. and Varney, M. (1977) *The Treatment of Football Injuries*, MacDonald James, London.

Hulme, C. (1994) Lone star. *Sports Magazine*, January, 31–6.

Ievleva, L. and Orlick, T. (1991) Mental links to enhanced healing: an exploratory study. *The Sports Psychologist*, **5,** 25–40.

Jackson, D.W., Jarrett, H., Bailey, D. *et al.* (1978) Injury prediction in the young athlete: a preliminary report. *American Journal of Sports Medicine*, **6,** 6–14.

Kahanov, L. and Fairchild, P.C. (1994) Discrepancies in perceptions held by injured athletes and athletic trainers during the initial injury evaluation. *Journal of Athletic Training*, **29,** 70–5.

Kleiber, D.A. and Brock, S.C. (1992) The effect of career-ending injuries on the subsequent well-being of elite college athletes. *Sociology of Sport Journal*, **9,** 70–5.

Kobasa, S.C. (1979) Stressful life events, personality and health: an inquiry into hardiness. *Journal of Personality and Social Psychology*, **37,** 1–11.

Kraus, J.F. and Gullen, W.H. (1969) An epidemiologic investigation of predictor variables associated with intramural touch football injuries. *American Journal of Public Health*, **59,** 2144–56.

Kubler-Ross, E. (1969) *On Death and Dying*, Macmillan, London.

Lampton, C.C., Lambert, M.E. and Yost, R. (1993) The effects of psychological factors in sports medicine rehabilitation adherence. *Journal of Sports Medicine and Physical Fitness*, **33,** 292–9.

Little, J.C. (1969) The player's neurosis–deprivation crisis. *Acta Psychiatrica Scandinavica*, **45,** 187–97.

Locke, E.A. and Latham, G.P. (1985) The application of goal-setting to sports. *Journal of Sports Psychology*, **7,** 205–22.

Lysens, R.L., Ostyn, M.S., Vanden Auweele, Y. *et al.* (1989) The accident-prone and overuse-prone profiles of the young player. *The American Journal of Sports Medicine*, **17,** 612–19.

May, J.R., Veach, T.L., Reed, M.W. and Griffey, M.S. (1985) A psychological study of health, injury and performance in players on the US Alpine Ski Team. *Physician and Sportsmedicine*, **13,** 111–15.

Mongillo, B.B. (1968) Psychological aspects in sports and psychosomatic problems in the athlete. *Rhode Island Medical Journal*, **51,** 339–43.

Moore, R.A. (1967) Injury in athletics, in *Motivation in Play, Games and Sports* (eds Charles C. Thomas, R. Slovenko and J.A. Knight), Springfield, Ill.

Nideffer, R.M. (1989) Anxiety, attention and performance in sports: theoretical and practical considerations, in *Anxiety in Sports: an International Perspective* (eds D. Hackfort and C.D. Spielberger), Hemisphere, New York, pp. 117–36.

O'Connell, J.K. and Price, J.H. (1982) Health locus of control of physical fitness-program participants. *Perceptual and Motor Skills*, **25,** 925–6.

Ogilvie, B. and Tutko, T.A. (1971) *Problem Athletes and How to Handle Them*, Pelham Books, London.

Passer, M.W. and Seese, M.D. (1983) Life stress and athletic injury: examination of positive versus negative events and three moderator variables. *Journal of Human Stress*, **9,** 11–16.

Reilly, T. (1975) An ergonomic evaluation of occupational stress in professional football. Unpublished PhD thesis, Liverpool Polytechnic.

Rotella, R.J. and Heyman, S.R. (1986) Stress, injury and the psychological rehabilitation of players. In *Applied Sports Psychology: Personal Growth to Peak Performance* (ed. J.M. Williams), Mayfield, Palo Alto, Ca.

Sanderson, F.H. (1981a) The psychology of the injury prone player, in *Fitness and Sports Injuries* (ed. T. Reilly), Faber and Faber, London, pp. 31–6.

Sanderson, F.H. (1981b) The psychological implications of injury, in *Fitness and Sports Injuries* (ed. T. Reilly), Faber and Faber, London, pp. 37–41.

Silva, J.M. and Hardy, C.J. (1991) The sport psychologist, in *Prevention of Athletic Injuries: the Role of the Sports Medicine Team* (eds F.O. Mueller and A.J. Ryan), F.A. Davis Co., Philadelphia, pp. 114–32.

Slenker, S.E., Price, J.H. and O'Connell, J.K. (1985) Health locus of control of joggers and non-exercisers. *Perceptual and Motor Skills,* **61,** 323–8.

Smith, A.M., Scott, S.G. and Wiese, D.M. (1990) The psychological effects of sports injuries: coping. *Sports Medicine*, **9,** 352–69.

Smith, R.E., Smoll, F.L. and Ptacek, J.T. (1990) Conjunctive moderator variables in vulnerability and resiliency research: life stress, social support and coping skills, and adolescent sport injuries. *Journal of Personality and Social Psychology*, **58,** 360–70.

Stein, C. (1962) Psychological implications of personal injuries. *Medical Trial Technique Quarterly*, **1,** 17–28.

Tunks, T. and Belissimo, A. (1988) Coping with the coping concept: a brief comment. *Pain*, **34,** 171–4.

Weiss, M.R. and Troxel, R.K. (1986) Psychology of the injured player. *Athletic Training*, **21,** 104–9.

Wichman, S. and Martin, D.R. (1992) Exercise excess: treatment patients addicted to fitness. *Physician and Sportsmedicine*, **20,** 193–200.

Wiese, D.M. and Weiss, M.R. (1987) Psychological rehabilitation and physical injury: implications for the sports medicine team. *The Sport Psychologist*, **1,** 318–30.

Worrel, T.W. (1992) The use of behavioral and cognitive techniques to facilitate achievement of rehabilitation goals. *Journal of Sport Rehabilitation*, **1,** 69–75.

Goal-setting during injury rehabilitation

12

David Gilbourne

Introduction

As a result of becoming injured soccer players may be unavailable to play in competitive matches. Sometimes injuries are serious and the rehabilitation period can be a matter of months rather than days. When this happens players face major changes to their daily routines and may suffer considerable trauma as a result. Doubts over the continuation of their careers together with long isolated periods of repetitive rehabilitation can lead to players becoming anxious and/or depressed. Psychological interventions can help athletes to cope successfully with rehabilitation and this chapter introduces the intervention technique of goal-setting and considers how it might be applied within a sports injury context.

Contemporary psychological theory, recent injury-based research and life development perspectives are linked in the formulation of a rehabilitation-specific approach to goal-setting. Finally, a three phase programme is outlined to provide broad guidelines for the successful initiation and development of a goal programme.

There are various situations in which Psychological Skills Training (PST) could be integrated into a soccer player's schedule. As an example, players could use stress management techniques to help them to cope with pre-match pressures or they could develop 'in play' strategies to

Science and Soccer. Edited by Thomas Reilly. Published in 1996 by E & FN Spon, London. ISBN 0 419 18880 0.

help maintain appropriate attentional focus. This chapter, however, concentrates on injured soccer players with specific emphasis on those undergoing long-term rehabilitation.

Interest in the potential of psychological skills to facilitate the rehabilitation process has increased in recent years. Vealey (1988) and Wiese and Weiss (1987) argued that injured athletes could use aspects of PST to help them return to required levels of fitness and form. Vealey (1988) also stressed how important it is for the sports medicine team to treat the 'whole athlete'. More recently a number of research publications and textbooks have been dedicated solely to the psychology of rehabilitation (Wiese *et al.*, 1991; Green, 1992; Heil, 1993; Pargman, 1993). These contributors have argued that goal-setting, imagery and relaxation skills may all prove to be helpful during rehabilitation.

12.1 GOAL-SETTING IN SPORT

Goal-setting involves the establishment of specific targets. These targets are seen to be more effective if they are stated in specific terms and are challenging in nature (for review see Beggs, 1990). Goal-setting was first used in industrial and organizational settings, and research considering the effectiveness of the process in these settings provided overwhelming support for the notion that the setting of hard, challenging goals improved performance. Within sports settings, however, research findings are less conclusive. Despite this, goal-setting remains a major component in most psychological skills programmes and is certainly popular within applied sport psychology circles.

12.1.1 Administration of a goal-setting programme

Goal-setting in competitive sports situations usually involves the coach and athlete working together as a team. In the context of sports injury rehabilitation the physiotherapist is often the 'administrator' central to the goal-setting process. The administrator has to identify appropriate goals for the player to work towards. It is also expected that the administrator of the programme would have the expertise to decide on suitable levels of goal difficulty and identify appropriate directions that an injured player's rehabilitation should take. For example, the physiotherapist would be in a position to ascertain the *suitability* of certain rehabilitation exercises and the appropriate *intensity* of the work required.

12.1.2 Structure of the programme

It is quite common for goal-setting programmes to differentiate between goals along temporal lines resulting in terms like short-, intermediate and long-term goals. What exactly constitutes a long-, intermediate or short-term goal is not clear. There is no definitive time frame into which these terms fit and they should always be viewed in a flexible manner. As a general guide, short-term goals tend to be set on a daily or weekly basis and act as the stepping stones that lead to the attainment of more distant intermediate and long-term goals. For example, a 'successful return to first team football' would not be an unusual long-term goal; how far ahead 'long-term' means in this instance will depend on the severity of the injury. An injury such as a tibial fracture will see the long-term goal set at anything between six months and a year. Furthermore, the need to adjust this goal may become apparent as rehabilitation goes particularly well or, conversely, the programme suffers a setback.

Setting goals at the correct level can be difficult. Difficult goals are perceived as having a greater impact than easy or 'do best' type instructions (Locke and Latham, 1985). The term 'difficult goal' can be ambiguous and practitioners from a range of settings, such as teaching, coaching and physiotherapy, have to judge the appropriate level of goal difficulty themselves. Common sense dictates what is challenging for one player may be easy or too difficult for another.

12.1.3 Goal-setting within injury rehabilitation settings

Authors in injury rehabilitation literature have argued that goal-setting is an ideal technique to use with injured athletes (Heil, 1993; Pargman, 1993). It has also been argued that it can be administered by physiotherapists and trainers (Wiese *et al.*, 1991).

The use of goal-setting is also a central theme in 'Life Development Intervention' (LDI), which perceives goal-setting to be a 'life skill' that can empower the injured athlete and assist him or her to overcome 'critical life events' (Danish *et al.*, 1993). The underlying themes within LDI can be adapted to help form a rehabilitation-specific 'goal-setting ethos'.

12.2 LDI AND GOAL-SETTING DURING REHABILITATION FROM INJURY

Two important factors can be highlighted within the injury rehabilitation experience. These refer to both place or location and time scale. Rehabilitation can often take place in a variety of settings with players undertaking prescribed rehabilitation sessions at their soccer club, at home, at local gymnasiums and so on. Some players typically move from situations of total rehabilitation support, say at the club, to relative isolation at home. This is similar to the way hospitalized patients may experience dramatically different environments upon discharge. It follows that the levels of guidance and support players receive may fluctuate in both a qualitative and quantitative manner. As

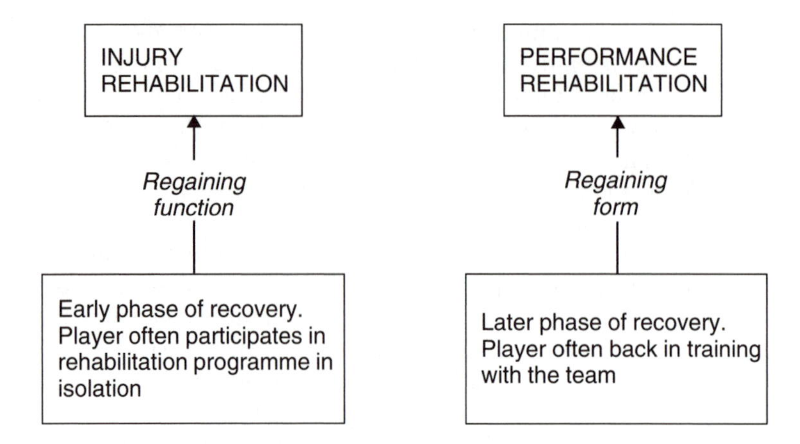

Figure 12.1 Injury rehabilitation: a longitudinal perspective.

different situations place unique demands on the player then any psychological skills employed would need to be flexible so they could be used across these different settings (Gilbourne and Taylor, 1995).

Secondly, rehabilitation can be considered from a longitudinal perspective (Figure 12.1). This view highlights a differentiation between what is termed injury and performance rehabilitation (Gilbourne and Taylor, 1995). The former relates to the early stages of rehabilitation when an injured joint or muscle group is the focus of attention; the latter refers to the last stages of recovery when the rediscovery of 'form' is the main consideration in the player's mind.

As an example, a professional soccer player in the early stages of recovery from knee reconstruction will focus on regaining strength in 'key' muscle groups; this is defined as **injury** rehabilitation. Later he will reach a stage where the knee has regained mobility and stability and is strong enough to undertake match-play movements. At this time gentle reintegration into training activities will begin. This phase heralds the onset of an **interface** between injury and performance rehabilitation. At such a time the player is likely to be reintroduced to old routines, yet, the injury can still be expected to be 'on his mind'. If progress is maintained a further stage will be reached where the player is focusing totally on rebuilding the technical or fitness facets of his game; this phase is defined as **performance** rather than injury rehabilitation.

The *situational* and *longitudinal* considerations inherent within an injured soccer player's experience of rehabilitation have implications for the way goal-setting is approached. The technique would need to be adaptable so it could be used in different situations. It also has to be perceived, by players, to be transferable over time. Players could then utilize the programme during *both* injury and performance rehabilitation.

When developing specific intervention strategies, it is important to consider *how* such interventions should be delivered. The LDI approach considers that interventions constitute 'life skills' and that these skills should be seen to be transferable across time and situations. Life Development Intervention encourages the demands of situational and longitudinal facets of injury rehabilitation to be met within a holistic framework of support and skills development. Furthermore, the LDI perspective stresses the opportunity for sport psychology interventions to 'enhance competence' and 'promote human development throughout the life span' (Danish *et al.*, 1993). Above all else, LDI principles are underpinned by a desire to 'empower' athletes with skills that can be used to overcome 'critical life events'.

Sports injury is highlighted within LDI as a typical 'critical life event' that invariably results in experiences of 'life change'. For example, the injured soccer player who faces surgical interventions may harbour concerns over the chance of full recovery and is likely, therefore, to perceive his career to be at a critical phase. Life changes may also be considerable; subsequent rehabilitation may be a lonely experience and the player may miss the daily routine of training with team-mates or preparing to play in a forthcoming competitive match.

The holistic perspective inherent in LDI promotes interest in the broader aspects of the injured player's daily existence. For example, the injury itself may prove to be a catalyst for problems in other aspects of his or her life.

> ... even in a single event other issues will intercede. Therefore, successful coping involves dealing with what is perceived as the primary event as well as other events. (Danish *et al.*, 1993, p. 361)

By emphasizing the need to consider other aspects of the player's life apart from sport, 'administrators' of a goal-setting programme could find themselves discussing goals that relate to personal relationships, or business engagements as well as those associated with the injury. For this to happen, goal-setting will need to be perceived by the players as something they can use *outside* the rehabilitation sessions to help them cope with other life demands. The notion of skills learned within a sporting context being applied by the players is associated with the enhancement of personal competence through the *teaching* of life skills (Danish *et al.*, 1993). The emphasis on teaching the skills to athletes is an important one. It allows goal-setting itself to be considered as something players can learn and then use on a lifespan basis.

12.2.1 The structure of goals within an LDI approach

Danish *et al.* (1993) stressed that goals refer to the *actions* undertaken to reach a desired end and should not be perceived to be the *ends* in themselves. They also emphasized that goals are different from results and should be seen as actions over which the player has some control. This view could be adapted to allow two types of goals to be set, **process** and **outcome** goals. Outcome goals are typically set 5–7 days ahead, are flexible in nature and fulfil the role of a slightly distant series of targets.

The issue of flexibility is important, particularly when setting goals in a rehabilitation setting. Sometimes progress is slow, whereas at other times unexpected strides can be taken. The goal-setting programme needs to be adjustable to accommodate such variances. The following example of outcome goals is taken from the 'goal-setting diary' of a professional soccer player.

By the end of the week I aim to:

(i) increase control in leg squat exercise;
(ii) introduce and establish an upper body weight training programme.

These 'outcome' goals reflect *general* rehabilitation targets. It is important, however, to note these goals alone will not tend to lead to players working more effectively on a daily basis. The outcome goals only serve to provide a forward staging post. Of utmost importance is the setting of process goals. Unlike outcome goals, these are set on a daily or session-by-session basis and reflect what the player has to do *today* in order to attain his outcome goals in several days time.

Process goals direct the daily steps taken by the rehabilitating player. They need to be *written down* in *specific* and *positive* terms. Players may initially balk at the thought of writing goals down, but committing goals to paper is important. All athletes engaged in a recent goal-setting exercise commented on the value of writing the goals down before the day began (Gilbourne and Taylor, 1995).

When writing goals down in some form of diary the player can think carefully about making the goals *specific*. This is broadly accepted as being helpful in focusing athletes into what exactly it is they wish to achieve. Danish *et al.* (1993) stressed the importance of goals being couched in *positive* language. Terms such as **might**, **try to**, or **don't** should be replaced by terms like **will** and **do**.

One note of caution should be raised regarding the use of language when goal-setting within a sports injury context. Sometimes goals will be set that aim to inhibit a player from doing too much too soon. A case could be made for these goals to be formulated in cautionary language – **be careful not to**, or **do not be tempted to**, for example.

Another factor that should influence the 'nature' of goals set is the establishment of a goal programme that encourages the player to focus on self-improvement and evaluate progress in a self-referent manner. Within sport psychology this particular issue is associated with **goal perspective** theory and requires the introduction and explanation of two important terms, **task** and **ego** involvement.

12.2.2 Setting task-oriented goals in both injury and performance phases

The terminology of task and ego involvement has emerged from a bank of motivation literature that has developed in the past 15 years (for review see Roberts, 1992). Being 'task involved' means that athletes are trying to improve on *their own* best performance and will assess progress by looking at how they did today against how they did yesterday. In contrast, '*ego*-involved' athletes display a tendency to compare their own performance against the performances of others.

Being 'task involved' is associated with positive behaviours such as persistence. Ego involvement is associated with more problematic behaviours, such as selecting tasks that are either too difficult or too easy.

Perceived competence is noted to play a mediating role in the way ego involvement influences behaviour. For example, when athletes perceive their level of competence to be high, ego involvement is not necessarily expected to lead to maladaptive behaviour. Greater concern is expressed when athletes, who do not deem themselves to be competent, compare themselves against others. They are likely to receive negative information and this in turn may result in a lack of persistence and a number of defensive behaviours that are not conducive to sound development.

Athletes might display a tendency to be highly task focused or highly ego focused. Just to confuse matters, they may also be high in both, high in one and low in the other, or low in both. The issue of what constitutes the *best* balance is still being debated.

In situations such as a training session at a soccer club coaches can be identified as emphasizing task or ego involvement. This 'situational orientation' is referred to as the **motivational climate** and is believed to influence behaviour (Ames, 1992). Levels of persistence and a tendency for participants to value the role of effort increase when the motivational climate is *task* focused.

The ability of physiotherapists or coaches to encourage a task-oriented climate by means of a goal-setting programme is of interest. This is possible during the early injury phase of the rehabilitation process. As players move into the performance phase of rehabilitation, it is increasingly likely they will find themselves in ego-focused situations where players compare themselves against each other. The following section considers how task-focused goal-setting could be used by the player through both injury and performance rehabilitation.

12.2.3 Task-focused goal-setting: a longitudinal perspective

The idea that within an injury rehabilitation setting goals should be task focused is supported by the view that an injured athlete is better advised to evaluate progress by focusing on his own development than to compare himself against others. In the case of soccer players, comparison with others

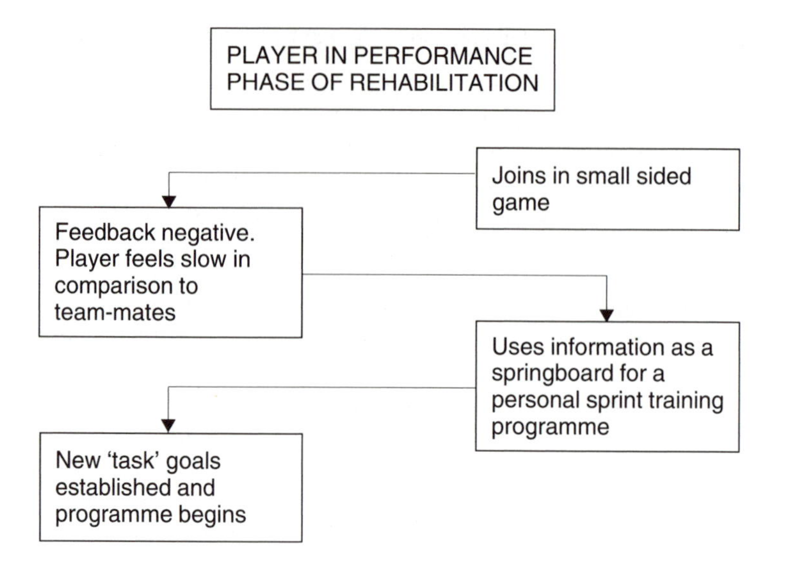

Figure 12.2 Using negative 'ego-oriented' feedback as a springboard for 'task-focused' goals.

might mean assessing progress against the performances of non-injured players or players with different injuries. This may be inappropriate, particularly in the early phase of injury rehabilitation.

A different case could be made for players who are in the performance phase of the rehabilitation process. They are likely to be training alongside non-injured players and it is only natural to compare oneself against others in such circumstances. In this situation, the notion of goal-setting being a life skill, transferable across time and situations, is again important. If the player had *learned* to use goal-setting in the injury phase of his rehabilitation and had been encouraged to set task-focused goals, then the skills would be in place for use in the performance phase. If, for example, a player compares himself to his team-mates during this latter stage of recovery and receives some negative information, say he has lost some of his ability to accelerate quickly, this information can then be used to set task-focused outcome and process goals that direct training and ensure progress (see Figure 12.2).

12.3 GOAL-SETTING IN PRACTICE

So far in this chapter goal-setting has been introduced and the topic developed by making reference to important theoretical constructs and research specific to rehabilitation. The final section suggests guidelines for the initiation, administration and longer-term monitoring of a goal-setting programme.

12.3.1 Phase 1: Introducing the programme

Players will need to attend some form of introductory session before the goal-setting programme can be initiated. This should provide them with basic information about the programme and allow them an opportunity to seek clarification on any aspect of the goal-setting process. A pre-goal-setting 'workshop' is a helpful way of introducing athletes to the programme. In these 'educative' meetings the most important points concerning goal-setting are presented. These 'key' factors (Figure 12.3) are important components that together form the basis for successful participation in a goal-setting programme. If players are able to understand these concepts, then the programme should start from firm foundations.

It can be helpful to provide players with written information that can be explored verbally in the workshop or given out as material to be taken away. Examples of 'informative' handouts are outlined towards the end of Phase 2 when goal-setting booklets are discussed. In the examples provided in Figures 12.5 and 12.6, (see below) goal-setting information is included in the booklets or diaries that players will use when they write down and assess their own goals. It is reasonable to utilize such booklets in the educative workshops at Phase 1, allow players to take the booklets away and request that they bring them along for the meeting at Phase 2.

12.3.2 Phase 2: Goal-setting baseline meeting

This meeting aims to encourage an exchange of views and clarify two perspectives, firstly the thoughts and feelings of the player and secondly those of the

Outcome goals relate to what players *want* to achieve by the end of the week or month

Process goals are set on a *daily* basis and act as stepping stones or action steps that lead to *outcome* goals being attained

It is important to write *down* goals in *specific* terms and to find time to write *down* thoughts on progress

In the *injury* phase of rehabilitation it is helpful to set *task-focused* goals

Goal-setting can be used to help organize and direct *other* facets of the player's life as well as helping with the injury rehabilitation

Figure 12.3 Important points for the player to understand.

physiotherapist. The meeting should also aim to conclude with the establishment of the first week's outcome goals and agreement on specific process goals for the day ahead.

If two people are to embark on a goal-setting programme, one as an administrator the other as a participant, then it is important to understand what each party is seeking to achieve and to establish common ground as soon as possible. Therefore, in the baseline meeting, a free exchange of views is important. The administrator would normally influence the format and atmosphere within the meeting and care needs to be taken to establish a friendly, informal climate.

Players may be reassured if the reason for the meeting is made clear. Very often explaining that setting goals is a two-way process and that their thoughts and feelings are of importance serves to reinforce the sense of involvement highlighted in the educational session. Although players are likely to benefit from talking about their injury, the meeting should generate important information for the administrator. For example, questions such as how they feel about the injury, are they concerned about any aspects of their rehabilitation and what are their long-term targets for full recovery are valuable in that they encourage the player to talk to the programme administrator about how *he or she* feels.

During the meeting the administrator should become increasingly aware of the player's concerns, hopes and personal targets. Knowledge of this nature helps to establish whether the player and administrator are viewing the rehabilitation in the same way. If, for example, the player was expecting to play in two months and the physiotherapist was thinking in terms of three to four months, then such inconsistencies could be discussed, explanations provided and a common direction established.

As well as the physiotherapist benefiting from listening to the player, the player similarly should benefit from hearing how the physiotherapist evaluates the situation. This exchange encourages the physiotherapist to talk about the injury and the rehabilitation process and *informs* the player about reasoning behind the proposed way forward. Setting goals for players without making it clear to them *why* this direction is required could lead to the outright rejection of goals or the demonstration of indifferent levels of 'goal acceptance'. This disappointing state of affairs may result in low adherence to rehabilitation goals. Conversely, if player and physiotherapist have exchanged views and established an agreed direction then goal acceptance and commitment are likely to increase.

It is also important to note that goals need not only deal with rehabilitation exercises. As a result of engaging in an open-ended discussion, it would not be surprising if the player made reference to issues that fall *outside* the injury rehabilitation programme. It is, however, likely that factors pertaining to the player's personal life will be indirectly related to the injury itself. As an example, a player may report being bored during the day as he or she is no

WEEKLY GOAL-SETTING DIARY
Goals for the week. By the end of the week I aim to:
DAILY OR SESSION GOALS GOALS GOAL ASSESSMENT Monday
Tuesday
Wednesday
Thursday
Friday
Overall evaluation of rehabilitation this week.

Figure 12.4 An example of a weekly goal-setting form.

longer mixing with the squad on a regular basis. Whilst it is often very helpful for players to talk about these concerns, goal-setting plays a protective role in organizing a player's life in a more global sense. In the above example, the player might establish goals that involve training with other injured players more often or visiting training sessions a few times a week in an attempt to maintain contact with the squad and experience team camaraderie.

This final point stresses the need for goal-setting to be perceived as a flexible tool and for the administrator to be encouraged to take a holistic interest in the player. The last stage of Phase 2 requires that the first goals be agreed and committed to paper. At this time it is helpful to provide the player with some kind of goal-setting form or diary. Figure 12.4 outlines a weekly goal sheet presently being used within an injury rehabilitation setting. The form was developed over a three-year period at the Lilleshall Sports Injury Centre and is presented in a bound booklet together with helpful information about goal-setting (see Figures 12.5 and 12.6). The form features process goals as 'daily' or 'session' goals and outcome goals as 'goals for the week'.

It may be helpful to use the informative material in the booklet during the educational workshop on goal-setting or to include the information as supportive material to reinforce these sessions. The goal-setting forms provide space for a number of weekly outcome goals and for specific daily process goals. Of great importance is the allocation of space for players to review or assess their goal progress on both a daily and weekly basis. The value of evaluating goal progress and maintaining a goal programme is considered in further detail in the discussion on Phase 3 of the programme.

12.3.3 Phase 3: Monitoring, maintenance and evaluation

Once Phase 1 and 2 have been successfully completed the player should be working on an established goal-setting programme. It is now important for the administrator of the programme to be vigilant and be available to talk about goal attainment and to check that the post goal-setting assessments are being written down. In these early stages the players are likely to need considerable assistance in writing goals down and our own experience indicates the physiotherapist tends to set most of the goals in the early stages of the programme (Gilbourne and Taylor, 1995). When players do set their own goals they may tend to be rather general. Any failure to write goals down in specific enough terms would reflect an on-going need for the administrator to monitor 'goal skill' development and to continue the educational input. Players who are ego oriented may also have difficulty establishing goals that reflect a task focus. This again would require some form of intervention from the administrator who may have to 'work harder' in convincing ego-oriented players that a task focus is in their interest. The above issues highlight the need to conceptualize goal-setting as a skill and therefore performance should improve with experience and practice.

A crucial phase in the early stages of goal-setting is reached at the end of the first week. At this point the players reflect on their rehabilitation through the week, decide how successful they were in attaining their goals and attempt to formulate thoughts on the way forward. Whenever possible it is recommended that a specific block of time is put aside to allow this review to take place. The player who is working alone should be encouraged to retire to a

GOAL-SETTING AS YOU COME BACK FROM INJURY

Goal-setting is a technique that encourages you to identify clear targets, assess your progress and become actively involved in the management of your rehabilitation. The ideas presented below have resulted from a long process of consultation with injured athletes and will hopefully help you in *your* rehabilitation from injury.

Goals. whenever possible, should be:

1. **SPECIFIC.** This means goals like 'work harder' need to be replaced by goals you can specifically focus on such as 'focus on quality when doing step ups'. This goal is more specific and you will be able to assess afterwards if you managed to attain the quality you wanted.

2. **CHALLENGING.** This means the goals are not too easy but set you something of a test.

3. **REALISTIC.** This relates to you having a chance of reaching your goals. If the goals are too easy then the challenge can go, too hard and you are unlikely to succeed. As you set goals more often you should find setting goals at a realistic level will become easier.

4. **MEASURABLE.** Some of the goals you set will allow you to measure progress, for example by assessing muscle development or increases in mobility. This is likely to involve your physiotherapist providing you with information. For other goals such as '**Stride out and keep balanced when running today**' or '**Concentrate during single leg squats**' then *you* will have to judge for yourself how well the running or leg squat exercise went.

WRITING THE GOALS DOWN

Writing down your goals for the day or forthcoming session encourages you to think through what it is you want to achieve. To begin with, at the start of the week, it is a good idea to write down the goals you want to attain by the *end* of that week. Again, your physiotherapist will be able to guide you in this exercise. When you have written down your *week's* goals, then you can write down goals for the day ahead or the rehabilitation session you are about to do *right now.*

MAKE THE GOALS 'SELF GOALS'

Rather than writing down goals that make you compare yourself against others we would encourage you to think about *yourself* and set goals that allow you to focus on your own personal progress. As an example '**Stride out more than last week when running today**' or '**Run the best I can in this session**' are both self goals.

EVALUATING GOALS

It is important to reflect on how your rehabilitation has gone both at the end of the day and, in a more overall sense, at the end of the week. When you have been working towards certain goals it is very useful to assess progress, write down your thoughts and then set fresh goals for the following week or the next day.

DIFFERENT TYPES OF GOAL

You can set goals to help you with almost any aspect of your daily life. Goals can help you remember to do certain chores, '**Make sure to phone the coach today**' or be clear about getting what you want from a meeting, '**Ask the physiotherapist what I should do this week**'. As we have already seen, goals can also help you go into a rehabilitation session with clear targets such as, '**Be positive today**' or '**Complete all the exercises with emphasis on quality**'. Goals can also help you avoid doing *too much too soon:* '**Be sensible, just do my exercises**'. Finally, remember any goals that relate specifically to your injury will need to be decided upon by you **and** your physiotherapist as you go through the week.

Figure 12.5 Informative material for players about to start goal-setting.

EXAMPLES OF GOALS SET BY INJURED ATHLETES

This section of the booklet has been provided to give you an idea of how injured athletes have, in the past, used goal-setting. All the goals presented here have been taken from 'goal-setting diaries' that injured athletes have used in their time at Lilleshall. The examples below demonstrate that goals can direct in many different ways and will hopefully help **you** see how you might use goals in the weeks ahead to help in your rehabilitation.

Examples of '**exercise specific**' week goals (goals set on Monday to be attained by Friday)

'Improve walking' 'Get the leg straighter' 'Have more control on my step ups'

'Gain more strength and control in my step ups'

'Get full knee extension without pain'

Notice *all* the above goals are **self goals** in that they focus the athlete's attention on *their* own progress and avoid focusing on others.

Examples of '**organizing**' week goals

'Get clear idea of rehabilitation programme for the summer'

'See what treatment facilities are available' 'Establish home routine'

Some of these weekly goals relate to organizing rehabilitation when away from Lilleshall. This ensures that the rehabilitation programme is maintained.

Examples of **daily** goals (goals set to attain on that specific day or in a session about to start)

'Last day, keep up quality of exercises' 'Phone coach'

'Do knee exercises when sat on the tube'

'Talk to physio. Find out her goal plan for me and tell her mine'

'Concentrate on slow stable movement' 'Stay cheerful'

'Improve step ups, very important to do them properly and with control'

The above examples show athletes setting **self goals** and using goal-setting to help them keep phone commitments, plan questions for meetings and focus on specific exercises within their rehabilitation sessions.

Finally, it may be useful to stress the links between your **week goals** and your **daily goals**. Once you have established what you want to achieve in that week, see your daily goals as the '**stepping stones**' or '**action steps**' that will help you reach those targets you have set for yourself for the end of the week. It is important, however, to stay **flexible** and understand that sometimes your week's goals will **not** be reached, whilst at other times you may surpass them. What is **crucial** is that you keep assessing progress, reflect on your day's work, write your thoughts down and set fresh goals.

Figure 12.6 Using examples from past goal-setting programmes as educational material.

quiet space in order to avoid being disturbed, and think through the good and the bad points from the past week. In these early stages of the goal-setting programme it is helpful if the programme administrator is available to discuss progress with the player and help him decide upon new outcome and process goals for the week ahead. This process should, whenever possible, be repeated throughout the course of the rehabilitation. Goals should be set, committed to, reviewed and new goals established.

The importance of injury rehabilitation being viewed from a longitudinal perspective has already been emphasized. This may have implications for the professional who oversees the programme. As the journey from injury through to performance rehabilitation takes place, the main administrator of the goal programme is likely to change. During the early stages of performance rehabilitation the physiotherapist would typically be expected to be the main point of contact. Later on, when the player is involved predominantly in performance rehabilitation, the coach would be an increasingly important figure.

This expected change of personnel implies that physiotherapists and coaches should be encouraged to liaise through the goal-setting process so that the transfer of the administrator's duties can take place with the minimum disruption. It should also be noted that as the player approaches the performance phase, he or she is likely to have become increasingly skilled in the use of goal-setting. In contrast to the considerable *educative* involvement required of the physiotherapist, the empowerment of the player may result in the coaching administrator providing only technical guidance.

The skill of formulating specific task-focused goals would have been learned during the injury rehabilitation phase. This may well result in the *role* of the administrator contrasting sharply between injury and performance phases of rehabilitation. It is still important for the player and coach to meet to review progress and plan goals for the weeks ahead. The process of shared review and two-way discussions on progress would be helpful to both player and coach. Such exchanges should still be sought by both parties as full recovery is approached.

Summary

Goal-setting can be applied within injury rehabilitation settings. In doing this, important considerations relate to situational and longitudinal features of the rehabilitation process. Finally, guidelines providing a step by step approach to the introduction and continued monitoring of a goal programme demonstrate how theoretical concepts can combine with research activity to improve the way physiotherapists and coaches work with injured players.

REFERENCES

Ames, C. (1992) Achievement goals, motivational climate and motivational processes, in *Motivation in Sport and Exercise* (ed. G.C. Roberts), Human Kinetics, Champaign, IL, pp. 161–77.

Beggs, W.D. (1990) Goal-setting in sport, in *Stress and Performance in Sport* (eds J.G. Jones and L. Hardy),Wiley, Chichester, pp. 135–70.

Danish, S.J., Petitpas, A.J. and Hale, B.D. (1993) Life development intervention for athletes: life skills through sports. *The Counselling Psychologist*, **3,** 352–85.

Gilbourne, D. and Taylor, A.H. (1995) Rehabilitation experiences of injured athletes and their perceptions of a task oriented goal-setting programme: the application of an action research design. *Journal of Sports Sciences*, **13,** 54–5.

Green, L.J. (1992) The use of imagery in the rehabilitation of injured athletes. *The Sport Psychologist*, **4,** 261–74.

Heil, J. (1993) *The Psychology of Sport Injury*, Human Kinetics, Champaign, IL.

Locke, E.A. and Latham, G.P. (1985) The application of goal-setting to sport. *Journal of Sport Psychology*, **7,** 205–22.

Pargman, D. (1993) *Psychological Bases of Sports Injuries*, Fitness Information Technology, Morgantown.

Roberts, G.C. (1992) *Motivation in Sport and Exercise*, Human Kinetics, Champaign, IL.

Vealey, R. (1988) Future directions in psychological skills training. *The Sports Psychologist*, **2,** 318–36.

Wiese, D.M. and Weiss, M.R. (1987) Psychological rehabilitation and physical injury: implications for the sports medicine team. *The Sport Psychologist*, **1,** 318–30.

Wiese, D.M., Weiss, M.R. and Yukelson, D.P. (1991) Sport psychology in the training room: a survey of athletic trainers. *The Sport Psychologist*, **5,** 15–24.

Environmental stress

13

Thomas Reilly

Introduction

Soccer is played worldwide and in highly varied environmental circumstances. In some instances the climatic conditions are too hostile or are temporarily unsuitable for playing and there is a lull in the competitive programme. This applies in northern climates in winter and in tropical countries during the rainy season. In the former it becomes impossible to maintain playing pitches and the weather is too cold to play in comfort. At another extreme is the stress imposed by a hot environment and the difficulty of coping with high heat and humidity. Usually the hottest part of the day is avoided and matches are timed for evening kick-offs. In highly competitive international tournaments this is not always feasible and some teams from temperate climates are obliged to compete in conditions to which they are unaccustomed.

Altitude constitutes another environmental variable that can make supra-normal demands on football teams. This has applied to those teams who have competed at the two World Cup finals in Mexico in 1970 and 1986. It applies also to teams playing friendly or international qualifying matches at moderate altitude. Additionally, training camps for top teams are sometimes located at altitude resorts and this constitutes a particular novel challenge to sea level dwellers.

The human body has mechanisms that allow it to acclimatize to some extent to environmental challenges. In the course of history it has evolved to match the environmental changes associated with the solar day. Consequently many physiological functions wax and wane in

Science and Soccer. Edited by Thomas Reilly. Published in 1996 by E & FN Spon, London. ISBN 0 419 18880 0.

harmony with cyclical changes in the environment every 24 hours. The sleep–wake cycle is dovetailed with alternation of darkness and light and the majority of the body's activities are controlled by biological clocks. These are disrupted when the body is forced to exercise at a time it is unused to, for example after crossing multiple time zones to compete overseas. It is also disrupted if sleep is disturbed.

In this chapter the major environmental variables that impinge on soccer play are considered. These include heat, cold, hypoxia, circadian rhythms and weather conditions. The biological background is provided prior to describing the consequences of environmental conditions for the soccer player.

13.1 TEMPERATURE

13.1.1 Thermoregulation

Human body temperature is regulated about a set point of 37 °C. This refers to temperature within the body's core and is measured usually as rectal temperature, tympanic or oesophageal temperature. Oral temperature tends to be a little lower than these and is less reliable since the temperature within the mouth can be influenced by drinking cold or hot fluids and by the temperature of the air inspired.

For thermoregulatory purposes the body can be conceived as consisting of a core and a shell. There is a gradient of about 4 °C from core to shell and so mean skin temperature is usually about 33 °C. The temperature of the shell is more variable than the core and responsive to changes in environmental temperatures. Usually there is a temperature gradient from skin to the air and this facilitates loss of heat to the environment.

The human exchanges heat with the environment in various ways to achieve an equilibrium. The heat balance equation is expressed as:

$$M - S = E \pm C \pm R \pm K$$

where

M = metabolic rate
S = heat storage
E = evaporation
C = convection
R = radiation
K = conduction

Thermal equilibrium is attained by a balance between heat loss and heat gain mechanisms (Figure 13.1). Heat is produced by metabolic processes, basal metabolic rate being about 1 kcal kg^{-1} h^{-1}. One kilocalorie (4.186 kJ) is the energy required to raise 1 kg water through 1 °C. Energy expenditure during

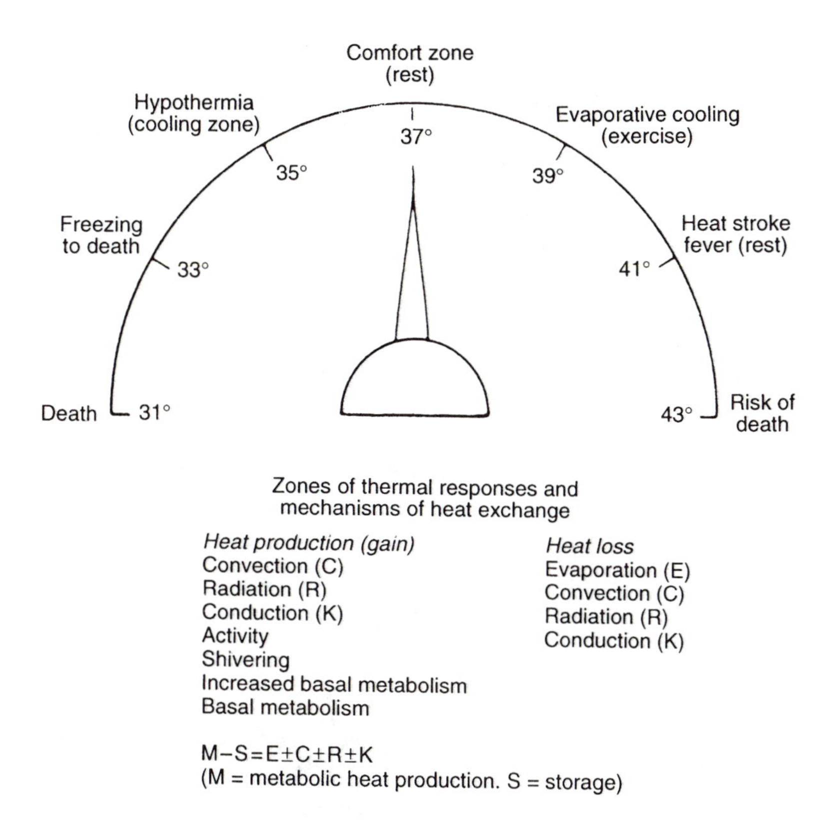

Figure 13.1 Heat loss and heat gain mechanisms.

soccer might increase this by a factor of 15, with maybe only 20–25% of the energy expended reflected in power output. The rest is dissipated as heat within the active tissues and as a result heat storage in the body increases. In order to avoid overheating, the body is equipped with mechanisms for losing heat. It also has built-in responses to safeguard the thermal state of the body in circumstances where heat might be lost too rapidly to the environment, for example in very cold conditions.

Body temperature is controlled by specialized nerve cells within the hypothalamic area of the brain. Neurones in the anterior hypothalamus respond to a rise in body temperature whilst a fall is registered by cells in the posterior hypothalamus. The neurones in the anterior portion constitute the heat loss centre since they trigger initiation of heat loss responses.

These are effected by a redistribution of blood to the skin where it can be cooled and stimulation of the sweat glands to secrete a solution on to the skin surface where evaporative cooling can take place.

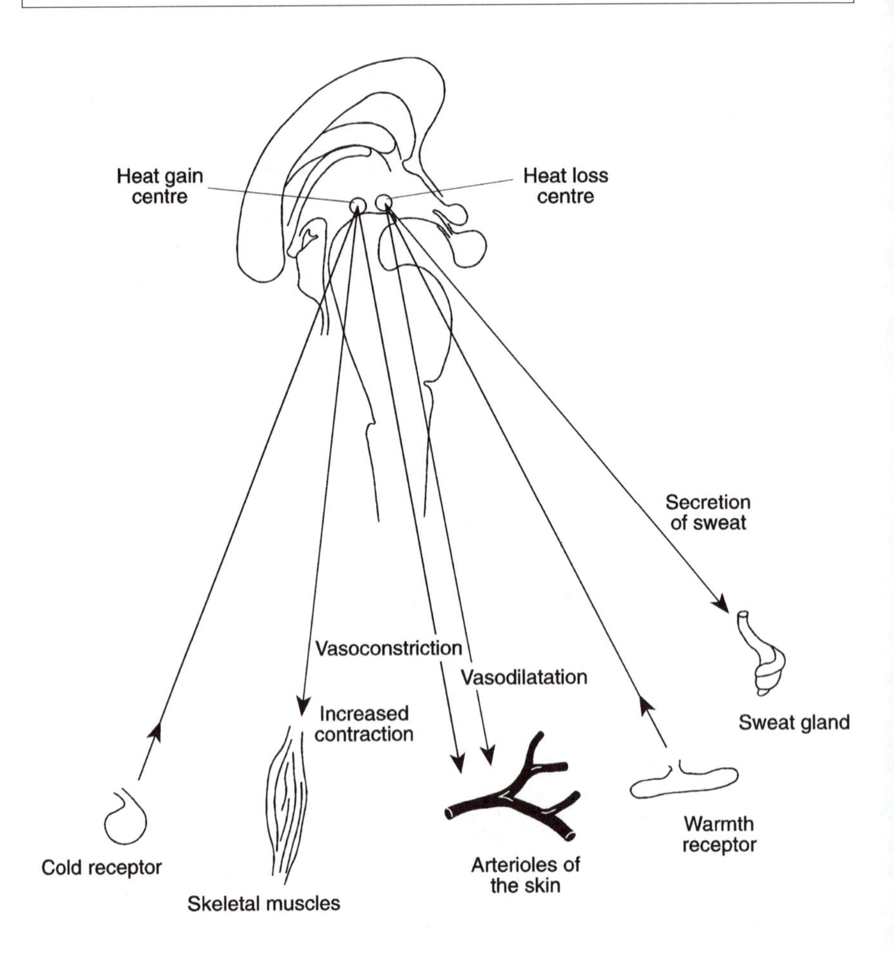

Figure 13.2 Control of human thermoregulation.

The hypothalamus is sensitive to the temperature of blood that bathes these cells controlling thermoregulatory responses. In addition to this direct information, the cells also receive signals from warmth and cold receptors located in the skin. In this way the heat loss and heat gain centres receive information about both the body's internal thermal state and environmental conditions (Figure 13.2).

13.1.2 Exercise in the heat

When exercise is performed, the temperature within the active muscles and core temperature rise. When the exercise is carried out in hot conditions the skin temperature is elevated, reflecting the external challenge to the body. The hypothalamic response is represented by a diversion of cardiac output to the skin: the body surface can lose heat to the environment (by convection and radiation) due to the warm blood now being shunted through its subcutaneous layers. In strenuous exercise, such as intense competitive soccer, the cardiac output may be maximal or near it and this increased cutaneous blood flow may compromise blood supply to the active muscles. In such instances the soccer player will have to lower the exercise intensity, perhaps by taking longer recovery periods than normal or by running 'off-the-ball' less.

There are indications from motion analysis of players of the extent to which high environmental temperature affects their work-rates during matches. Ekblom (1986) reported that the distance covered in high-intensity running during match-play at an ambient temperature of 30 °C was 500 m compared to 900 m when the temperature was 20 °C. It is likely that this lowered work-rate reflects changes in the overall pace of the game. The exercise intensity and the level of play affect the magnitude of the rise in core temperature. Rectal temperatures averaging 39.5 °C have been reported for Swedish First Division players in ambient temperatures of 20–25 °C. The corresponding average for players of lower divisions was 39.1 °C (Ekblom, 1986).

The distribution of blood to the skin is effected by means of dilatation of the peripheral blood vessels. There is a limit to the vasodilatation that occurs in thermoregulation. This is because increased vasodilatation reduces peripheral resistance and so causes a fall in blood pressure. The kidney hormone renin stimulates angiotensin which is a powerful vasoconstrictor and this response corrects a drop in blood pressure. The blood pressure decline is more of a risk in marathon running than in soccer play: whilst both entail an average loading of 75–80% $\dot{V}O_2max$, the greater duration of marathon racing provides extra stress, probably with consequences for body water stores.

The sweat glands are stimulated when core temperature rises, loss of sweat by evaporation being the major avenue by which heat is lost to the environment during intense exercise. The glands respond to stimulation by noradrenaline and secrete a dilute solution containing electrolytes and trace elements. Heat is lost only when the fluid is vaporized on the surface of the body, no heat being exchanged if sweat drips off or is wiped away. When heat is combined with high humidity, the possibilities of losing heat by evaporation are reduced since the air is already highly saturated with water vapour. Consequently hot humid conditions are detrimental to performance and increase the risk of heat injury.

Soccer players may lose 3 litres or more of fluid during 90 minutes of play in the heat. This is an average figure which varies with the climatic conditions and also between individuals. Some players may sweat little and will be at risk when competing in the heat due to hyperthermia. Those who sweat profusely may be dehydrated near the end of the game. A fluid loss of 3.1% body mass has been reported during a match at 33 °C and 40% relative humidity. A similar fluid loss occurred when the ambient temperature was 26.3 °C but humidity was 78% (Mustafa and Mahmoud, 1979).

It is important that players are adequately hydrated prior to playing and training in the heat. Water is lost through sweat at a faster rate than it can be restored through drinking and subsequent absorption through the small intestine. Thirst is not a very sensitive indicator of the level of dehydration. Consequently players should be encouraged to drink regularly, about 200 ml every 15–20 min when training in the heat. The primary need is for water as sweat is hypotonic. Electrolyte and carbohydrate solutions can be more effective than water in enhancing intestinal absorption (see Chapter 6).

One important consequence of sweating is that body water stores are reduced. The body water present in the cells, in the interstices and in plasma seems to fall in roughly equal proportions. The reduction in plasma volume compromises the supply of blood available to the active muscles and to the skin for cooling. Whilst the kidneys and endocrine glands attempt to conserve body water and electrolytes, the needs of thermoregulation override these mechanisms and the athlete may become dangerously dehydrated through continued sweating. The main hormones involved in attempting to protect against dehydration are vasopressin, produced by the pituitary gland, and aldosterone secreted by the adrenal cortex which stimulates the kidneys to conserve sodium.

Many components of soccer performance will be adversely affected once core temperature rises above an optimal level. This is probably around a body temperature of 38.3–38.5 °C (Åstrand and Rodahl, 1986). Performance also deteriorates with progressive levels of dehydration. This drop in performance can be offset to some degree by fluid replacement. This includes cognitive as well as physical and psychomotor aspects of skill. The data shown in Figure 13.3 show how decision-making was best maintained when an energy drink was provided to subjects compared with water only, which itself was superior to a trial when no fluid was provided (Reilly and Lewis, 1985).

The thermal strain on the individual player is a function of the relative exercise intensity (% $\dot{V}O_2max$) rather than the absolute work-load. Therefore the higher the maximal aerobic power ($\dot{V}O_2max$) and cardiac output, the lower the thermal strain on the player. A well-trained individual has a highly developed cardiovascular system to cope with the dual roles of thermoregulation and exercise. The highly trained individual will also acclimatize more quickly than one who is unfit. Training also improves exercise tolerance in the heat but does not eliminate the necessity of heat acclimatization.

13.1.3 Heat acclimatization

Acclimatization refers to reactions to the natural climate. The term **acclimation** is used to refer to physiological changes which occur in response to experimentally induced changes in one particular climatic factor (Nielsen, 1994).

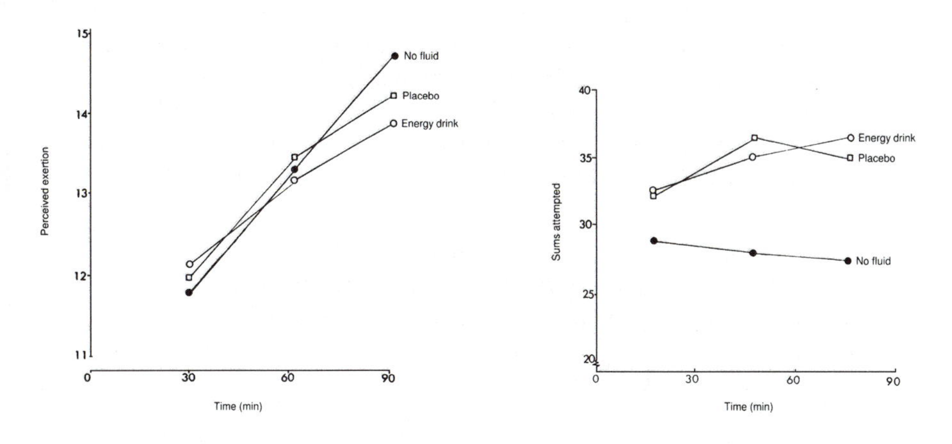

Figure 13.3 Rating of perceived exertion and the speed of adding under three experimental conditions: no fluid, a placebo and an energy drink.

The main features of heat acclimatization are an earlier onset of sweating (sweat produced at a lower rise in body temperature) and a more dilute solution from the sweat glands. The heat-acclimatized individual sweats more than an unacclimatized counterpart at a given exercise intensity. There is also a better distribution of blood to the skin for more effective cooling after a period of acclimatization, although the acclimatized player depends more on evaporative sweat loss than on distribution of blood flow.

Heat acclimatization occurs relatively quickly and a good degree of adaptation takes place within 10–14 days of the initial exposure. Further adaptations will enhance the athlete's capability of performing well in heat stress conditions (Nielsen, 1994). Ideally, therefore, the athlete or team should be exposed to the climate of the host country for at least two weeks before the event. An alternative strategy is to have an acclimatization period of two weeks or so well before the event with subsequent shorter exposures nearer the contest. If these are not practicable, attempts should be made at some degree of heat acclimatization before the athlete leaves for the host country. This may be achieved by pre-acclimatization.

1. Exposure to hot and humid environments, the players seeking out the hottest time of day to train at home.

2. If the conditions at home are too cool, players may seek access to an environmental chamber individually for periodic bouts of heat exposure. It is important that the players exercise rather than rest under such conditions. Repeated exposure to a sauna or Turkish bath is only partially effective. About 3 hours per week exercising in an environmental chamber should provide a good degree of acclimatization.
3. The microclimate next to the skin may be kept hot by wearing heavy sweat suits or windbreakers. This will add to the heat load imposed under cool environmental conditions and induce a degree of adaptation to thermal strain.

On first exposure to a hot climate the players should be encouraged to drink copiously to maintain a pale straw-coloured rather than dark urine. They should drink much more fluid than they think they need since thirst is often a very poor indicator of real need. When they arrive in the hot country they should be discouraged strongly from sunbathing as this itself does not help acclimatization except by the development of a suntan which will eventually protect the skin from damage via solar radiation. This is a long-term process and is not beneficial in the short term, but negative effects of a sunburn can cause severe discomfort and a decline in performance. Players should therefore be protected with an adequate sunscreen if they are likely to be exposed to sunburn.

Initially training should be undertaken in the cooler parts of the day so that an adequate work-load can be achieved and adequate fluid must be taken regularly. If sleeping is difficult, arrangements should be made to sleep in an air-conditioned environment but to achieve adequate acclimatization the rest of the day should be spent exposed to the ambient temperature other than in air-conditioned rooms. Although sweating will increase as a result of acclimatization, there should be no need to take salt tablets, provided adequate amounts of salt are taken with normal food.

In the period of acclimatization the players should regularly monitor body weight and try to compensate for weight loss with adequate fluid intake. Alcohol is inappropriate for rehydration purposes since it acts as a diuretic and increases urine output. Players are reminded to check that the volume of urine is as large as usual and that it is a pale straw colour rather than dark (de Looy *et al.*, 1988).

13.1.4 Heat injury

Hyperthermia (overheating) and loss of body water (hypohydration) lead to abnormalities that are referred to as heat injury. Progressively they may be manifest as muscle cramps, heat exhaustion and heat stroke. They are observed more frequently in individual events such as distance running and cycling than in soccer but can occur in soccer matches or training sessions in the heat.

Heat cramps are associated with loss of body fluid, particularly in games players competing in intense heat. Although the body loses electrolytes in sweat, such losses cannot adequately account for the occurrence of cramps. These seem to coincide with low energy stores as well as reduced body water levels. Generally the muscles employed in the exercise are affected, but most vulnerable are the leg (upper or lower) and abdominal muscles. The cramp can usually be stopped by stretching the involved muscle, and sometimes massage is effective.

Heat exhaustion is characterized by a core temperature of about 40 °C. Associated with this is a feeling of extreme tiredness, dizziness, breathlessness and tachycardia (increased heart rate). The symptoms may coincide with a reduced sweat loss but usually arise because the skin blood vessels are so dilated that blood flow to vital organs is reduced.

Heat stroke is a true medical emergency. It is characterized by core temperatures of 41 °C or higher. Hypohydration – due to loss of body water in sweat and associated with a high core temperature – can be driven so far as to threaten life. Heat stroke is characterized by cessation of sweating, total confusion or loss of consciousness. In such cases treatment is urgently needed to reduce body temperature. There may also be circulatory instability and loss of vasomotor tone as the regulation of blood pressure begins to fail.

13.1.5 Competing in cold

Soccer in countries such as the United Kingdom is a winter sport and is often played in near-freezing conditions. Core temperature and muscle temperature may fall and exercise performance will be increasingly affected. Muscle power output is reduced by 5% for every 1 °C fall in muscle temperature below normal levels (Bergh and Ekblom, 1979). A fall in core temperature to hypothermic levels is life-threatening and the body's heat gain mechanisms are designed to arrest the decline.

Among the responses to cold initiated by the posterior hypothalamus is a generalized vasoconstriction of the cutaneous circulation. This is mediated by the sympathetic nervous system. Blood is displaced centrally away from the peripheral circulation and this increases the temperature gradient between core and shell. The reduction in skin temperature in turn decreases the gradient between the skin and the environment which protects against a massive loss of heat from the body. Superficial veins are also affected in that blood returning from the limbs is diverted from them to the vena comitantes that lie adjacent to the main arteries. In this way the arterial blood is cooled by the venous return almost immediately it enters the limb by means of counter-current heat exchange.

One of the consequences of the fall in limb temperature is that motor performance is adversely affected. In addition to the drop in muscular strength and power output as the temperature in the muscle falls, there is impairment of conduction velocity of nerve impulses to the muscles. The sensitivity of muscle spindles also declines and there is a loss of manual dexterity. For these reasons it is important to preserve limb temperatures in soccer players during competition. The goalkeeper in particular must maintain manual dexterity for handling the ball. Indeed much of the activity of the goalkeeper is spontaneous rather than directly imposed by demands of the game. The goalkeeper must remain alert during the periods when he is not directly involved in anticipation of the parts of the game when he is called upon.

Shivering is a response of the body's automatic nervous system to the fall in core temperature. It constitutes involuntary activity of skeletal muscles in order to generate metabolic heat. Shivering tends to be intermittent and may persist during exercise if the intensity is insufficient to maintain core temperature. It may be evident during stoppages in play, especially when cold conditions are compounded by sleet.

Early symptoms of hypothermia include shivering, fatigue, loss of strength and coordination and an inability to sustain work-rate. Once fatigue develops, shivering may decrease and the condition worsens. Later symptoms include collapse, stupor and loss of consciousness. This risk applies more to recreational rather than professional soccer as some players might not be able to sustain a work-rate to keep them warm in extreme cold. In such events the referee would be expected to abandon play before conditions became critical.

Cold is less of a problem than heat in that the body may be protected against exposure to the ambient environmental conditions. The important climate is the microclimate next to the skin and this may be maintained by appropriate choice of clothing. Behaviourally players might respond to cold conditions by maintaining a high work-rate. Alternatively, they may be spared exposure to the cold by conducting training sessions in indoor facilities where these are available.

Clothing of natural fibre (cotton or wool) is preferable to synthetic material in cold and in cold–wet conditions. The clothing should allow sweat produced during exercise in these conditions to flow through the garment. The best material will allow sweat to flow out through the cells of the garment whilst preventing water droplets from penetrating the clothing from the outside. If the fabric becomes saturated with water or sweat, it loses its insulation and in cold–wet conditions the body temperature may quickly drop.

Players training in the cold should ensure that the trunk area of the body is well insulated. The use of warm undergarments beneath a full tracksuit may be needed. Dressing in layers is well advised: the outer layers can be discarded as body temperature rises and if ambient temperature gets warmer.

When layers of clothing are worn the outer layer should be capable of resisting both wind and rain. The inner layer should provide insulation and should also wick moisture from the skin to promote heat loss by evaporation. Polypropylene and cotton fishnet thermal underwear has good insulation and wicking properties and so is suitable to wear next to the skin.

Immediately prior to competing in the cold, players should endeavour to stay as warm as possible. A thorough warm-up regime (performed indoors if possible) is recommended. It is thought that cold conditions increase the risk of muscle injury in sports involving intense anaerobic efforts; warm-up exercises may afford some protection in this respect. Competitors may need to wear more clothing than they normally do during matches.

Aerobic fitness does not directly offer protection against cold. Nevertheless it will enable games players to keep more active when not directly involved in play and not increase the level of fatigue. Outfield players with a high level of aerobic fitness will also be able to maintain activity at a satisfactory level to achieve heat balance. On the other hand the individual with poor endurance may be at risk of hypothermia if the pace of activity falls dramatically. Shivering during activity signals the onset of danger.

13.2 ALTITUDE

13.2.1 Physiological adjustments to altitude

As altitude increases the barometric pressure falls. At sea level the normal pressure is 760 mmHg, at 1000 m it is 680 mmHg, at 3000 m it is about 540 mmHg. High altitude conditions are referred to as hypobaric or low pressure and the main problem associated with this environment is hypoxia or a relative lack of oxygen.

Normally, the proportion of oxygen in the air is 20.93%, and the partial pressure of oxygen at sea level is 159 mmHg. The partial pressure of oxygen decreases with increasing altitude: this corresponds to the fall in ambient pressure since the proportion of oxygen in the air is constant. As a result there are fewer oxygen molecules in the air at altitude for a given volume of air. Less oxygen is inspired for a given inspired volume and this ultimately means a reduction in the amount of oxygen delivered to the active tissues.

As far as the uptake of oxygen into the body through the lungs is concerned, the important factor is the tension of oxygen in the alveoli (PO_2). Here the water vapour pressure is relatively constant at 47 mmHg as is the PCO_2 of 35–40 mmHg. The result of the fall in ambient pressure and consequently alveolar tension is that the gradient across the pulmonary capillaries for transferring oxygen into the blood becomes less favourable. Exercise that depends on oxygen transport mechanisms will be impaired at about 1200 m once desaturation occurs. This refers to the oxygen association curve of haemoglobin (Hb) which is sigmoid-shaped and is affected by pressure. Normally the red blood cells are 97% saturated with O_2 but this figure falls when PO_2 levels drop at a point corresponding to this altitude (1200 m). The O_2–Hb curve is little affected for the first 1000–1500 m of altitude because of the flatness at its top. As the pressure drops further to reach the steep part of the curve, the supply of oxygen to the body's tissues is increasingly impaired. Nevertheless, at an altitude of 3000 m the arterial saturation is about 90%.

The immediate physiological compensation for hyoxia is an increase in ventilation. This is represented by an increased tidal volume (depth of breathing) and an increased breathing frequency. A consequence of this hyperventilation is that there is an increase in the CO_2 blown off from blood passing through the lungs. The elimination of CO_2 leaves the blood more alkaline than normal due to an excess of bicarbonate ions, CO_2 being a weak acid in solution in body fluid. Over several days the kidneys compensate by excreting excess bicarbonate, so returning the blood to the normal pH level. However, the body's alkaline reserve is decreased and so the blood has a poorer buffering capacity for tolerating additional acids (such as lactic acid diffusing from muscle to blood during exercise).

Once at altitude, there is an increased production of the substance 2,3-BPG (bisphosphoglycerate) by the red blood cells. This is beneficial in that it aids in unloading oxygen from the red blood cells at the tissues.

The oxygen-carrying capacity of the blood is enhanced by an increase in the number of red blood cells. This process begins within a few days at altitude and is stimulated by the kidney hormone erythropoietin. This causes the bone marrow to increase red blood cell production: this requires that the body's iron stores are adequate and may indeed mean supplementation of iron intake prior to and during the stay at altitude. There is an apparent increase in haemoglobin in the first few days at altitude which reflects haemoconcentration due to a drop in plasma volume. Nevertheless there is a gradual true rise in haemoglobin which may take 10–12 weeks to be optimized. Even after a year or more at altitude the increases in total body haemoglobin and red cell count do not attain values observed in high altitude natives. As a result sea level natives will never be able to compete in aerobic events (including soccer) at altitude on equal terms with those born at altitude. They have to devise strategies to allow them to demonstrate their superior skills as well as prepare physiologically by acclimatizing.

13.2.2 Exercise at altitude

Soccer players will experience more difficulty in exercising at altitude compared with sea level in spite of the physiological adjustments to hypoxia that take place. Changes in maximum cardiac output and in the oxygen transport system lead to a fall in maximal oxygen uptake ($\dot{V}O_2max$). At an altitude of 2300 m, corresponding roughly to Mexico City, the initial decline in $\dot{V}O_2max$ is about 15%. After 4 weeks at this altitude there is an improvement in $\dot{V}O_2max$ but it still remains about 9% below its sea level value. For sea level dwellers the initial decline in $\dot{V}O_2max$ is 1–2% for every 100 m above 1500 m (Åstrand and Rodahl, 1986).

Soccer play is mostly at submaximal intensity, although periodically there are short maximal efforts. Maintaining a fixed submaximal exercise intensity is more stressful at altitude than at sea level. The highest level of endurance exercise that can be sustained is determined by the intensity above which lactate accumulates progressively in the blood. This 'lactate threshold' is lowered at altitude although the percentage of $\dot{V}O_2max$ at which it occurs is unaltered. In order to cope with the lack of oxygen, the active muscles rely more on anaerobic processes and so soccer players will need longer low- intensity recovery periods during match-play, following from their bouts of all-out high-intensity efforts.

Heart rate, ventilation and perceived exertion are all elevated beyond the normal sea level responses at any given submaximal exercise intensity. As a result the pace of tolerable exercise is reduced. Soccer players should be prepared to pace their efforts more selectively during matches at altitude. They will also need to accept a lower intensity during training sessions. This is especially important in the first few days at altitude.

Different individuals will be affected to varying degrees depending on factors such as level of aerobic fitness, prior acclimatization, previous experience of altitude. Physiological factors such as pulmonary diffusing capacity, total body haemoglobin, iron stores, nutritional state and so on may also determine why some individuals will suffer more at altitude than others.

Successful adaptation to altitude results in a decreased tachycardia in response to submaximal exercise compared with the heart rate on initial exposure. The heart rate response may approach sea level values after three to four weeks of exposure. Adaptations of skeletal muscles occur to aid their struggle against hypoxia. Improvements in maximum blood flow capacity and oxidative metabolism require a sojourn of many months at altitude. These long-term adaptations will not be of benefit to anaerobic processes. Anaerobic efforts such as sprinting may in fact be improved at altitude due to the reduced air resistance against which the body moves. Such conditions may in fact be favourable for improving running speed. The buffering capacity of muscle is improved with a prolonged stay at altitude and this complements the adaptations that occur in oxygen transport mechanisms.

13.2.3 Altitude sickness

The immediate and short-term adjustments which help the body adapt to altitude can have adverse side-effects. The most common problem is referred to as acute mountain sickness. This is characterized by headaches, nausea, vomiting, loss of appetite, sleep disturbances and irritability. These problems can be encountered at altitudes above 2000–2500 m. The syndrome develops progressively, reaching a peak within about 48 hours of initial exposure, and then disappears with adaptation. The problems are related to changes in the pH of the cerebrospinal fluid consequent to respiratory alkalosis and also increases in cerebral blood flow stimulated by hypoxia. Onset of acute mountain sickness may be sudden when ascent is rapid and exercise may increase the likelihood of its development. For this reason intense training is discouraged for three to four days on going to altitude.

Individuals with low body stores of iron may experience difficulties at altitude once red blood cell production is stimulated by erythropoietin. This can be accentuated if appetite is affected by acute mountain sickness. Careful attention to diet is needed when going to altitude and during the immediate period of adaptation.

Attention is also directed towards adequate hydration. The air at altitude tends to be drier than at sea level. More fluid is lost by means of evaporation from the moist mucous membrane of the respiratory tract. This is accentuated by the hyperventilation response to hypoxia. The nose and throat get dry and irritable and this can cause discomfort. It is important to drink more than normal to counteract the fluid loss. Indeed a rigorous regime of drinking fluids has been shown to help offset the fall in plasma volume that is a characteristic response to altitude (Ingjer and Myhre, 1992).

It should be mentioned that the ambient temperature drops by about 1°C in every 300 m ascent. As a result some of the problems linked with cold environments are relevant considerations at altitude.

13.2.4 Soccer strategies

Acclimatization is imperative for a soccer team scheduled to compete at altitude. Major international tournaments have taken place at altitudes where aerobic processes are compromised. These have included two World Cups at Mexico City, the Olympic Games soccer tournament in Mexico in 1968 and the World Students Games in 1979 at Mexico. Other countries play their home matches at altitude, including Colombia and Bolivia. In qualifying for the 1994 World Cup finals, Bolivia played its home matches at 2800 m, a factor that bestowed a considerable advantage to its players.

Teams playing at altitude may need to redistribute work-loads among players so that individuals can take longer recovery periods than normal. They may also need to time their offensive moves more effectively and concede possession to the opponents for longer than customary. Teams that rely on a high all-round work-rate from players, particularly in putting pressure on opposition players in possession of the ball, will need to modify their usual style of play. Occasionally the direct style of play in quickly transferring the ball from defence to attack with long passes might prove effective.

The lowered air resistance at altitude alters ball-flight characteristics. Consequently long kicks will travel further and shots at goal will travel faster. It seems important that all players should experience these conditions before actually competing in matches at altitude. This would be especially important for the goalkeeper and the strikers. There is no real method of simulating these conditions at sea level and so players have to accustom themselves to this aspect of skill at altitude.

Multi-venue soccer tournaments may entail qualifying matches or early rounds at different altitudes. This happened to some teams playing in the 1986 World Cup in Mexico, a number of matches being scheduled close to sea level. In such circumstances it is difficult for the team management to make plans and generally they prepare for the worst possible eventualities. Some flexibility is available in the choice of altitude for living accommodation and the team may descend to a lower altitude for specific strenuous training sessions. In this way the players can maintain a high standard of training stimulus and achieve a measure of acclimatization to altitude.

Many Olympic athletes use altitude training camps in the belief that the adaptations that occur will benefit subsequent performance at sea level. There are advantages and disadvantages to the practice. This ploy is unlikely to be of much help to soccer players whose competitive season tends to leave little room for such manoeuvres.

13.2.5 Preparing for altitude

Players scheduled to compete at altitude must consider the physiological consequences of such an engagement. Detailed preparatory recommendations have been outlined by de Looy *et al.* (1988) and much of the advice is appropriate to soccer teams.

It is not advisable to do strenuous training for at least two to three days until the stage of vulnerability to acute mountain sickness has passed. After that, prolonged training sessions should be reduced in intensity to the same perceptual load as at sea level; full work-outs are not advisable for seven to ten days after arrival. Recovery periods between intense short-term efforts should be lengthened when intermittent exercise is being performed: this applies both to conditioning work and to games practice.

Rehydration following training at altitude must be complete as more fluids than normal may be lost during exercise. The diet should contain a greater than normal proportion of carbohydrate, especially in the first days at altitude. This will compensate both for the increased reliance on glycogen as a fuel for exercise and for the fall in the tension of CO_2 in the blood consequent to hyperventilation.

About 14 days should be allowed before competition for acclimatization to altitudes of 1500–2000 m and 21 days before matches at 2000–2500 m. These periods may be shortened if the players have had previous exposures to altitude in their build-up to the tournament. Unacclimatized individuals need about one month to adapt to altitudes above 2500 m and may lose match fitness in the process. Fortunately, soccer play at this altitude is uncommon for sea level dwellers.

If it is impractical to stay at altitude for a long period before a competition, some degree of acclimatization may be achieved by frequent exposures to simulated altitude in an environmental chamber. Continuous exercise of 60–90 minutes, or 45–60 minutes of intermittent exercise performed four or five times a week at simulated altitude of 2300 m has shown good results in three to four weeks (Terrados *et al.*, 1988).

Portable simulators that induce hypoxia are available for wear as a back-pack. These lower the inspired-oxygen tension and accentuate exercise stress but also increase the resistance to breathing. There is no convincing evidence that they promote the kind of adaptations that are experienced at altitude or that result from prolonged exercise in a hypobaric chamber. Nevertheless they may have psychological benefits for players in accustoming them to hypoxia. Portable simulators were used by the Danish soccer team, along with exercise tests in a hypobaric chamber (Bangsbo *et al.*, 1988) in preparation for the 1986 World Cup in Mexico.

13.3 CIRCADIAN VARIATION

13.3.1 Circadian rhythm

Circadian rhythms refer to cyclical changes within the body that recur around the solar day. An example is the rhythm in core temperature (Figure 13.4) which shows a cycle every 24 hours. A cosine wave can be fitted to the observations in rectal temperature and the time of peak occurrence identified. This is referred to as the acrophase and is usually found between 1700 and 1800 hours. Many measures of human performance follow closely this curve in body temperature (Reilly *et al.*, 1993). These would include components of motor performance (such as muscular strength, reaction time, jumping performance) that are important in soccer play.

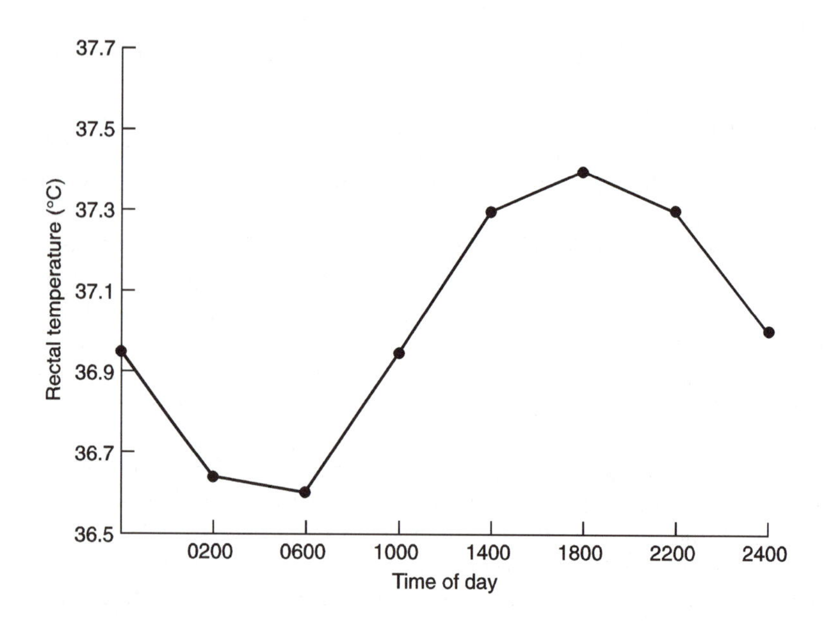

Figure 13.4 Circadian rhythm in rectal temperature.

The other biological rhythm of major importance is the sleep–wake cycle. This is linked with the pattern of habitual activity, i.e. sleeping during the hours of darkness and working or staying awake during daylight. Thus there is a sharp contrast in arousal states between night time and day time. Arousal tends to peak just after mid-day at the time that circulating levels of adrenaline are at their highest. A team forced to compete at a time of day it would normally be inactive would not be well equipped, biologically or psychologically, to do so.

Circadian rhythms are described as endogenous or exogenous depending on the degree to which they are governed by environmental signals. These include natural and artificial light, temperature, type and timing of meals, social and physical activity. Endogenous rhythms imply internal body clocks, the suprachiasmatic nucleus cells of the hypothalamus being thought to be the site of control of circadian rhythms. Timekeeping functions have also been attributed to the pineal gland, its hormone melatonin and related substrates such as serotonin. Local timekeepers have also been isolated in both cardiac and skeletal muscle. It is likely that there is a family of clocks within the body which control a host of circadian rhythms and which are organized in a hierarchy. The most relevant of these for sports performance seem to be the body temperature curve and the sleep–wake cycle.

13.3.2 Sleep

Sleep is an enigma in the sense that it has never been conclusively explained why it is needed. One school of thought relates sleep to the restitution of the body's tissues. An alternative view is that the need for sleep is specific to nerve cells – the so-called brain restitution theory of sleep. Nevertheless it is obvious that sleep is essential and this need is most apparent when sleep is deprived or disturbed.

Whilst the average sleep of a 20–30-year-old is about 7 hours each night, there is a large variation between individuals both in the need for sleep and in the amount of sleep taken. Some athletes feel uneasy unless they sleep soundly for 8–9 hours and place a priority on their sleeping arrangements. Brain states may be monitored during sleep by electroencephalography (EEG). Traces from EEGs demonstrate cycles of about 90 minutes, each cycle containing stages known as REM (rapid eye movement) and non-REM sleep. Non-REM sleep is further classified into stages 1 to 4. It is easy to awaken individuals from REM sleep but more difficult during non-REM sleep. Consequently, players who doze in the morning may slip into a further 90 minute cycle of sleep which in all probability does them little good.

Professional soccer players tend to get adequate sleep when a complete week is considered. However, it is often inconsistent, players staying up late especially after an evening match. In such events it is difficult to get to sleep since catecholamine levels are elevated above normal and players still rethink details from the match completed earlier. Relaxation following a game may require a conscious behavioural strategy which differs from individual to individual.

Pre-match anxiety can disrupt sleep the night before playing. Complaints from players that they were unable to sleep are generally untrue as short periods asleep, albeit snatched unwittingly during the night, do provide a restorative function. In such cases a short nap during the day could be encouraged. A brief afternoon nap prior to an evening kick-off can promote a release from pre-match anxiety. There is a natural tendency to drowse in the mid-afternoon which is generally referred to as a 'post-lunch dip'.

Muscular performance may be unaffected by sleep loss, at least as shown in experiments of partial sleep deprivation where subjects are permitted only 2.5–3 hours sleep a night (Reilly and Deykin, 1983). Complex tasks and decision-making, especially if demanded over a prolonged period such as 90 minutes, deteriorate with the duration of the task. Thus concentration during a soccer match following nights of disrupted sleep requires a distinct motivational drive from the player concerned.

It has been possible to play soccer for days without sleep, although this was done indoors and at a low intensity. For over 91 hours the level of play showed a cyclic change that corresponded to the circadian rhythm in core temperature of the players (Reilly and Walsh, 1981). The appearance of psychotic-like symptoms, particularly following the second night without sleep, makes meaningful play difficult under such circumstances. Clearly such a regime is not conducive towards serious soccer performance.

13.3.3 Training and time of day

The majority of motor performance measures demonstrate a peak in performance that occurs close to the acrophase of the circadian rhythm in body temperature. On this basis the ideal time for soccer play would be about 1700–1800 hours, assuming the environmental temperature is comfortable. There is probably a window of some hours during the day when maximal performance can be achieved. This can be realized with appropriate warm-up and physical and mental preparation. Consequently kick-offs at 1500 and 1930 hours do not necessarily entail sub-optimal performance, particularly as muscle and core temperatures rise during the course of match-play. Particular consideration may be needed in late kick-offs, say 2000 hours in cold conditions.

There is often a mis-match between the time of training and the time at which matches are played. The majority of professional soccer teams train in the morning, starting at 1000 or 1100 hours. Strenuous physical conditioning exercise is best conducted in the early evening, the time at which many amateur teams train. Joint stiffness is greatest in the morning and so special attention should be given to flexibility exercises in warming-up prior to morning training sessions.

Skills may be best acquired in mid-day sessions just as the curve in arousal approaches and reaches its high point. Consequently there is a case for young professionals to have their skills work at light intensity in morning sessions. The more intense exercise can be retained for a later session following lunch and a rest for recovery.

13.3.4 Jet lag

Footballers are sometimes called upon to travel vast distances to play in international team or club contests. They may also participate in closed-season tournaments or friendly matches overseas. Such engagements are made possible by the speed of contemporary air flight. Although international travel is commonplace nowadays it is not without attendant problems for the travelling footballer, which the team management and back-up staff should recognize in advance.

In the course of travelling abroad players encounter disruption of their regular routine. They may be particularly excited about the trip or may have had worries associated with planning for the departure. Vaccinations may be required, according to the country to be visited. The majority of top teams have arrangements overseen by their administrative and medical staff, as far as possible, in order to avoid otherwise inevitable embarrassments.

There are still routines to be faced – travelling to the airport of departure, checking in and going through security controls, possibly taking advantage of duty-free facilities or coping with the frustrations of delayed flights. Experienced management staff try to shield the players from such irritations. There is also a possibility of protracted disembarkation procedures on arrival and mix-ups with ground travel and accommodation. Having arrived safely at the destination, the player may be suffering travel fatigue, loss of sleep perhaps (depending on flight times), and a cluster of symptoms which have come to be known as jet lag.

Jet lag refers to the feelings of disorientation, light-headedness, impatience, lack of energy and general discomfort that follow travelling across time zones. These feelings are not experienced with travelling directly northwards or southwards within the same time zone when the passenger simply becomes tired from the journey or stiff as a result of a long stay in a cramped posture. The feelings associated with jet lag may linger for several days after arrival and may be accompanied by loss of appetite, difficulty in sleeping, constipation and dizziness. In extreme cases the individual may even burst into tears when facing unanticipated difficulties. On the other hand some people claim they never experience any problems and deny that the phenomenon of jet lag exists. Although there are undoubtedly individual differences in the severity of symptoms, many people may simply fail to recognize how they themselves are affected, especially in tasks requiring concentration and complex co-ordination.

Following a journey across multiple time zones the body's circadian rhythms at first retain the characteristics of their point of departure. However, the new environment forces new influences on these cycles, mainly the time of sunrise and onset of darkness. The body attempts to adjust to this new context but core temperature is relatively sluggish in doing so. As a rough guide it takes about one day for each time zone crossed for body temperature to adapt completely. The individual may have difficulty in sleeping for a few days but activity and social contact during the day help in the adaptations of the arousal rhythm. Thus arousal adjusts more quickly than does body temperature to the new time zone. Until the whole spectrum of biological rhythms adjusts to the new local time, thereby becoming re-synchronized, the performance of the footballer may be below par.

Allowing for individual differences, the severity of jet lag is affected by various factors. In general, the greater the number of time zones travelled, the more difficult it is to cope. A 2-hour phase shift may have marginal significance but a 3-hour shift (e.g. British or Irish teams travelling to play European football matches in Russia or Turkey or teams within the USA travelling coast to coast) will entail desynchronization to a substantial degree. In such cases the flight times – time of departure and time of arrival – may determine how severe are the symptoms of jet lag that occur. It may also be wise to alter training times to take the direction of travel into account. Such a ploy was shown to be successful in American football teams travelling across time zones within the USA and scheduled to play at different times of day (Jehue *et al.*, 1993).

The severity of symptoms may be worse two or three days after arrival than on the day immediately following disembarkation. Symptoms then gradually abate, but may still be acute at particular times of day. There will be a window of time during the day when the time of high arousal associated with the time zone departed from and the new local time overlap. This window may be predicted in advance and should be utilized for timing of training practices in the first few days at the destination. Observations on footballers travelling from Britain to Oceania (Australia, New Zealand and Papua New Guinea) indicate that morning training sessions suit players best over the first few days (Reilly and Mellor, 1988).

The direction of travel also affects the severity of jet lag. It is easier to cope with flying in a westward direction compared to flying eastward. In flying westward the cycle is lengthened and body rhythms can extend in line with their natural free-wheeling period of about 27 hours and thus catch up. Our observations on travelling to Korea (9 hours in advance of British Summer Time) and Malaysia (7 hours in advance of British Summer Time) were that periods of nine and six days respectively were inadequate for jet lag symptoms to disappear. In contrast re-adaptation was more rapid on returning to Britain. However, when time zone shifts approach near maximal values – the maximum is a 12-hour change – there may be little difference between eastward and westward travel.

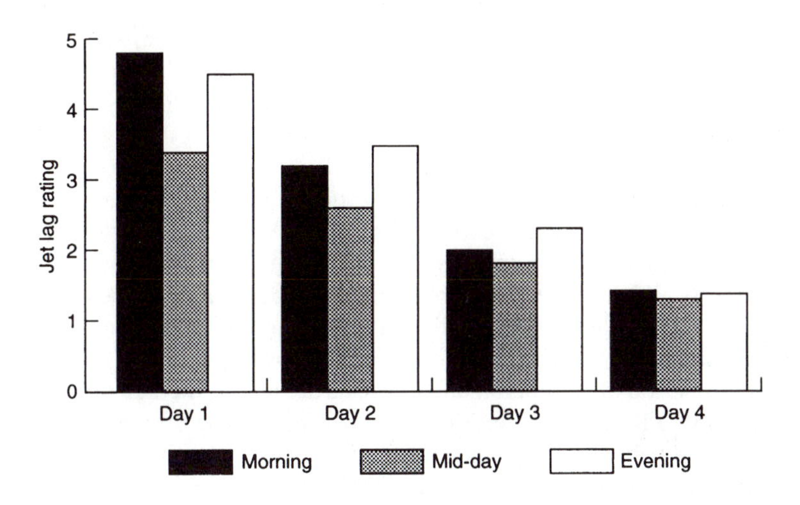

Figure 13.5 Mean jet lag ratings (scale 0 to 10) of soccer players after returning to England from Japan. Values had returned to zero on Day 5. (Reproduced with permission from Reilly, 1993.)

In the past, English League teams travelling to Japan to compete in the international club final have failed to allow time for jet lag symptoms to decay. Not only did these representatives (Liverpool, Aston Villa, Nottingham Forest) lose but their performances in the week following the return home were clearly compromised (Figure 13.5). Understanding the physiological processes involved in jet lag and the consequences of their disruption should encourage the use of practical measures to help the traveller to cope better.

British football teams travelling to Australia have used sleeping pills to induce sleep whilst on board. Although such drugs as benzodiazepines are effective in getting people to sleep, they do not guarantee a prolonged period asleep. Besides, they have not been satisfactorily tested for subsequent residual effects on motor performances such as soccer skills. They may also be counter-productive if administered at the incorrect time. A prolonged nap at the time the individual feels drowsy (presumably at the time that he would have been asleep in the time zone he departed from) simply anchors the rhythms at their former phases and so resists the adaptations to the new time zone.

On reaching the destination a key factor is to fit in immediately with the phase characteristics of the new environment. Players will already have worked out the local time for their disembarkation. There may be other environmental factors to consider such as heat, humidity or even altitude.

Having travelled westward players may be allowed to retire to bed early. Early onset of sleep will be less likely after an eastward flight. In this case a light training session on that evening would be helpful in instilling local cues into the rhythms. There is some evidence that exercise does speed up the adaptation to a new time zone. Observations on professional players returning to England from the Far East showed that a light training session on the afternoon of arriving home was successful in alleviating jet lag symptoms (Reilly, 1993).

For the first few days in the new time zone, training sessions should not be all-out. Skills requiring fine coordination are likely to be impaired and this might lead to accidents or injuries if sessions with the ball are conducted too strenuously. Where a series of tournament engagements is scheduled, it is useful to have at least one friendly match during the initial period, that is, before the end of the first week in the overseas country.

In this period of adaptation a few caveats are noted. Alcohol taken late in the evening is likely to disrupt sleep and so is not advised. An alternation of feasting and fasting as recommended for commercial travellers in the USA is unlikely to gain acceptance among footballers. Nevertheless, they could benefit from biasing the macronutrients in their evening meal largely towards carbohydrates. These would include vegetables with a choice of chipped, roast or baked potatoes, pasta dishes, rice and bread. These should include sufficient fibre to safeguard against constipation.

In the early days in the new country the players should be discouraged from taking prolonged naps. A nap at the time they would have been asleep had they stayed at home would make subsequent sleep more difficult and retard the adjustment of the major biological clocks to the new regimes.

By preparing for time zone transitions and the disturbances they impose on the body's rhythms, the severity of jet lag symptoms may be reduced. There has been little success in attempting to predict good and poor adaptors to long haul flights. The fact that an individual escapes lightly on one occasion is no guarantee that that same individual will do so again on the next visit. The disturbances in mental performance and cognitive functions have consequences not only for players but also for management and medical staff travelling with the team, who by no means have immunity against jet lag symptoms.

13.4 ENVIRONMENTAL MONITORING

The environment in which the soccer player trains and competes has implications not just for performance but also for health and safety. The quality of the playing surface, for example, forces a choice of appropriate footwear so that performance can be executed without increased risk. Playing conditions sometimes exceed the bounds of safety and the match referee is entitled to declare the pitch unplayable.

A similar decision may be made in cases of air pollution (ozone, CO, SO_2, or Pb). Ozone concentration may exceed acceptable limits in some of the world's major cities (Mexico City, Seoul, Los Angeles) but generally is not a major problem for European players. The air is impurified in foggy conditions, but in such cases matches and training are usually curtailed for reasons of visibility.

Preparations for hot conditions entail choice of appropriate clothing. Light, loose clothing helps in creating convective air currents to cool the skin. Clothing of natural fibre such as cotton (or at least a cotton-polyester mix) is desirable under warm and radiant environmental conditions. In contrast, clothing with good insulation and preferably in layers helps maintain a warm microclimate next to the skin in cold conditions.

Calculating the risk of heat injury requires accurate assessment of environmental variables. The main factors to be considered are the dry bulb temperature, relative humidity, radiant temperature, air velocity and cloud cover. The most widely used index in sporting contexts is the wet bulb temperature, which takes both ambient temperature and humidity into account. The wind-chill index is employed in determining risk in cold conditions. Quite apart from the chilling effect of the wind, blustery conditions make ball flight more difficult to anticipate and skills become more erratic as a consequence.

The novel environmental challenge – hypoxia, temperature, travel, weather – calls for preparation on behalf of team management. Conditions may even change dramatically during the course of play. An awareness of the dynamic biological adjustments that the body makes means that adverse effects and discomfort associated with environmental variables can be countered to a large degree.

Summary

Soccer is played in a variety of challenging environments. Stresses may include heat, cold, altitude or disruption of the circadian body clock. Some account must be taken of the environment in which competition is scheduled. This may involve physiological preparation, and changes in tactics may also be needed to enable the players to cope.

REFERENCES

Åstrand, P.O. and Rodahl, K. (1986) *Textbook of Work Physiology*, McGraw-Hill, New York.

Bangsbo, J., Klausen, K., Bro-Rasmusen, T. and Larson, J. (1988) Physiological responses to acute moderate hypoxia in elite soccer players, in *Science and Football* (eds T. Reilly, A. Lees, K. Davids and W.J. Murphy), E. & F.N. Spon, London, pp. 257–64.

Bergh, U. and Ekblom, B. (1979) Effect of muscle temperature on maximal muscle strength and power in human skeletal muscles. *Acta Physiologica Scandinavica*, **107,** 33–7.

Ekblom, B. (1986) Applied physiology of soccer. *Sports Medicine*, **3,** 50–60.

Ingjer, F. and Myhre, K. (1992) Physiological effects of altitude training on elite male cross-country skiers. *Journal of Sports Sciences*, **10,** 37–47.

Jehue, R., Street, D. and Huizengar, R. (1993) Effect of time zone and game time on team performance: National Football League. *Medicine and Science in Sports and Exercise*, **25,** 127–31.

de Looy, A., Minors, D., Waterhouse, J. *et al.* (1988) *The Coach's Guide to Competing Abroad*, National Coaching Foundation, Leeds.

Mustafa, K.Y. and Mahmoud, E.D.A. (1979) Evaporative water loss in African soccer players. *Journal of Sports Medicine and Physical Fitness*, **19,** 181–3.

Nielsen, B. (1994) Heat stress and acclimation. *Ergonomics*, **37,** 49–58.

Reilly, T. (1993) Science and football: an introduction, in *Science and Football II* (eds T. Reilly, J. Clarys and A. Stibbe), E. & F.N. Spon, London, 3–11.

Reilly, T. and Deykin, T. (1983) Effects of partial sleep loss on subjective states, psychomotor and physical, performance tests. *Journal of Human Movement Studies*, **9,** 157–70.

Reilly, T. and Lewis, W. (1985) Effects of carbohydrate feeding on mental functions during sustained physical work, in *Ergonomics International 85* (eds I.D. Brown, R. Goldsmith, K. Coombes and M.A. Sinclair), Taylor and Francis, London, pp. 700–2.

Reilly, T. and Mellor, S. (1988) Jet lag in student Rugby League players following a near-maximal time zone shift, in *Science and Football* (eds T. Reilly, A. Lees, K. Davids and W.J. Murphy), E & F.N. Spon, London, pp. 249–56.

Reilly, T. and Walsh, T. (1981) Physiological, psychological and performance measures during an endurance record for 5-a-side soccer play. *British Journal of Sports Medicine*, **15,** 122–8.

Reilly, T., Atkinson, G. and Collwells, A. (1993) The relevance to exercise performance of the circadian rhythms in body temperature and arousal. *Biology of Sport*, **10,** 203–16.

Terrados, N., Melichna, J., Sylven, C. *et al.* (1988) Effects of training at simulated altitude on performance and muscle metabolic capacity in competitive road cyclists. *European Journal of Applied Physiology*, **57,** 203–9.

PART THREE

Behavioural science and soccer

Soccer skills practice

14

Dick Bate

Introduction

Beginners do not fully understand or play a game in its complete sense during their early encounters with it. Games involving large numbers of participants tend to be complex in nature, making them difficult to understand and to participate in fully. Many hours of practice and experience of a variety of situations are required before a player can be recognized as a competent or advanced performer. The rate of progress towards competence depends on factors such as ability, fitness level, knowledge and experience of similar events, age, time and degree of involvement in practice and playing. The coach's knowledge, understanding, efficiency and skill in devising and conducting effective practice situations are also relevant. Introducing young beginners to a highly complex full game is likely to end in failure and frustration. In order to achieve total mental and physical integration into the full game, a series of steps must be incorporated in the overall progress to competence. Learning takes place through guidance, tuition, experience and repetitive practice. It is practice, mental and physical, that moves performers forward and both player and coach have responsibilities here, more so the coach. It is human nature to play – to practise can be to interrupt the natural desire to 'play' and can be seen as an interference in the pleasure of playing the game. Consequently a coach's understanding of practice – how to devise, how to present and how to organize – is important in producing highly competent games players. Without demanding, challenging, interesting, relevant and effective practice, players are unlikely

Science and Soccer. Edited by Thomas Reilly. Published in 1996 by E & FN Spon, London. ISBN 0 419 18880 0.

to progress at satisfactory rates. The coach must analyse performance, establish practice objectives, implement effective coaching programmes and evaluate subsequent skill progression (see the coaching process model, Chapter 15).

14.1 SOCCER SKILL

All sports involve the application of skill of some kind either cognitive or intellectual, perceptual or motor. Football involves all three skill types operating simultaneously in a rapidly changing environment. Skills have been classified as 'open' or 'closed' (Knapp, 1974). **Open** skills are those which are dictated by and are varied according to external situations, for example, opponents, support players, movement, weather, ground conditions and so on – how a player acts according to what he or she sees going on in the game. **Closed** skills are pre-learned sequences of movements, little affected by the environment, and are well timed and coordinated. Soccer is an 'open-skilled' game requiring rapid responses to unpredictable situations but with some 'closed skill' events also evident, such as free kicks, corner kicks, etc.

Soccer skill involves making correct decisions and then executing that which has been decided upon. The coach has to educate players to make correct decisions and to equip players with the necessary skills and techniques to carry out these decisions.

Technique refers to the relationship and harmony a player demonstrates with the ball and describes the performance of a solitary action in isolation from the game, e.g. a shot or a pass. A technical practice involves players working in isolation on the various aspects of the game such as shooting, passing and ball control. In technique practice, decision-making is minimal and is usually concerned with *how* to perform the technique or required action. To practise these techniques in isolation from the game or game-type situation is unproductive for gifted players but relevant for beginners of the game. Technical mastery of the ball is essential but ultimately technical practice must take place in an environment where tactical decision-making is also required.

The execution of a technique or soccer action such as passing, dribbling or shooting is a part of skilled performance, essential but relatively valueless as a lone facet. Players are judged to be truly skilful in the game of soccer when they can make the best decisions about where and when to play the ball and then perform the skill accurately.

Knapp (1974) defined skill as:

> the learned ability to bring about predetermined results with maximum certainty and minimum outlay of time and energy.

In terms of soccer skill and the context in which we are using the word, 'skill' could be defined as:

the learned ability to be able to select and then perform the correct technique as determined by situational demands.

An important point that is highlighted within this definition is that if 'skill' in this context is a *learned* ability then it can be taught and by implication it can be improved.

The quality of skill performance is determined by many factors, both physical and mental, but within the context of this chapter skill can be said to be dependent upon three sequential processes. The three processes are **reception of information**, **analysis** of that information and **execution of the selected response**. Players make decisions based on the perceptual stimuli (what they observe and notice to be of importance to their performance) that are presented to them, especially movements of the ball, opponents and support players. On the basis of this perceived information players assess the situation and the possible responses prior to selecting the most appropriate response. Once a response has been selected, the players have to execute the response in the desired manner. Errors may be due to problems with perception, analysis or execution of a technique. The most important point for coaches to grasp is that all of these processes can be developed through appropriately structured training.

14.2 LEARNING

Learning is the relatively permanent change in behaviour as a result of experience. It includes also the modification and refining of behaviour through experience, training or practice. Skilled performance is the result of learning and the coach is responsible for creating the learning situation in soccer. His work content and method of operation must wherever possible be interesting, stimulating and excite the performers. This does not always occur with hardened and experienced professionals! One of the chief roles of any coach is to motivate players to learn in spite of many setbacks that may occur along the way. Without interest, incentive or stimulation any learning situation may be active in the physical sense but is unlikely to be educational in benefit. Getting the players to understand and agree to the need for practice is a further role of any coach and the players' attitudes to training and learning often determine how fast learning takes place. Performers are more likely to learn if they are interested in what is going on and any success often has the effect of generating further interest. One of the chief factors for any coach to carry in his mind when devising coaching programmes is what the players are interested in achieving at this stage.

The coach should remember that all players differ in age, ability, stage and rates of development, physique and attitude. Consequently teaching method and approach should be modified to suit group needs. One of the arts of the coach is to decide at what level to start the work and how to present it to suit the levels and abilities of those under his charge. Many times coaches work way above or way below the levels of challenge necessary for the players. The coach must at all times match his methods, language, coaching levels and content to the players, remembering key words such as 'challenging', 'interesting', 'educating' if he is involved in teaching and developing players' abilities and team understanding. The aim must be to assist individual players, groups within a team and the team itself to achieve its potential as completely and efficiently as possible.

Learning is the development of patterns of behaviour to be stored in a memory bank for later recall. True learning causes a permanent and lasting change in behaviour. In the development of players and teams, all these factors are major considerations for the coach in planning practice sessions.

14.3 PRACTICE

Practice involves rehearsal for whatever is required by the game, a player's positional functions and their technical and decision-making responsibilities as they play the game. Practice can only be deemed to be effective when what is being rehearsed is improved in accuracy, consistency, efficiency and control. This can be indicated by the degree to which players become less concerned with the mechanics of performance and more assured of the certainty of the performance. A skilled player does not have to be consciously aware of every movement he makes, as much of what he does is instinctive.

There are essentially three phases in the acquisition of skills (Fitts, 1964). The first is the cognitive stage where the player must understand what is required of them from an analysis of what is happening around them. Next is the intermediate stage where responses are learned, errors are gradually eliminated and new movement patterns begin to emerge. The autonomous stage is where the skill no longer requires conscious control, uncertainty is eliminated and skills require less information processing. The most stringent test comes in the game situation. If performance in games does not improve as a result of practice, then practice has been ineffective.

A coach's aim in practice would be to organize and control players' learning in an attempt to perfect the most relevant and efficient techniques and skills for the game (Worthington, 1974). Whilst doing this, the coach should be working for a stable performance, especially where distractions are similar to those found in competitive games. To develop the correct mechanical actions of, say, passing the ball, without combining them with decisions concerning choice of receiver, target area for the pass, and timing of release, would be meaningless. So in practice the coach should be operating in circumstances as close to match conditions as possible.

The choice and application of those skills determine the success and skilled nature of performance. As a guideline, practice situations for developing soccer should follow these procedures.

1. Develop each technique/skill in a 'closed' situation and in an order and manner which permit no interference with mechanical performance. Practise singly and with concentration on 'how' actions are performed. This policy would be especially applicable to young performers who are building up their memory patterns of behaviour or for the introduction of a new technique unrelated to any others.
2. The technique/skill is then exposed to performance in a changing situation. Opposition and support players, a direction, a target, a spatial restriction, are introduced and carefully controlled by the coach so that success is possible. In this situation a player has to decide when, where, how and whether to use the designated skill in a constantly changing environment. Equally important is for the learner to appreciate when not to employ that skill and to make more appropriate and effective choices of action. Only by performing in an 'open', variable and rapidly changing practice situation does a young performer develop an understanding of the game and when and where to choose and implement appropriate skills.
3. From this simple 'open' practice, a performer should be placed in a game or game-type situation with an increasing number of options offered to their decision-making processes and variable physical and mental pressures being applied by opponents and situations in the game. With experience, the player can fit into the game with all its requirements. The player learns what will and will not work, what is expedient and what is not. Trial and error learning plays some part, but trial and success are all-important factors in learning and the development of a player.

Identifying the problems of players and teams from game situations, then devising practice situations that transfer effectively into performance are perhaps key roles of any coach. The process of effectively transferring practice into playing performance is closely related to transfer and specificity of training. Specificity of practice means simply that *what* you do in practice corresponds to what you experience in competition. The principle of specificity of practice can be summed up by the following statements.

1. Players will react and perform in a competitive game situation relative to what and how they have been practising.
2. The more closely you simulate game situations in practice the better game performance is likely to be.

In the early stages of learning, practice needs to be simple so that concentrated and focused learning may take place. Distances, speeds, conditions and situations in which players operate should still be as close as possible to a game situation, but with all interference removed that could confuse and adversely affect performance of the skill. Players should be encouraged to think and to concentrate on only one thing, and if necessary to reduce the speed of action in order to develop 'correct' performance of the task. As quickly as appropriate, speed of performance should be allied to accuracy and graded challenges introduced.

As the player progresses, specificity of practice is essential for developing players who can understand and operate in a competitive game. Players must learn, and familiarize themselves with stressors introduced by the game. These might include opponents, opponents' movements and proximity, presence of support players, speed of operation, and spatial limitations. Only by doing this will players learn appropriate and timely responses to game conditions and operations. The coach must make certain that what is practised and how it is practised is required by match situations.

Knapp (1974), commenting on transfer of training, indicated that transfer 'can best be explained on the grounds that it occurs to the extent that the two situations are similar'. In a game, the player is involved in **assessment, judgement** and **action**. If only one of these three processes is faulty then a player will not succeed. Practice must involve players in the same actions as the game. The question for the coach is how to achieve this.

The game involves three major elements highly relevant for practice. These are **support players** to work with, **opponents** to play against, and **targets** (direction and limits). By incorporating these elements into practice players will be involved in making decisions and taking action to succeed. The ability to make correct decisions is arguably the most important factor in developing a skilled player.

The role of the coach is twofold: (1) to educate players to make correct decisions, and (2) to equip players with the necessary skills to carry out these decisions. Players do not improve just because they play the game often,

Table 14.1 Principles for devising effective soccer practice situations

1. Players must accept the need for improvement, preferably completely and enthusiastically. Without a player's total agreement improvement may be only marginal or non-existent
2. The coach must devise relevant and effective practice situations based on actual match performance and its requirements
3. Players must know what they are trying to achieve. There should be an objective and a purpose known to all who participate in the sessions
4. Practice should be of short duration, repeated regularly and frequently with occasional periods of long duration practice
5. Performance must be evaluated fairly, objectively if possible and players should be given feedback as quickly as possible. Information that becomes available as learning advances is essential to continued learning. The more varied, accurate and better the quality of feedback the more productive the practice
6. All practice sessions must aim for and demand high quality performance
7. All practice sessions should finish on a successful note
8. There must be the highest degree of similarity between the practice situation and the real situation in which skill improvement will be measured, bearing in mind the players' age, development stage, physique and ability levels

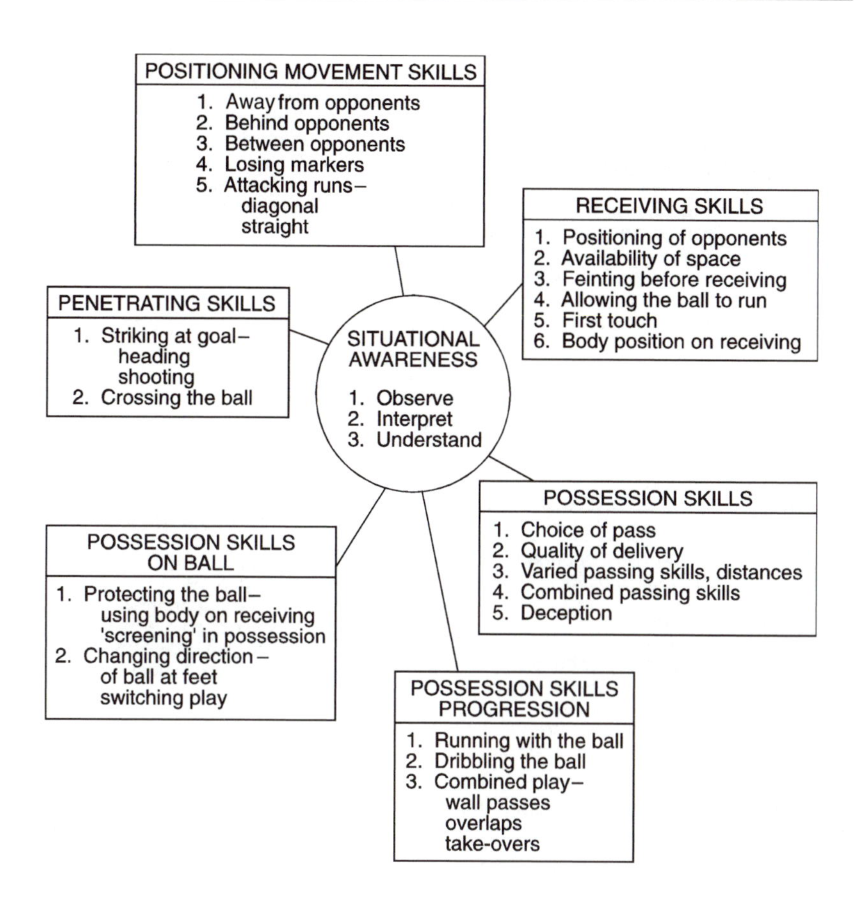

Figure 14.1 Categorization of attacking skills for youth coaching programmes.

although this helps. They need to be educated, trained and coached in meaningful practice situations. Coaches should follow the principles outlined in Table 14.1 for devising effective soccer practices.

Individual learning and progress develop at varying rates. A successful and interested player is more likely to persist in practice. Success governs both progress and interest. Practice will bring about some change. If players are allowed to practise less than the highest quality of execution, they will learn less than the highest quality of execution.

Key characteristics in any coaching and practice sessions would be: quality, duration, frequency, intensity and specificity. Repetitive, frequent practice is an important component in the acquisition of soccer skill and must entail rehearsal in the fullest sense of the word.

'What' to practise and develop is a problem for some soccer coaches. Team coaching is influenced by the adopted style of play, either imposed by the coach

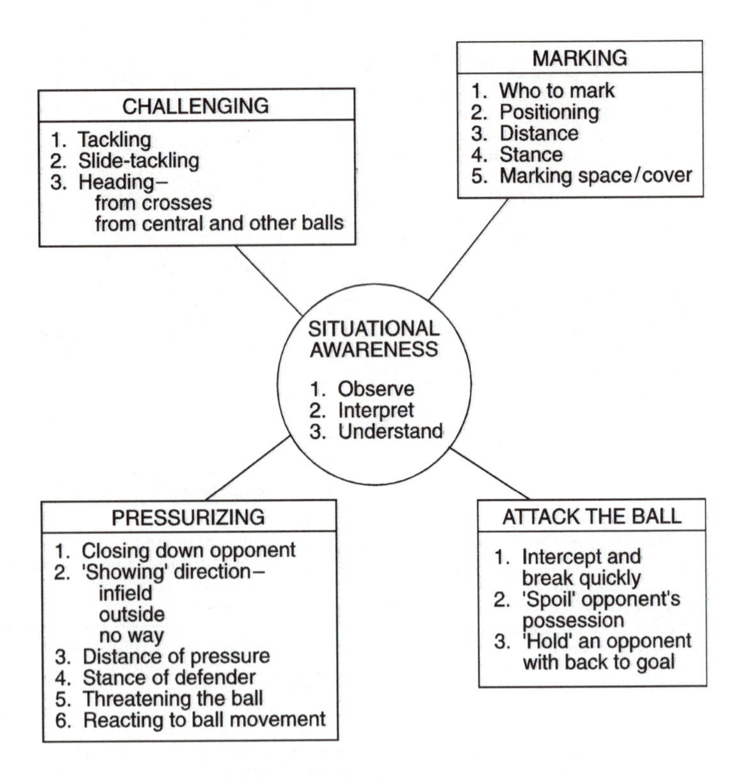

Figure 14.2 Categorization of defending skills for youth coaching programmes.

or determined by the players with the approval of the coach. Style refers to the 'attitudes' of a team, how it chooses to operate technically to achieve its end, its manner of playing and the predetermined or developed tactics used to accomplish its objectives.

But what do younger, developing players need to practise regularly and frequently? What should they strive to understand and to become competent in performing? No tactics, style or strategies have yet been imposed, so what are the skills required by any style of play and any game? The ingredients necessary for the development of young, youth age players are detailed in Figure 14.1 for attacking skills, and Figure 14.2 for defending skills.

14.4 ORGANIZATION OF COACHING AND PRACTICE SESSIONS

14.4.1 Purpose of the session

All coaching sessions need a purpose. Sessions may be dual-purpose in that part of the session is devoted to development of fitness and part to the development of, say, team understanding. Also, some of the time may be devoted to group or individual player improvement. The coach and the players need to be in no doubt whatsoever as to '*the training message for the day*'. The coach needs to spend time in thought and preparation of sessions, knowing precisely what he intends to achieve. The coach will have an aim to achieve during his working time with the players. Coaching sessions should never be *ad hoc* in that players and coach arrive and participate in purposeless, ill-thought-out and ill-planned sessions merely to occupy time and attention.

14.4.2 Location of the session

Where is the session best conducted? Is the use of a full pitch or half a pitch or even a more restricted area most suitable to conduct the session? Fitness training may be best conducted in woodland or on an undulating surface or even on a beach. Shooting, crossing and goalkeeping practices should be conducted on the pitch and in front of goal. *All sessions should be located in the most relevant and appropriate areas of the soccer pitch.* Carefully marked-out and correct areas should be used with the appropriate numbers of players for both the activity and the area being used.

14.4.3 Duration of the session

Depending on the time available, each section of the coaching session should be carefully planned in terms of time allocation. Lengthy and irrelevant warm-ups devour valuable coaching time. Sometimes, what the coach intends to achieve can be done quickly and efficiently through good planning, organization and thorough teaching. Sometimes a coach must persevere and take time to explain clearly to the players what is necessary, and so time allocation has to be waived. A contingency time plan should be decided upon but a degree of flexibility should be built into this plan. A coach should always allow more time than is thought necessary for any unforeseen circumstances that may arise to prolong parts of the session.

14.4.4 Numbers of players available

In order to conduct a coaching session, a coach always needs to know the numbers of players available. The coach should meticulously check the players' name list and ensure everyone has been included *in a meaningful and realistic role*. The coach should allocate players to realistic and purposeful roles during the session and ensure that players are practising those skills and tactics that they employ during the match in those areas of the pitch where they are expected to perform them. Consequently, practice will be meaningful and closely related to the game situation. *Practice then becomes game-rehearsal, which is what it should be*. Only essential players should be included when working tactically, until the coach needs to enlarge the practice in terms of areas, numbers and objectives. Non-essential players quickly become bored and can undermine the quality of a coaching session because of a lack of realistic involvement and lapses in concentration. They can be gainfully directed to training drills until needed. Consequently, only the essential, relevant players should be included until all are eventually needed for the final coaching progression into an 11 versus 11 game.

14.4.5 Equipment needed

Sessions should always be aided by the effective use of training aids. Video recorders and a TV set can be used for instructional purposes if available.

A blackboard/wiper-board and appropriate writing materials can help a coach to explain aims to players. Bibs, footballs, markers are also necessary for field use and when needed areas and targets should be clearly marked out with the efficient use of markers. More than one set of bibs is useful so that neutral players and any player in particular can be identified from others in the squad.

14.4.6 Age, experience, quality, ability of players

The coach should always *consider the realistic abilities of the players being coached.* Setting players unrealistic objectives and targets for their abilities is a wasteful exercise. Abilities and understanding should be developed and stretched beyond present levels but sensibly so. Advancing too far, too quickly and beyond players' current abilities is sure to be met with frustration, disappointment and resentment. The aims and objectives and consequent organizational structure of the session should always reflect players' levels of operating. To challenge and interest players should be the target of coaching sessions. This can be done by designing progression in practice and demanding high quality work from individuals and groups in the pursuit of even higher levels of attainment consistent with realism in the setting of goals and incentives.

14.4.7 Assistance available

Where a coach needs to work with a small group of players, others may not need to be involved in that particular aspect of the game. They should be usefully employed in working on either their individual abilities or in group work, such as shooting or crossing. Players need to be supervised and assisted at coaching and training sessions wherever possible. A good coach can always assist a player in some aspect of that individual's game, but a coach cannot be working in two places simultaneously and so needs assistance to ensure the efficient organization and conduct of the session. When planning, the availability of assistance needs to be considered and not only the availability but also the quality of that assistance. The coach must ask the question '*What are the assistants capable of coaching?*' An able assistant, briefed well before any coaching session, should be able to undertake any work detailed by the head coach. An assistant who is not well qualified, but is learning the trade, should only be assigned work that he or she can capably supervise and conduct. Discussions should have taken place beforehand on how the work is to be carried out.

14.4.8 Coaching method

The art of coaching entails putting into practice the conceptual aspects of the session. The coach must know exactly the cause of the problem to be rectified or the ingredients of success in developing an aspect of play. The coach must know if the problem is technical, tactical or one of understanding, so that an appropriate practice situation can be devised in which to educate the players. That situation may be a simple but realistic technical practice or a complex 11 versus 11 game. The coach must design an efficient and effective teaching vehicle to explain the important points to the players. *Designing and conducting effective coaching sessions are probably the most important of all coaching functions.* Progressing to incorporate ideas, players, and strategies into the full-game situation is the ultimate and final coaching step that a coach must manage. Knowing if, when and how to make this progression, according to the abilities and stages of development of individuals and groups, is the key to developing team success.

Choosing the coaching method is influenced by many factors. Players available, numbers of players, area to work in, purpose of the session and other factors all influence the coach's decision as to the type of practice situation to use.

In the overall planning of any sessions, the main factors can be summarized as follows.

1. What needs to be practised?
2. When will it be practised?
3. Where will it be practised?
4. Who is involved in practice?
5. How will it be practised?

Underlying these questions is why the practice is being carried out.

14.4.9 Other considerations

On occasions it may be necessary to hold a theory session conducted in a classroom or lecture theatre. A 'split-session' of football and fitness development may involve a change of practice location from pitch to woodland. In this instance, the whole practice period and the time for travel should be incorporated in planning.

14.5 A TYPICAL COACHING/TRAINING SESSION (Figure 14.3)

14.5.1 Warm-up (10–15 min)

A purposeful and effective warm-up is necessary prior to any coaching or training session. The warm-up prepares the body to manage the work planned for the remainder of the session. Warm-up should include raising the heart rate and metabolic rate by jogging and running. Whole-body exercise for agility and specific flexibility work should also be included. Warm-ups can be conducted under the guidance of a coach or if players are experienced and trustworthy, can be carried out individually or in small groups. Over-lengthy warm-ups are not necessary and can be completed efficiently in approximately 15 minutes. A ball can be used in warm-ups but in a controlled and careful manner; injury can occur if players suddenly overstretch or accelerate quickly before flexibility work has been completed. Warm-ups should be conducted progressively and logically, exercising and stretching major muscle groups before smaller groups.

14.5.2 Ball work (15–20 min)

Ball work can be technical or tactical in nature. Simple practices to develop passing techniques, or controlling techniques, can be used to increase the intensity of the session. The coach can use this part of the session to develop techniques and skills relevant to the style of play adopted by the team. Games for retaining possession of the ball can be used to develop technical and related tactical abilities and can also increase the physical output in the session. Again, ball work can be made relative to the selected playing style, using, for instance,

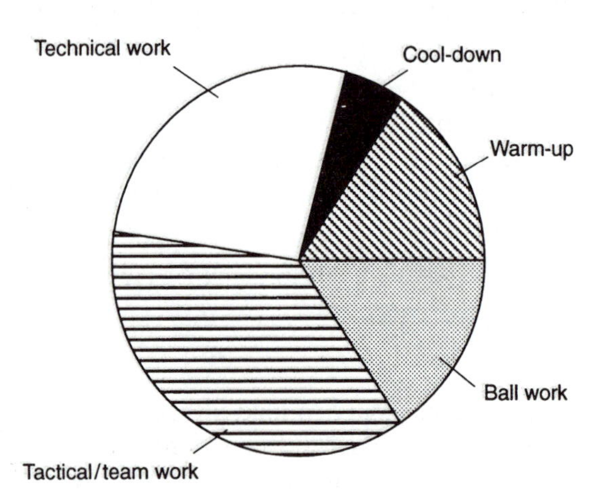

Figure 14.3 Structure of a typical training/coaching session.

wall passes, through passes or combination play incorporated in possession games of equal or unequal numbers.

14.5.3 Tactical team work (30–45 min)

This part of the session is used for either rectifying or developing group or team play. Introducing a new tactic or refining an existing tactic can be included. Groups of players or the whole team should be involved in practice and it is here that the group members need to be educated as to their function at certain stages of the game. The group session can be developed into an 11 versus 11 coached game; this part of the whole coaching session is probably the most crucial in developing team understanding and team play.

Simply playing a free game of 11 v. 11 with little or no coaching input has limited value unless the coach uses it for observation purposes or to decide if playing tactics or policies are fully understood.

14.5.4 Technical work (20–30 min)

This section of the overall session may be used for free time, maybe 6 v. 6, 8 v. 8 or even 11 v. 11, with the intention being that players use this period for free expression, or to try out previously learnt skills or strategies. This period may be used to practise the fundamental aspects of the game that are crucial to successful play. Practising crossing, finishing, defending 1 v. 1 or any other underlying concepts of successful soccer should be conducted here. Again, choosing skills relevant to the style of play of the team should be uppermost in the coach's mind.

14.5.5 Cool-down (5–10 min)

There are sound physiological reasons why players should cool down after training sessions. Removal of blood lactate, for example, is accelerated if an active recovery follows strenuous training.

Light jogging and stretching should be used to finish off any coaching/training session so that players can mentally relax after what could have been a highly intensive training session. This should leave them in as beneficial a condition as possible to continue with the next session whenever detailed by the coach.

14.5.6 Duration of session

The session outlined can be conducted in approximately 90 minutes. The appropriate time proportion would be:

1. Warm-up 10–15 min

2. Ball work 15–20 min
3. Tactical team work 30–45 min
4. Technical work 20–30 min
5. Cool-down 5–10 min

14.5.7 Individual training

Individual players may need to be helped with certain weaknesses in their game. This work should be undertaken before or after the training session as other players need not be involved, unless a small number are held back to assist.

14.5.8 Fitness training sessions

This aspect of player or team development can be incorporated in any training session or may be conducted singly. For instance, to include fitness training in a session as outlined above the distribution could be as indicated in Figure 14.4.

Fitness work should be conducted after any ball work, especially ball work closely related to tactical understanding and decision-making. Generally ball work should precede fitness work. Any strength development as part of the fitness programme should be conducted at the end of the fitness session and should be followed by a cool-down also.

An efficient and relevant fitness-related session can be conducted in 45 minutes. Consequently it can be included as part of any overall coaching/training session.

A time proportion of such a session lasting approximately 90 min would be:

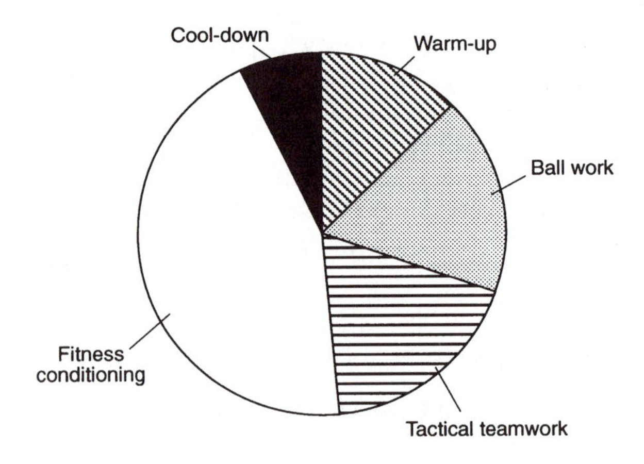

Figure 14.4 Structure of a training session including a fitness conditioning programme.

1. Warm-up 10–15 min
2. Ball work 15–20 min
3. Tactical team work 15–20 min
4. Fitness/conditioning 40–45 min
5. Cool-down 5–10 min

Summary

The Oxford Advanced Learner's Dictionary defines 'practice' as 'regularly repeated exercise in order to improve one's skill'. Soccer coaches should be concerned with ensuring that any 'regularly repeated exercise' is relevant, is based on match-play and transfers effectively from practice to match situation. Not all transfer enhances learning.

Positive transfer occurs when performance is improved by previous experience and learning. In order to promote positive transfer coaches should study the relevant tasks and select an appropriate teaching environment and method. The most important tasks of any coach, after identifying a problem and its key elements, are to ascertain and devise how to transfer effective practice into effective performance. Assessing the players' level of skill and understanding and their operating function is the key to deciding how and where to conduct practice situations. Learning progresses more quickly when the learner concentrates, devotes maximum attention and effort to the task, when feedback is accurate and motivation to succeed is high. Knowing this, coaches should design situations where learners have an opportunity to succeed through relevant, regular and challenging practices. Timely injection of necessary coaching information can accelerate the learning process. The whole learning and practice process is a combination of inputs from coach, player and game.

REFERENCES

Fitts, P. (1964) *Perceptual Motor Skill Learning*, Categories of Human Learning, Academic Press, New York.

Knapp, B. (1974) *Skill in Sport*, Routledge and Kegan Paul, London.

Worthington, E. (1974) *Teaching Soccer Skill*, Physical Recreation Series, Pelham, London.

Coaching science

15

Andy Borrie

Introduction

Everyone has at least a vague understanding of the term coaching and the kind of activities that coaching entails. However, even within coaching, the field of coaching science is, at best, poorly understood. The first issue that should therefore be addressed in this chapter is the question of what the term 'coaching science' really means.

Coaching science is concerned primarily with analysing and understanding the role of the modern coach and the problems faced by practising coaches. Coaching science should provide empirically based information that the coach can use to enhance effectiveness. The disciplines that can be considered to contribute to coaching science are almost as diverse as the roles of the coach. Woodman (1993) cited the disciplines of exercise physiology, biochemistry, biomechanics, sports medicine, psychology, sociology and pedagogy as all contributing to the subject of coaching science. These disciplines all develop knowledge that can either be used to understand coaching or utilized directly by the coach.

It is important at this stage to make one other key point concerning coaching science. Whilst coaching science is an important factor in developing coaching, it would be wrong for coaches to assume that coaching science can ever provide a proven set of steps to universal coaching success. Ultimately coaching is an art that requires creative input from the coach; what science can provide is knowledge that coaches can utilize in their own individual styles. It could be argued that

Science and Soccer. Edited by Thomas Reilly. Published in 1996 by E & FN Spon, London. ISBN 0 419 18880 0.

the art of coaching is actually the **personalized** application of the science.

The value of adopting a scientific approach to the analysis of coaching method is best appreciated by considering the complexity of coaching. Any review of the general coaching literature shows quite clearly that all authors agree that coaching requires the coach to engage in a wide range of roles (Sabock, 1985; National Coaching Foundation, 1986; Fouss and Troppman, 1987). Coaches can be involved in the establishment of basic skills in beginners, the provision of sound technical and tactical advice to intermediate performers or the planning and implementation of long-term training programmes with elite performers. The variety of tasks, challenges and problems facing coaches in these different situations is immense. Many also believe that the role of the modern coach has now expanded far beyond direction of practice sessions (Pyke, 1992; Woodman, 1993). The expanded coaching role involves taking responsibility for the performer outside of the practice/competition environment and being aware of the performer's overall social and psychological development. Coaches are now frequently expected to take on almost any task that creates a better working environment for the performer or for the coach. In response to the increasing demands, modern coaches have had to develop a wide range of technical, interpersonal and managerial skills in order to function effectively.

The complexity of coaching is indisputable, consequently *only* a systematic and objective assessment of the subject will allow scientists and coaches to develop a clear understanding of coaching behaviour. Scientific analysis of coaching behaviour is fundamental to the continued provision and development of high quality coaches.

15.1 THE COACHING PROCESS

The tasks involved in developing sports skills may range from simply correcting a child's attempt to kick a football through to developing cohesive attacking play with an international forward line. Regardless of the differences in these activities, many authors accept that the process of successfully coaching both the child and the international is essentially the same (Fairs, 1987; Woodman, 1993). In this context, the term process refers to the series of stages that a coach has to go through in order to help the player learn and improve in a particular skill. Models of the coaching process have been proposed by several authors (Franks *et al.*, 1986; Fairs, 1987). A schematic view of the stages within the coaching process is presented in Figure 15.1. This model of the coaching process is a simplistic one that focuses on the understanding of performance and the development of skill. Coaching science has in fact made a significant contribution to the understanding of several elements of the process.

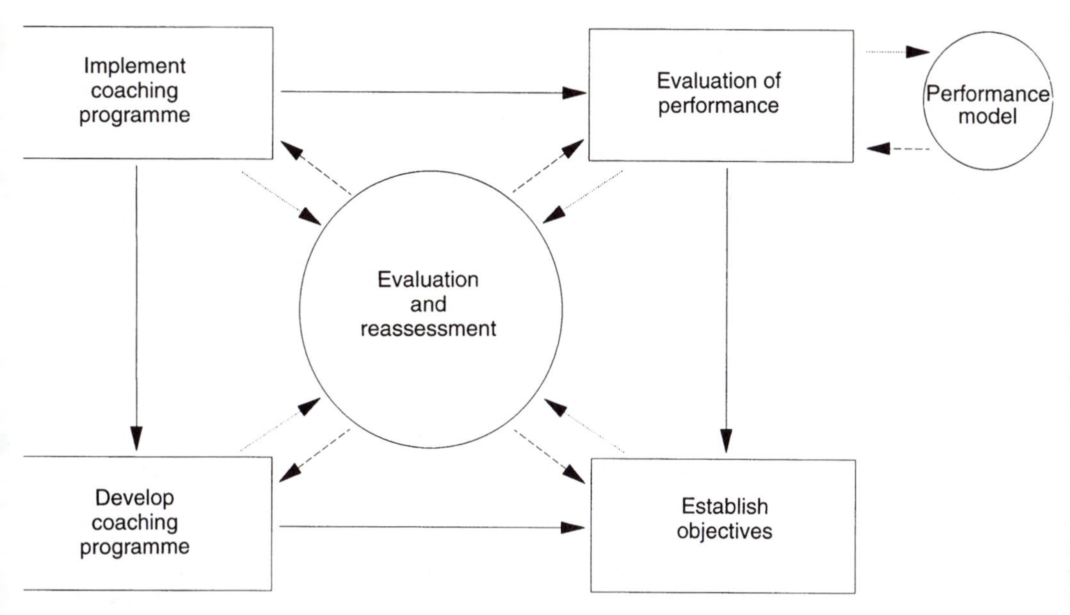

Figure 15.1 Schematic representation of the coaching process. (Adapted from work by Fairs, 1987.)

- Evaluation of performance – notational analysis research (see Chapter 20).
- Establishing objectives – goal-setting research (for a comprehensive review see Gould, 1993).
- Developing coaching programmes – motor skill research (for a comprehensive review see Schmidt, 1991).

The contribution of coaching science to the development and understanding of elements in the coaching process is clear; however, it must be realized that the model contained in Figure 15.1 does not fully express any of the roles of the coach. The effectiveness of the implementation of the coaching process will be dependent upon the quality of the interaction between player and coach. Unless coaches can establish good relationships with their players then it is unlikely that they will ever fully achieve their objectives regardless of the quality of either their analysis, the subsequent performance goals or the coaching programme. Analysis of coach–player interaction and factors which affect it are therefore of prime importance in helping to build models of effective coaching behaviour and understanding how best the coaching process can be pursued. Consequently, the remainder of this chapter will review current concepts of coaching behaviour and attempt to highlight the implications of this research for the soccer coach.

15.2 COACHING BEHAVIOUR ANALYSIS

For many years scientists and coaches have been concerned with trying to identify the 'perfect' coach or the 'ideal' coaching style. The early approaches to coaching analysis stemmed from work on leadership in general settings outside of sport, e.g. business and military settings. The nature of the task facing coaches was deemed to be essentially the same as for any other individual filling a leadership role, with leadership commonly being defined as the behavioural process of influencing the activities of individuals or groups towards specific goals (Stogdill, 1974). The research approaches adopted for leadership research in general were subsequently used for the study of sports coaches.

The earliest research attempted to assess whether coaches could be distinguished from the general population on the basis of their personality profile (Ogilvie and Tutko, 1966; Hendry, 1974; Sage, 1975). The results of these studies were remarkably inconclusive and it is now widely accepted that there is no single personality profile that is ideal for coaching. A simple comparison of the personality characteristics exhibited by top coaches/managers within soccer will show that there are marked differences between individuals. Graham Taylor, Brian Clough and Jack Charlton would all appear to have different personality profiles but all have been successful managers and coaches at one time or another. This may seem a rather negative start to a discussion of the factors that determine coaching effectiveness. However, this research has emphasized one crucial point with respect to coaching. The fact that there is no one universal personality that all effective coaches possess means that prospective soccer coaches should not be discouraged because their personality profile does not match that of more experienced coaches. The research shows that a range of personality types can be successful in coaching, therefore coaches should be encouraged to be true to themselves and not try just to imitate recognized successful coaches.

When coaches could not be distinguished on the basis of their personality researchers then started to consider behaviour patterns. This research continued with a universal theme with studies directed towards trying to identify a particular set of behaviours that were exhibited by all successful leaders. The earliest behavioural studies were conducted at the Ohio State and Michigan Universities (Stogdill, 1974). Although the findings from the two research groups differed in certain respects both groups did identify two general dimensions of leadership behaviour that related to group effectiveness. Whilst the terminology differed the two research groups described essentially the same general behavioural patterns (Table 15.1). The leadership behaviour dimensions that were identified have formed a conceptual basis for much of the subsequent research into leadership. It must be remembered though that these early studies were conducted outside the sports domain. Within the sport domain studies of leadership behaviour have identified dimensions of behavi-

Table 15.1 General leadership behaviour dimensions

Relationship behaviour
Refers to leader behaviour that is characterized by friendship, mutual respect, understanding, trust and good interpersonal communication between the leader (coach) and follower (player)

Task behaviour
Refers to leader behaviour that is characterized by emphasis on setting and achieving goals, establishing patterns of organization and hard work

our similar to those proposed by the general leadership research (Danielson *et al.*, 1975; Chelladurai and Saleh, 1980). Chelladurai and Saleh's study identified five categories of coaching behaviour which have now been studied extensively. The categories that they defined are summarized in Table 15.2.

The dimensions identified by Chelladurai and Saleh have clear similarities to the concepts of task-oriented and relationship-oriented behaviour. It would appear that, however expressed conceptually, concern for the individual and the organized drive towards achieving sporting goals are important aspects of coaching behaviour.

The most important question for the coaching community is how much of each dimension of behaviour should a coach exhibit in order to optimize their effectiveness? Unfortunately the early general research was not successful in

Table 15.2 Dimensions of coaching behaviour identified by Chelladurai and Saleh (1980)

Training and instruction behaviour
Behaviour aimed at improving the performance of athletes by emphasizing and facilitating hard and strenuous training, clarifying relationships amongst group members through structuring and coordinating team activities

Democratic behaviour
Behaviour of the coach that allows greater participation by athletes in decisions relating to the establishment of group goals, practice methods and strategies, etc.

Autocratic behaviour
Behaviour of the coach that allows little athlete participation in decision-making. All decisions pertaining to team affairs are handled by the coach alone

Social support behaviour
Behaviour of the coach that indicates a genuine concern for all athletes and their welfare. Also behaviour that fosters a positive group atmosphere and interpersonal relationships between team members

Rewarding behaviour
Behaviour of the coach that provides positive and rewarding reinforcements for athletes through recognition of good performance

answering this question. The failure of the universal approaches to coaching analysis prompted researchers to start looking for other factors that interact with the coach's behaviour pattern to determine whether the coaching is going to be effective. The many variables that influence coaching effectiveness will now be considered.

15.3 ELEMENTS IN OPTIMAL COACHING BEHAVIOUR PATTERNS

It is now accepted that there is no universal behaviour pattern that produces optimal coaching effectiveness in all situations. The majority of researchers now agree that three factors interact in determining what pattern of coaching behaviour will produce the best results. These interacting variables are **the coach, the player** who is being coached and **the situation** in which the coaching is done. All these have an influence on the effectiveness of various coaching styles in producing good competitive performance and satisfied players. In the following discussion each of the three main variables is isolated and the factors associated with it are discussed more fully. It must be remembered though that in reality these variables are not isolated and are continually interacting within the overall coaching situation. Where possible, data are drawn from soccer-specific research and the implications for the soccer coach are highlighted.

15.3.1 Player characteristics

Research has shown quite clearly that characteristics of the player such as age, maturity, playing experience and ability all influence the kind of coaching behaviour that is preferred by performers. The influence of maturity on coaching behaviour has been assessed by numerous researchers and several models of the inter-relationship between these two variables have been proposed (Hersey and Blanchard, 1969; Chelladurai and Carron, 1983; Case, 1987). Maturity has been defined as:

> the relative mastery of skill and knowledge in sport, the development of attitudes appropriate to sport, and experience and the capacity to set high but attainable goals. (Chelladurai and Carron, 1983, p. 372)

These models are referred to as situational models of leadership and all related maturity to the two basic dimensions of task and relationship-oriented behaviour. Whilst it seems clear that maturity is an important mediating factor in determining coaching effectiveness the literature is not in full agreement on the exact relationship between player maturity and coaching behaviour. In most of the studies that have been conducted maturity has been operationalized by studying performers of increasing age and ability level.

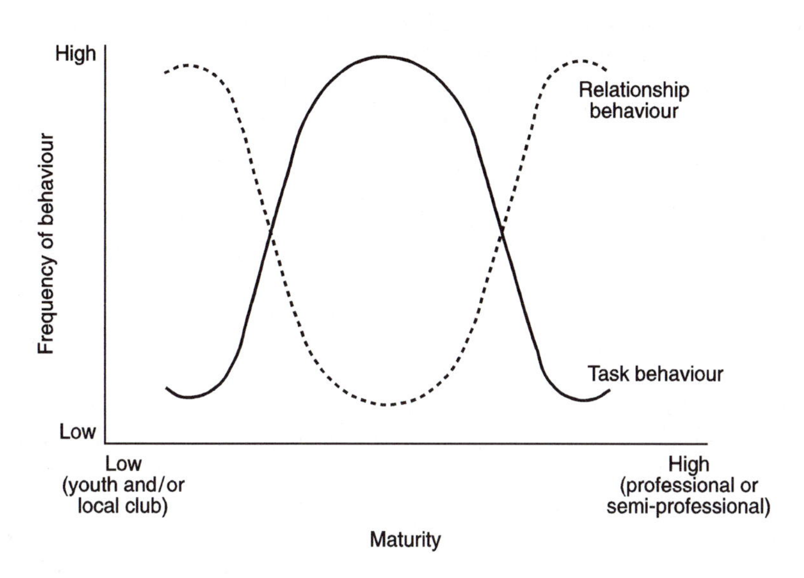

Figure 15.2 Case's (1987) situational model of leadership presented with respect to soccer.

All the situational models that have been developed show that the coach needs to vary behaviour according to the maturity of the group being coached (Hersey and Blanchard, 1969; Chelladurai and Carron, 1983; Case, 1987). Additional research has also supported the concept that coaching behaviour has to alter with reference to maturity (Terry, 1984). However, the proposed models have all differed in the type of coaching behaviour deemed to be most effective at the various stages in player development. Consequently, because there is no universal agreement in this area, it is difficult to be more prescriptive other than stating that coaches should be aware of the possible influence of maturity on preferred coaching behaviour. The problem is further compounded in soccer because, as yet, none of the research has included a sample of soccer players and coaches. However, there is indirect evidence to support the general applicability of the most recent situational model, proposed by Case (1987), and also evidence that parts of the Case model can be applied to soccer. Some authors have also favoured the Case model because, 'intuitively, the model is appealing because it is somewhat consistent with expectations' (Cox, 1992, p. 397). For these reasons we shall look at the Case model in a little more detail.

The model (Figure 15.2) suggests that at low levels of age/maturity, when young players are developing rapidly psychologically, socially and physically, coaching behaviour needs to be relationship-oriented, i.e. showing concern for the player as an individual. Within youth soccer this would mean that effective coaching should emphasize personal development and that high levels of social support should be provided. This concept is supported by the results of numerous studies of youth sports coaching.

In the late 1970s a series of studies (Smith *et al.*, 1977, 1978, 1979) showed that young players responded favourably to coaches who created a more socially supportive sporting environment. This was found to be independent of the team's win/loss record, indicating that for young sports participants winning is not that important a part of their sport experience. This research has been replicated recently with identical results (Smoll and Smith, 1993). Coaches who improved their level of social supportive and rewarding behaviour received more favourable responses from their players, with the coaches being perceived as being better teachers. The players also reported having more fun in their sport and having a higher level of attraction for team-mates. Children with low self-esteem found that self-esteem improved and all children reported lower levels of competitive anxiety when taught by coaches with high levels of social support behaviour (Smoll and Smith, 1993).

The importance of relationship-oriented behaviour with young players is also supported by two soccer-specific studies. Dubois (1981), in a study of youth soccer coaches, found a positive relationship between supportive coaching behaviour and successful team performance. The higher the level of supportive behaviour the greater the level of team cohesion and performance. Wandzilak *et al.* (1988) in a similar study did not find a relationship between winning and supportive behaviour but did identify a positive correlation between supportive behaviour and indices of player satisfaction. The soccer coaches studied used a high percentage of positive remarks and encouraging comments in both practice and game situations. The young players' responses indicated that the coaches who had higher ratings for positive behaviour were more popular with the players and that these players exhibited a greater liking of soccer.

The data from these studies would suggest that coaches who emphasize support for the young player are more likely to have:

1. Players who derive high levels of satisfaction from their soccer.
2. Players who want to stay involved in the game.
3. Players with lower anxiety.
4. Players with higher self-esteem.
5. Teams with greater cohesion.
6. Higher levels of performance.

It is also worth noting that in all of these studies it was found that coaches are weak at perceiving their own behaviour patterns and that the young soccer players were more accurate judges of the behaviour being exhibited by their coaches. Consequently coaches should play close attention to feedback from the young players they are coaching.

The approach that Case (1987) suggests for effective coaching of intermediate level players is more task oriented. The assumption is that players will have perhaps developed greater levels of personal self-confidence and therefore need less external support in the form of relationship-oriented coaching behaviour. As player development takes place the coach needs to provide a more structured training environment with the focus on technical and tactical instruction. At this stage of their development, players need to have more specific performance goals and a clearly defined coaching programme. Unfortunately there is no support for this concept outside of the original research by Case, who looked at successful basketball coaches.

At the elite end of the playing spectrum the experienced professional has already attained a high level of personal performance. At this stage of development the Case model suggests that players' needs revert to more relationship-oriented behaviour. The professional/experienced player perhaps needs personal guidance and advice rather than direct instruction. The coach's primary focus may shift from developing personal performance to more team- or group-related tasks. This is not to suggest that coaching of technique/skill does not take place but that the nature of the coaching should change, with the coach drawing more on the players' experiences in developing the coaching programme. There is no research directly related to soccer but it has been shown that elite performers in other sports do prefer higher levels of social support than intermediate performers (Chelladurai and Carron, 1983; Terry 1984). In his comparative study of club and elite performers of comparable age Terry (1984) found that elite performers preferred higher levels of democratic and social support behaviour, thus supporting the high maturity aspect of the Case model.

In discussing the influence of maturity on coaching behaviour it must be emphasized that it is not just age alone that determines preference for particular coaching behaviour. The original concept of maturity combined factors of age, experience and ability and it would appear that it is a combination of these factors that is important in determining behavioural preferences.

It is not possible to state that the Case model describes the relationship between maturity and coaching behaviour within soccer; however the indirect support suggests that it might provide a useful conceptual base for future research. Currently it is impossible to prescribe exactly the type of behaviour that will be most successful when coaching at different levels within soccer. All coaches must still be aware of team and player characteristics when deciding upon the behaviour pattern that they are going to adopt in their coaching. What is indicated quite clearly by the research is that in order to be effective across a range of levels within soccer, coaches need to be able to vary their coaching behaviour to suit the group being coached.

A further factor that might be influential in terms of player preference for particular coaching behaviour is the cultural background of the individual. Two studies by groups headed by Chelladurai (Chelladurai *et al.*, 1987, 1988b) found that cultural differences exist in preferences for coaching behaviour. It must be emphasized though that differences appear to be related to culture, not necessarily to nationality. Terry (1984) found no differences in preference for coaching when subjects were grouped by nationality. It was reported that all of the sample came from backgrounds with shared cultural and sporting ideologies. This suggests that it is the cultural difference not the differences in nationality *per se* that influence preference for coaching behaviour.

This issue may be of real significance in professional soccer where there is inter-continental movement of players and coaches. The importing of foreign players into all European leagues is commonplace. Coaches need to be aware that imported players may well have different expectations and needs with respect to their coaching. Unfortunately Chelladurai and his colleagues only looked at two cultures, Japanese and Canadian; therefore it is impossible to be more specific about the differences between the major cultures within world soccer, e.g. South America versus Western Europe.

Overall, the research into player characteristics that influence effective coaching behaviour has highlighted several influential factors, but it is still impossible to be prescriptive in terms of the 'correct' coaching behaviour for a particular soccer team. If coaches are simply aware of the characteristics that have been outlined in this section, it should help them to provide more effective coaching behaviour.

15.3.2 The situation

Although the characteristics of the players being coached are clearly part of the overall 'situation' in which the coach operates, there are also other situational factors that affect the player/coach interaction. In fact, the nature of the game itself may have an influence on effective coaching behaviour although the research is equivocal.

Soccer is a sport with high task dependency in that all players are dependent upon each other for successful completion of their task, which is to score goals. Sports such as gymnastics, golf, etc. have minimal task dependence and are classified as independent, whereas a team sport like soccer is classified as interdependent.

Research into task dependency has shown that performers from interdependent sports prefer a greater emphasis on training and instruction behaviour in comparison with independent sport performers (Terry, 1984). This view is supported by Gordon (1986), who looked at the coaching perceptions and preferences of 161 Canadian University soccer players. The results of the study showed that the players rated emphasis on hard training as being the most important factor in their perception of the effectiveness of their coach.

These findings would suggest that a soccer coach should perhaps spend more time structuring the training environment than coaches from independent sports. The results have also to be viewed in comparison to the work on player maturity where emphasis on training behaviour is suggested with intermediate performers. It may be that preference for emphasis on training is consistent across age/ability groups. Young soccer players may prefer higher levels of instruction than their peers in independent sports but still require less instruction than older soccer players. However the relationship here is unclear and needs further clarification.

Another possible variable that may affect preference for coaching behaviour particularly in professional sports is that of power structure. This concept was first proposed by Fiedler (1967) in his contingency model of leadership behaviour. Although the model is non-sporting and relatively old, the concept of power structure as an influential variable in coaching is not without merit, particularly in soccer where the coach/manager has significant authority.

The contingency model suggested that the level of authority that the leader, in this case a coach/manager, has over an individual is influential in determining the most effective type of coaching behaviour in a given situation. The contingency model has not received much empirical support but the concept that power may be a mediating variable in determining coaching preferences has an intuitive appeal in professional sports such as soccer. It might be argued that the relationship between a coach and team within an amateur club will be fundamentally different to the same relationship in a professional club. The level of power and control that the coach/manager can exert over many aspects of the players' environment might change the nature of the coach/player interaction. Coaches need to recognize that the power structure within a given situation may be a further mediating factor in determining optimal coaching behaviour. Unfortunately this is yet another factor that needs to be more rigorously assessed empirically before firm conclusions can be made.

15.3.3 Coach characteristics

The final variable that needs to be considered is the element within the coach's control that has not yet been considered, that is, decision-making style.

Chelladurai *et al.* (1987) identified autocratic and democratic styles of decision-making as being important dimensions of coaching behaviour. The traditional didactic method of teaching sports skills is rooted firmly in the autocratic decision-making style, i.e. the coach decides what needs to be taught, the coach designs the practices, the coach organizes and runs the practices. In this scenario responsibility for decision-making is left solely in the hands of the coach. The autocratic style can be split into wholly autocratic decisions, where the coach gathers all information and makes the decision, and consultative, where players are involved in gathering information and discussing the problem but the coach takes the decision.

In recent years the suitability of this type of coaching is becoming increasingly questioned. Many general coaching texts suggest that if a coach adopts a more democratic style of decision-making, this will yield more positive results (National Coaching Foundation, 1986; Martens, 1990).

The democratic decision style obviously allows the player greater input into the coaching process and this is felt to be advantageous for three reasons. First, the involvement of players in the decision-making process means that there will be more information available for generating alternative solutions to problems. It is therefore probable that any decision taken will be of a higher quality. The decision reached is also more likely to generate a course of action that meets the players' needs. Secondly, involvement in making a decision gives the player ownership of the resulting action and should result in increased acceptance of the decision. The research on goal-setting discussed in Chapter 12 emphasizes the importance of ownership of decisions/goals in promoting effort. Finally, increased participation in decision-making can also contribute to improved feelings of self-worth and self-confidence within performers. There are also disadvantages to a democratic decision style that are not found with autocratic decision-making.

The nature of democratic decision-making, i.e. with several interested parties involved in the process, means that it is inherently more time-consuming than the autocratic process. In allowing players to contribute to any decision the coach cannot dictate the flow of group discussion. Consequently control of a debate may be lost and tangential discussions arise that slow the decision-making process. Autocratic decision-making is free from this problem. The greater the degree of autocracy that is exercised the quicker a decision can be made since discussion is reduced to a minimum.

The second major disadvantage centres around the degree of integration within the group and also has a bearing upon the previous point. If a group is characterized by internal divisions, rivalry and a general lack of cohesion then a group decision may prove problematical. The main problem is in getting universal acceptance of the group decision. In a divided group a decision reached 'jointly' may further split the group in that one party feels that they have lost the argument and are unwilling to accept the final decision. This will make it difficult to implement the decision and pursue the chosen course of action. The alternative scenario occurs when factions within an overall group reach a compromise decision. This might maintain the credibility of all concerned but may not result in an optimal decision for promoting team effectiveness. Autocracy avoids these problems because the group is not involved in the decision. Autocracy also has the advantage of taking responsibility for a decision away from players who may not want, or be ready for, the responsibility inherent in decision-making.

The question that we need to consider concerns the correct balance between the different decision styles. This has been investigated in both soccer-specific and non-soccer situations. Models of decision-making that have been developed propose that there are seven factors that influence the style of decision best suited to a particular situation.

1. **Time restrictions** – the less time available for a decision then the lower the number of players who can be involved in the discussion.

2. **Quality requirement** – sometimes only one answer is acceptable; on other occasions any acceptable alternative will suffice.
3. **Amount of information** – who has the most information relative to the problem.
4. **Problem complexity** – complex problems are often more easily solved by one individual if they have sufficient knowledge.
5. **Group acceptance** – if a decision is to be successfully implemented then the decision must be accepted by the team/group.
6. **Coach power** – if the coach has control over rewards and punishments, or is held in high esteem, players will be more likely to accept their decisions.
7. **Inter-personal group relations** – if a group is highly cohesive then a group discussion is more likely to reach a sound decision. A fragmented team is less likely to reach decisions that are accepted by all players.

Gordon (1988) investigated the decision-making preferences of both players and coaches involved in university soccer. All of the players and coaches were presented with 15 different soccer-specific situations where a decision had to be made. The 15 situations included different combinations of the factors outlined in Table 15.3. The players and coaches were asked to nominate their preferred decision-making style in each situation. The results showed that both players and coaches preferred the coach to make autocratic decisions in the majority of situations (Table 15.3). It can be seen that the players preferred the coach to take decisions in approximately 79.6% of all situations. This same preference for decision-making has also been noted in studies of other university sports such as basketball (Chelladurai *et al.*, 1988a).

These findings appear to contradict the assertions in the general coaching literature that the democratic style is preferred by players. Gordon (1988) suggested that socialization factors and soccer behavioural norms may have contributed to the finding. The players were also relatively young and inexperienced in the situation in which they were studied, consequently they may have recognized the greater experience of the coach and readily acknowledged the traditional role of the coach. This acknowledgement was then expressed in a desire for the coach to take responsibility for decisions in soccer-specific situations. The study suggests that autocratic decision-making is of great value in soccer; however the results have to be seen in the context of the samples that was studied. In this case the subjects were young and inexperienced and it was highlighted in the section on player characteristics that age and experience have a strong influence on players' preferences. If Gordon's study (1988) was replicated with either youth players or professionals it would clarify the impact of age and experience upon preferences for decision style.

The results of this type of study also have to be weighed against the players' experiences of democratic decision-making and the skills of the coaches in structuring participation in the decision process. Some coaches may not find it easy to engage in the democratic process. The skills required in making

Table 15.3 Percentages of decision styles preferred and perceived by players and their coaches

Group	*% Ao.*	*% Do.*
Coach (perceived)	79.6	18.5
Player (perceived)	82.6	15.4
Player (preferred)	73.1	24.9

Source: Gordon, 1988.

democratic decisions are different from those required for making effective autocratic ones. In allowing the player greater freedom in developing the coaching plan there is inevitably a loss of control of the situation on the part of the coach. Coaches may find this difficult to come to terms with if they have only ever experienced wholly autocratic coaching methods in the past. Equally some players may find it difficult to assume a role in the decision-making process when faced with the responsibility that is an integral component of the decision-making process. It may well be that soccer coaches do not provide good examples of democratic decision-making, therefore the players do not appreciate the benefits of this style.

The main implication of Gordon's study is that no single decision-making style was universally preferred by players or coaches. Coaches should not use one decision style but vary their style in relation to a specific situation. It was suggested that *situations* should be labelled as autocratic or democratic, not coaches. The most effective coaches will be those who are sensitive to the situation and can be dynamic in terms of their use of decision styles.

Summary

Modern soccer coaching is a complex task that requires the coach to operate effectively in many different roles in support of their players. Coaching science has already contributed extensively to the understanding of many aspects of the coaching process such as goal-setting or learning skills. It must be recognized that the effective implementation of a coaching programme is dependent upon the quality of the interaction between coach and player. Coaching science has also developed our understanding of coaching behaviour and its impact upon coach/player interaction.

The previous discussion has highlighted the range of variables that influence effective coaching behaviour. The empirical evidence does not, as yet, allow coaching scientists to be very prescriptive in terms of optimal coaching behaviour for specific situations. It can be stated that there is no universal personality type or set of behaviours that guarantees success in coaching. The research indicates quite clearly that the effectiveness of coaching behaviour is influenced by player characteristics (age, ability, culture), situation characteristics (task dependence, power structure) and coach characteristics (decision-making style). These variables dynamically interact to create an infinite number of subtly different coaching environments. Coaches can best optimize their coaching by being aware of these influencing variables and being dynamic in terms of the behaviour pattern that they adopt.

Whilst science has supported and developed an understanding of the coach and the coaching process in general, there is still much that we do not understand with respect to specific sports, i.e. soccer. What is required now is for soccer coaches and scientists to work in partnership to identify the aspects of coaching behaviour that are most important within soccer. Once the most important aspects have been identified then soccer can work towards developing coach education programmes that allow coaches to develop effective coaching behaviour patterns.

REFERENCES

Case, B. (1987) Leadership behaviour in sport: a field test of the situational leadership theory. *International Journal of Sports Psychology*, **18,** 256–68.

Chelladurai, P. and Carron, A.V. (1983) Athletic maturity and preferred leadership. *Journal of Sports Psychology*, **5,** 371–80.

Chelladurai, P. and Saleh, S.D. (1980) Preferred leadership in sports: development of a leadership scale. *Journal of Sports Psychology*, **2,** 34–45.

Chelladurai, P., Haggarty, T.R. and Baxter, P.R. (1988a) Decision styles of basketball coaches and players. *Journal of Sports and Exercise Psychology*, **11,** 201–15.

Chelladurai, P., Imamura, H., Yamaguchi, Y. *et al.* (1988b) Sport leadership in a cross-national setting: the case of Japanese and Canadian university athletes. *Journal of Sport and Exercise Psychology*, **10,** 374–89.

Chelladurai, P., Malloy, D., Imamura, H. and Yamaguchi, Y. (1987) A cross-cultural study of preferred leadership in sports. *Canadian Journal of Sports Sciences*, **12,** 106–110.

Cox, R.H. (1992) *Sports Psychology: Concepts and Applications*, WCM Publishers, Dubuque.

Danielson, R.R., Zelhart, P.F. and Drake, D.J. (1975) Multi-dimensional scaling and factor analysis of coaching behaviour as perceived by high school hockey players. *Research Quarterly*, **46,** 323–34.

Dubois, P.E. (1981) The youth sport coach as an agent of socialization: an exploratory study. *Journal of Sport Behaviour*, **4,** 95–107.

Fairs, J. (1987) The coaching process: the essence of coaching. *Sports Coach*, **11(1),** 17–19.

Fiedler, F. (1967) *A Theory of Leadership Effectiveness*, McGraw-Hill, New York.

Fouss, D.E. and Troppman, R.J. (1987) *Effective Coaching: a Psychological Approach*, Collier Macmillan, London.

Franks, I.M., Sinclair, G.D., Thomson, W. and Goodman, D. (1986) Analysis of the coaching process. *Sports Science Periodical on Research and Technology in Sport*, January.

Gordon, S. (1986) Behavioural determinants of coaching effectiveness, in *Coach Education: Preparation for a Profession*. Proceedings of the VIII Commonwealth and International Conference on Sport, Physical Education, Dance, Recreation and Health, E. & F.N. Spon, London, pp. 92–6.

Gordon, S. (1988) Decision-making styles and coaching effectiveness in university soccer. *Canadian Journal of Sports Sciences*, **13(1),** 56–65.

Gould, D. (1993) Goal-setting for peak performance, in *Applied Sports Psychology: Personal Growth to Peak Performance* (ed. J.M. Williams), Mayfield Publishing, Mountain View, pp. 158–69.

Hendry, L. (1974) Human factors in sport systems. *Human Factors*, **16,** 528–44.

Hersey, P. and Blanchard, K.H. (1969) Lifestyle theory of leadership. *Training and Development Journal*, May, 26–34.

Martens, R. (1990) *Successful Coaching*, Human Kinetics, Champaign, IL.

National Coaching Foundation (1986) *Coach in Action,* Springfeld Books, Leeds.

Ogilvie, B. and Tutko, T. (1966) *Problem Athletes and How to Handle Them*, Pelham Books, New York.

Pyke, F. (1992) The expanding role of the modern coach. *The Pinnacle*, **9,** 3.

Sabock, R.J. (1985) *The Coach*, Human Kinetics, Champaign, IL.

Sage, G.H. (1975) An occupational analysis of the college coach, in *Sport and Social Order* (eds D. Ball and L. Loy), Addison-Wesley, Reading, Mass.

Schmidt, R.A. (1991) *Motor Learning and Performance: From Principles to Practice*, Human Kinetics, Champaign, IL.

Smith, R.E., Smoll, F.L. and Curtis, B. (1978) Coaching behaviours in Little League baseball, in *Psychological Perspectives* in *Youth Sports* (eds F.L. Smoll and R.E. Smith), Hemisphere, Washington DC, pp. 173–201.

Smith, R.E., Smoll, F.L. and Curtis, B. (1979) Coach effectiveness training: a cognitive-behavioural approach to enhancing relationship skills in youth sport coaches. *Journal of Sport Psychology*, **1,** 59–75.

Smith, R.E., Smoll, F.L. and Hunt, E.B. (1977) A system for the behavioural assessment of athletic coaches. *Research Quarterly*, **48,** 401–7.

Smoll, F.L. and Smith, R.E. (1993) Educating youth sport coaches: an applied sport psychology perspective, in *Applied Sports Psychology: Personal Growth to Peak Performance* (ed. J.M. Williams), Mayfield Publishing, Mountain View, pp. 36–50.

Stogdill, R. (1974) *Handbook of Leadership: A Survey of Theory and Research*, The Free Press, New York.

Terry, P. (1984) The coaching preferences of athletes. *Canadian Journal of Applied Sports Sciences*, **9,** 188–93.

Wandzilak, T., Ansorge, C.J. and Potter, G. (1988) Comparison between selected practice and game behaviours of youth soccer coaches. *Journal of Sport Behaviour*, **11,** 78–88.

Woodman, L. (1993) Coaching: a science, an art, an emerging profession. *Sports Science Review*, **2,** 1–13.

The science of soccer management

16

Malcolm Cook

Introduction

There have been a number of sound, developmental areas where soccer has made quite dramatic changes in the United Kingdom. The introduction of the Youth Training Scheme (YTS) for young 'apprentice' footballers breathed life into clubs, as have the schemes of soccer excellence which the great majority of clubs have implemented. The Taylor Report (1990) has greatly accelerated the growth and development of new, modern stadia, whilst the advent of television and satellite stations has brought additional cash into the game. Many clubs have taken advantage of this trend.

Clubs now need to promote themselves much more dynamically, to grow or even, in some instances, survive. This has meant huge changes from the traditional ways that the game was presented and played. One area that has not seen many changes, however, is soccer management. There have been over 2000 managers sacked from their clubs in the United Kingdom since the Second World War, without any institution strong enough to represent them or look after their interests. The job of soccer manager has certainly widened and become much more difficult and stressful. This is reflected in the physical and psychological illnesses of many famous managers, some of whom have died prematurely. Others such as Johann Cryuff, Kenny Dalglish, Graeme Souness, Arthur

Science and Soccer. Edited by Thomas Reilly. Published in 1996 by E & FN Spon, London. ISBN 0 419 18880 0.

Cox and Don Howe have seemingly experienced stress-related reactions.

There is need of a complete overhaul of the management scheme in the United Kingdom and we need to improve the status of managers, give them positive support and training and have a strong, creditable organization to represent them properly. A sound management structure is required to train managers effectively; the present 'part-time' training does not prepare them for this most demanding occupation. Comparisons with continental training systems such as those in France, Germany and Holland, who all have accredited management schemes over a number of years of intensive training, demonstrate some of the reasons for their success at club and international level and put managers from the United Kingdom at a serious disadvantage in the performance of their jobs.

16.1 CATEGORIES OF SOCCER MANAGEMENT

There are no definitive and standardized titles, roles or demarcation lines for duties and responsibilities of a soccer manager. There has been a gradual evolution in the game but each country and club has its own interpretation of the scope for management within its own organization. Two clubs can have a team manager, but the role, responsibilities and scope going with the post can differ greatly.

The categories include national team consultant, general manager, team manager, player, coach, assistant and youth manager. The titles bestowed on managers usually denote their apparent areas of duty. There are two major areas that the manager will be responsible for, namely the club and the team. More administrative skills are associated with the former while more personal, practical skills are connected with the latter.

The major difference between a club and **national team manager** is that club managers usually deal with their players on a daily basis, and national team managers only see their players on a few occasions each season. Much of the latter's time will be spent in travelling to observe club teams in order to keep up to date with current and potential international players' form as well as visiting other countries who will provide opposition at a later date. Some national governing bodies also employ their national manager in a dual role as the director of coaching responsible for the development of players in their country. Much of the job will be concerned with good team selection, motivation of players and a sound, tactical judgement to prepare the team for the different styles and systems that they will encounter at home and abroad. The manager will need to become something of a diplomat and communicator to meet the ever-increasing demands by the communication media on his time. Andy Roxborough (Scotland) and Jack Charlton (Republic of Ireland) have both led their countries to European and World Cup final stages and developed their own strategies for handling meetings with media representatives.

Some clubs who have appointed young managers (often a player manager) also retain the services, albeit on a part-time basis, of the older retiring managers to act as advisers to their younger counterparts who are 'learning the trade'. The **consultant manager** has normally been attached to the club for a long time, has had some success and is well respected by all. The consultant is the 'elder statesman', who has the experience but not the energy to withstand the pressure of the job. On the surface, the concept seems sensible, but in practice it does not always turn out to be effective. Both managers will have different, individual ways of doing things which can create conflicts of interest and philosophies. The consultant manager may provide more pressure for the acting manager, albeit unintentionally, by being the one the new manager has to follow and emulate. Successful consultant managers in English League soccer have included Paisley (Liverpool), Mee (Watford) and McMenemy (Southampton).

The **general manager** will cooperate closely with the team manager to bridge the gap between the board of directors and the management at the club. This job title could be 'manager' or 'executive manager/director', as it is at some clubs; however, this type of manager is essentially concerned with the day-to-day running of the club outside the team affairs. Such managers will oversee the wider aspects of the club, including commercial, secretarial, financial, ticket sales and ground development and so on. Some successful club general managers have been Jimmy Hill (Coventry City), Don Revie (Leeds United) and Matt Busby (Manchester United).

The **team manager** has basically the same role in the United Kingdom as the coach abroad. The responsibilities of this position are mostly concerned with the team and include organizing the training and coaching programme, acquiring training venues, match preparation, team selection, player recruitment, staff training and recruitment. The team manager has the responsibility for the running of the first, reserve and youth teams at the club. Some effective team managers have been Terry Venables (Tottenham Hotspur), Kenny Dalglish (Liverpool), Joe Royle (Oldham Athletic), Walter Smith (Glasgow Rangers) and Kevin Keegan (Newcastle United).

The category of '**player manager**' has been in vogue for some time and involves the person being employed by the club both as a player and the manager. Normally such individuals are approaching the end of their playing career and have a good reputation and name in the game. Also they can influence success by their playing expertise, knowledge and influence on the pitch. Often player managers have the assistance of an older, more experienced assistant to help them to learn their 'managerial ropes', by allowing them to become more involved in team affairs. They use each other as a sounding board for their ideas. Quite often, at the initial stage, player managers have proved successful at clubs but, inevitably, as they get older, and the overall demands of the job increase, it becomes much more difficult to do justice to the job. They have to try to train daily with the other players so as to be fit enough to last the physical rigours of playing throughout the season, with the added mental strain of knowing that their playing performance will be assessed much more critically than anyone else in the team. On top of this, there is still a need to do some administrative, everyday duties, such as attending board meetings, telephoning other managers regarding possible signings or selling players. They find that they cannot split themselves to do the myriad things that the manager needs to do and eventually retire from the playing side. They will then need to rely on the more traditional management skills and lose the influence they carried by being a player. Some effective player managers were Dalglish (Liverpool), Souness (Glasgow Rangers) and Hoddle (Chelsea).

The **coach manager** is essentially a coach who believes that most of the team success can be derived from the work done on the training ground (Tutko and Richards, 1974). Unlike team managers, who delegate much of the training work to the coach, coach managers do this themselves, as they are effective in this area. Otherwise they are like team managers, responsible for the team selection, player recruitment and tactics. Some successful coach managers have been Malcolm Allison (Manchester City) and Howard Wilkinson (Leeds United).

Many clubs, particularly the bigger ones, have found that the everyday running of the organization has become too much for one person. Consequently the position of **assistant manager** has evolved and is very much in vogue in the United Kingdom. Once again, clubs differ in how this person is used, some delegating the assistant manager to the more administrative work whilst the team manager attends to team affairs. Others may do the reverse. Essentially, the assistant manager is there to support and take the burden off the manager, to liaise in the most effective way for the good of the club and team. The personalities of the two and their philosophies of how the game should be played and players handled, to name but two aspects of management, are very important to the success of this pairing. Ideally, such trust should be built up that they can, at times, alternate when the manager is unavailable. Examples of these situations include talking to the media, taking responsibility for the team preparation before a match whilst the manager watches another player with a view to making a signing, or actually signing players themselves. Some people do not relish the strain of being the leader and prefer to take a more low-key role. This can be very effective in areas where the manager is not very strong and so the two are complementary to each other. Some effective partnerships involving assistant managers have been Clough and Taylor (Nottingham Forest), Mercer and Allison (Manchester City), Nicholson and Bailey (Tottenham Hotspur) and Mee and Howe (Arsenal).

In the United Kingdom one management position may be called the **youth development officer** or **youth coach**. This person will be responsible for the youth squad of apprentices plus school-boy signings and the younger members of centres of excellence at the club. Youth coaches will be involved in the youth scouting network, liaising with their scouts to ensure that they pick up the young talent that is available. They will also need considerable public relations skills to make contact with local schools and junior club representatives plus the ability to convince parents of promising young players that they should sign for that particular club. Youth managers will not only plan and execute the training and coaching programme with the management of the team and prepare for its competitive schedule of matches, but will also look after the education and general welfare of players under their charge.

The job involves both administrative and practical duties. The occupant needs to take a wider view of the development of youngsters rather than winning as the main aspect. Clubs employ youth coaches who have teaching experience with children but little professional playing experience. Alternatively they may use older coaches who have been involved as coaches at an older level for some years, or are just coming to the end of their playing career.

16.2 SOCCER MANAGER'S ROLES

The soccer manager's roles have widened considerably in the United Kingdom since the 1980s. As the requirements of the job became clearer and the need for greater professionalism was apparent, the modern manager has found that it requires multifaceted skills to operate successfully in the game. Delegation of some of the more onerous jobs is essential, to others who have the time and maybe are more suitable to do them. However, the manager needs feedback from them over what is going on in what is the manager's domain, whether those jobs are carried out personally by the manager or not.

The major roles of the manager are **leadership**, **organization** and **motivation**. The manager must also function as a **businessman** and **tactician**. Managers need to have the necessary leadership qualities (strong personality, knowledge of the game and associated factors, ability to make good decisions under pressure and solve problems), and motivate all functions of the club towards planned objectives. They need the ability to take people, staff, players and fans along with them even when things may go wrong for a period of time.

Managers should have the ability to plan effectively and organize the resources of workforce, time, facilities, money and so on towards achieving success. They need to pay attention to detail when planning and must set out individuals' roles and responsibilities within the club. They are also obliged to monitor progress and development continuously.

The morale of many people at the club's offices and grounds will depend to a large extent on the manager's skill at motivation which will come from that individual's personality, attitudes, planning and methods of persuading staff and players alike to work for the benefit of the club. Players tend to lean towards the winning of games with scant regard for human relationships in the reaching of this objective. The manager needs to find out whether the majority of players in the squad are task or player-oriented and keep this fact in mind when looking at how to maintain motivation over longish periods of time.

The modern game has become very business-minded because of the prevailing economic circumstances in which soccer suffers, like every other industry. Paradoxically, in the United Kingdom there is a much greater pressure on the major clubs in particular to gain success and this promotes the spending of more and more money. For example, Blackburn Rovers, Leeds United, Arsenal, Manchester United, Liverpool and others have all committed many millions of pounds for expensive ground developments and the purchase of players during the 1990s, whilst a number of lower League clubs have fought for financial survival. The amounts spent in England pale when compared with spending at top European clubs such as Barcelona (Spain), AC Milan (Italy) and Marseilles (France). These clubs have extraordinary stadia and pay staggering amounts on the transfer market for players who also acquire massive salaries and bonuses for joining the club. At many clubs the manager will be at least partly responsible for the finances of the club including dealing with 'selling' the club to potential signings, transfer arrangements, contracts, ground developments and generally looking after the cash flow and balance at the club.

Although the coach is with the team for the majority of the time on the training field, the manager will be responsible for the team selection, system of play and tactics employed for each game. Many managers are good coaches too and take some time in training sessions to work with the players to get their tactical ideas across more forcibly. The manager needs to be able to put out the strongest combination of players, the most fluid way of playing to suit these players and to formalize the tactics to give the team the best chance of getting a positive result. Tactical application is largely problem-solving and the manager does as much spontaneously during the game as before it at the planning stage. The ability to develop a sound strategy for the players and team is an important requirement for the manager.

16.3 SOCCER MANAGERIAL STYLES

There is a range of management styles with success coming from varying personalities and methods (Cook, 1982). Some managers have been successful at one club, yet not at others. Some seem to be 'right' for a club, arrive at the right time and become an immediate success whilst others, who have proved successful in the past, do nothing at the new club (e.g. Shankly at Liverpool was in the first category and Clough at Leeds in the second). The management

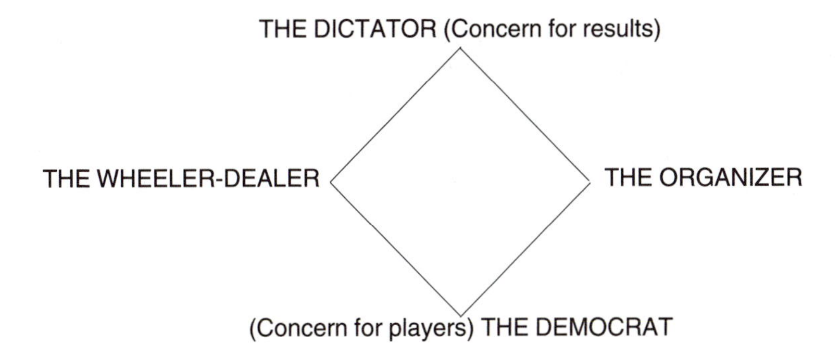

Figure 16.1 Soccer managerial grid.

styles seem to fit the club and their needs at the time. Factors which are important include the team's make-up of players, style of play, League position, expectations from the board of directors and the fans.

Every manager displays their own unique style which has some advantages and disadvantages. Many are too rigid and cannot develop more flexibility in their style which would give them better performances and success. The model shown in Figure 16.1 has been modified from the Blake Managerial Grid (Blake and Mouton, 1985) to examine extreme styles.

The 'Dictator' accepts only one way, their own way. Such individuals rarely, if ever, listen to or trust anyone. They think they have all the answers but are insecure. They do not allow players to grow by themselves; they direct them. 'Dictators' come in two varieties: the 'Sergeant Major', who tends to be a bully and verbally assaults players constantly, or the 'Hard Man' who is quieter but tough and who can frighten and intimidate with their looks, demeanour and general manners. 'Dictators' can be sarcastic in front of the players and physically strong with a hard reputation, but often have little 'personality'. They do not like to be questioned about their decisions, and see this as a challenge to their authority.

Advantages of the 'Dictator' type are:

1. they plan and prepare players well so they know the team tactics thoroughly;
2. their teams are usually mentally tough, resilient and determined;
3. the players are always well prepared physically and disciplined;
4. if the team is on a winning run, the team spirit can be very good.

Disadvantages linked with the 'Dictator' are:

1. morale and discipline can suffer when there is too much pressure or if the team is on a losing run;
2. such managers are feared by players who cannot perform; they treat them all the same and lack flexibility to change even when it is obvious that they should;

3. if they show lack of sportsmanship and do not practise what they preach, they can lose the respect of some players.

A second type of manager can be described as the 'Organizer'. Managers of this type are full of new theories to deal with and read a lot. They keep the players up-to-date but forget to treat the players as human beings, not machines. They have a computerized mind which rarely shuts off. They are a little paranoid about their schedules and preparation, which must be followed to the letter.

Advantages of this type are:

1. the team is always well prepared tactically: match preparation is thorough with nothing left to chance and great attention is paid to detail;
2. such managers are always up-to-date with new rules, tactics and knowledge, and can gain advantage over less well prepared teams;
3. the team is normally well disciplined and organized and rarely caught out with any surprises that the opposition can present.

Limitations of this type are:

1. the manager can be too rigid and inflexible with problems;
2. they will tend to blame players when things go wrong, rather than their own plans (their view is that if it works on paper it should work on the field);
3. they are not very concerned with players – only with the task in hand;
4. they can confuse players with technical jargon when simple ideas or words would communicate the message more effectively;
5. the predictability of the training and preparation means that it soon becomes too monotonous and players lose motivation;
6. players cannot express themselves in this system.

A third type is referred to as the 'Wheeler-dealer'. This character is often shrewd but personable; they live off their wits and 'hunches'. When these come off, the players, board of directors and media think the manager has special gifts. These managers often ride their luck and can make good transfer deals on the market. However, the players can easily become unsettled by their behaviour and lack of direction over a period of time.

Advantages associated with the 'Wheeler-dealer' are:

1. they often have a 'charismatic' personality which attracts people to them; they exude confidence;
2. their high profile breeds confidence in the team and can attract interest to the club which otherwise would not be there;
3. their unpredictability can keep players on the alert: players don't know what to make of them and this serves to motivate them;
4. on occasions, such managers can take the pressure from their players by their high profile: the media personnel can concentrate on the manager

and avoid the players, which can be positive for the team, especially if confidence is low at the time.

Disadvantages of this approach are:

1. their unpredictability and double-standards can lose such managers the respect of their players and cause team disruption;
2. teams can feel that more attention is given to the managers and their personality than to the team and its performance;
3. the team lacks preparation and planning and comes unstuck at times with tactical problems from opposing teams;
4. the team's physical conditioning can also be suspect as the manager lacks organization, preparation and objectivity.

A further type can be classified as the 'Democrat'. These individuals are normally seen as the 'nice guy' who wants to build teamwork through friendship. They do not like conflicts within the team and are not always good at handling players' disagreements.

Advantages of this style are:

1. the team spirit can be high if things are going well on the field;
2. the more relaxed, positive atmosphere can encourage players to take more risks in the game, and let themselves go and be more creative;
3. better communication between coach, manager and team inspires more trust and leads to less conflict in the dressing room;
4. these managers' open-door policy can generate greater respect for themselves and create a good atmosphere.

Disadvantages are:

1. this type of manager is seen by some in the team as weak, especially 'con-men' who can cause problems;
2. sometimes they are seen as not being committed enough to winning, being too concerned about players' and others' feelings and over-sensitive to the needs of some players to win as the priority;
3. they may not handle stress well when it is presented or prepare players and team satisfactorily; the discipline may drop as a result.

It is possible to attach famous managers to a specific style. One practical exercise is to put the following famous managers on the grid in Figure 16.1:

Alex Ferguson
Kevin Keegan
Johann Cryuff
Kenny Dalglish
Jack Charlton
Joe Royle

Figure 16.2 Alex Ferguson of Manchester United (*left alongside author*). Where does he fit within the range of managerial styles?

Some managers almost defy description with their style which is unorthodox but often successful. What about the charismatic Bill Shankly and Brian Clough?

It should be accepted that there is no perfect style. The currently successful manager develops the ability to 'transfer' his style at the appropriate time with particular players and situations.

16.4 STRESS IN SOCCER MANAGEMENT

The job for soccer managers is notoriously insecure and the pressures on managers have become intense with the fear of losing their position very strong indeed. It was reported in *Goal* magazine in November 1991 that the average rate for a manager getting the sack over the previous 25 years was one every two weeks, a grand total of 750. The lower divisions suffered most, with an average change of club manager at least once every four years. The stress on managers has greatly intensified since this time, with well-known personalities such as Souness, Dalglish, Clough and Cryuff suffering mentally and physically when trying to cope with their club's insatiable appetite for success. There

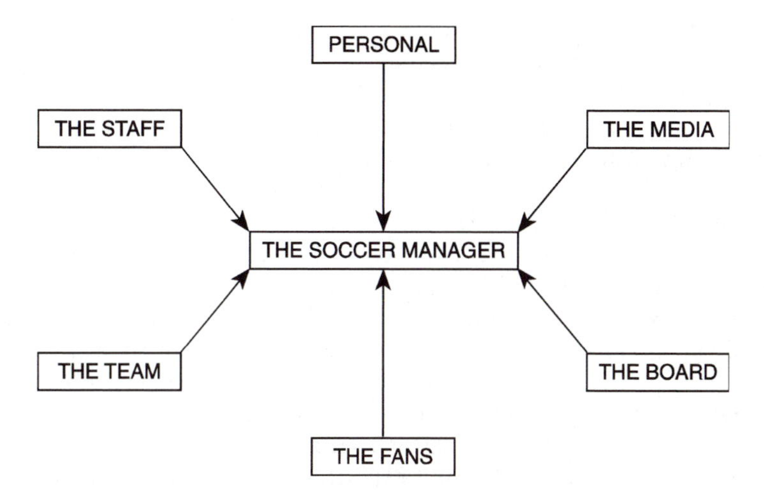

Figure 16.3 Sources of stress for the soccer manager.

are many stresses on the modern manager which need to be identified and coped with in the everyday running of the club (Figure 16.3).

The communication media's interest in soccer is truly staggering. There is a constant diet of soccer for the public via the television, radio, newspapers, magazines and many other sources of information such as club programmes. The communication media as a whole have become driven to sensationalism and instant answers to, at times, complex problems. Often the managers feel they are unfairly treated by the press/media. There is little doubt that the continuous exposure via the scrutiny of the television, radio and newspaper journalists, which often intrudes into all areas of the occupation in an unsympathetic manner, causes much anxiety to most managers.

The chairman and members of the board of directors can be a source of stress or support depending on the relationship that exists with the manager and the current situation at the club. The manager is the link between the board and the team and, as such, needs to be able to communicate and mediate effectively. There has been a lack of clarity, vision and direction by boards of directors at clubs in the past which has undoubtedly contributed to the large numbers of managers who have been sacked, though, in more than a few cases, they have done very well with the meagre resources available. The best of managers will encounter at some time a slump in the team's performance which often needs the patience of the members of the board rather than the infamous 'vote of confidence' which normally precedes the end of the manager's employment at the club. Many managers feel that the directors, in essence, do not know enough about the game and the business of soccer and treat it as a part-time hobby. The emergence of managers such as Jim McClean (Dundee United), Terry Venables (England), Jimmy Hill (ex-Coventry) and others may start to change the system and allow a closer and more professional working relationship with the team managers.

A new type of supporter has emerged over the past decade who, unlike his predecessors, is not content to just watch his team once or twice a week and return to his everyday life until their next game. The modern fan wants a great involvement in the club and vehemently articulates this clearly via the local news media. The enthusiastic football supporter also expresses his views through fanzines, magazines, club AGMs and other meetings and through supporters' associations. Some fans are very demanding and vociferous towards the board and often the manager, which can only add to the pressures already upon him in a very direct way – to have 20 000 or so fans vent their displeasure at the team and manager for a poor performance cannot be dismissed easily, and must be faced by the manager.

Working with people day by day in continually competitive circumstances, where so much is demanded and expected of the team and the manager, creates much tension and anxiety. There are conflicts with players for a host of reasons which can include grievances over being dropped from the team, substitution, wage or bonus demands and personality clashes. The better the human management skills, the more the manager can avoid potential problems and maintain motivation and an effective working atmosphere. There are always potential problems around the players in the various squads of a personal, domestic or competitive nature to test the manager's resolve and judgement.

The manager's staff can provide problems for the manager inadvertently and otherwise, which affect his actions. It is essential that the manager carefully vets and assesses the coach's and assistant's philosophy, style and character before selecting them to work at the club. Managers opt for the assistants who will not threaten their jobs, such is the mistrust and fear in the game. On occasions, rather than relieving the stress, this can actually aggravate it, as the person appointed has neither the personality, ability nor character to do the job which they have been assigned to perform. Sometimes the assistant is found to be over-ambitious and will work, albeit subtly, against the manager. It is most important that the staff and the manager provide good leadership for the team in the quest for success by working along the same lines.

The personality of the manager, his management style and how he perceives the everyday problems in running a soccer club are all important in the handling of stress. Most managers are heavily committed to the game and to their club, and work long hours in stressful conditions in the attempt to bring their club success. Fear of losing, or experiencing a sequence of losses at the beginning, middle or end of a season, are very potent stresses (Teipel, 1993). Many, however, fail to understand or see what factors are causing stress and take avoiding action. Brian Clough was one of the first managers to take a mid-season break by going abroad for a few days to 'recharge his batteries'.

For the great majority of managers, this is seen to be out of the question, but they all need some 'time out' a few times each season.

16.4.1 Handling stress

Managers are instantly recognizable by their style of presentation. Some are sour and seem to have little sense of humour whilst others, like Ron Atkinson (Coventry City), express a sense of balance and seem to take things in their stride. Managers need to come to terms with any personal or domestic problems, as these can affect their performance as managers. They need to learn from experience, education and self-examination how to use their strengths and hide their weaknesses.

Summary

There is no single system that applies to soccer management in the United Kingdom. It is not surprising therefore that there is variation in management styles. A common factor is the psychological stress that professional soccer managers experience. This may be linked with the high rate of turn-over of managers in the professional clubs and the associated job insecurity. An ability to tolerate occupationally related and personal stress is a requirement of a good manager.

REFERENCES

Blake, R. and Mouton, J. (1985) *The Managerial Grid*, 3rd edn, Gulf Publishing Company, Houston, Tx.
Cook, M. (1982) *Soccer Coaching and Team Management*, A & C Black Ltd, London.
Taylor, The Rt Hon. Lord Justice (1990) *The Hillsborough Stadium Disaster: Final Report*, HMSO, London.
Teipel, D. (1993) Analysis of stress in soccer coaches, in *Science and Football II* (eds T. Reilly, J. Clarys and A. Stibbe), E. & F.N. Spon, London, pp. 445–9.
Tutko, T.A. and Richards, T. (1974) *The Psychology of Coaching*, Allyn and Bacon, London.

Psychology

17

Frank Sanderson

Introduction

For the soccer player, such factors as physical conditioning, skill and match experience are fundamentally important in determining level of success. However, there are many psychological factors which are equally important to consider if the player is to achieve his or her potential, and this chapter focuses on them.

In discussing personality and performance, the importance for the coach and manager of having insight into the player's personality is emphasized. The key concept of motivation is then explored, together with an examination of the extent to which players take personal responsibility for outcomes such as defeat or victory.

As soccer, particularly at the professional level, can be a cause of considerable psychological stress, the concepts of stress and anxiety are examined. Guidelines on achieving optimum performance through, for example, relaxation procedures and goal-setting are provided.

Finally, the social psychology of soccer is outlined. The main research evidence relating to aggression, team cohesion, crowd effects and leadership is presented.

17.1 PERSONALITY AND THE PLAYER

A player's personality is as critical in determining success in the sports arena as physical ability. True, a seven-stone weakling is unlikely to become an

Science and Soccer. Edited by Thomas Reilly. Published in 1996 by E & FN Spon, London. ISBN 0 419 18880 0.

effective central defender, regardless of personality, but, equally, the well-proportioned and skilful athlete will not succeed without such attributes of personality as determination and will to win. It is therefore useful for those who interact with the player, i.e. coaches, managers, physiotherapists, doctors, to have some general understanding of personality and to appreciate how it might relate to performance, whether it be in a match, in the training room or, for that matter, on the recovery bench.

The word 'personality' derives from the Greek word 'persona' meaning 'mask'. Anyone involved with helping the player should take time to get behind the mask and deal with the real person.

Some of the main points to note about personality are:

- the personality of each individual is unique;
- the role-playing behaviour or 'persona' of the individual is superficial and an unreliable indicator of personality;
- we assess an individual's personality from the way he or she typically responds, i.e. a person who tends to display nervousness and apprehension in a wide variety of situations can be described as having an anxious personality. But we cannot safely conclude this about a young player who displays anxiety during a first-team debut. We would need much more evidence.

17.1.1 Traits and states

Adjectives descriptive of behaviour, such as anxious, outgoing, aggressive and so on, can refer to stable, enduring characteristics (**traits**) or to transient fluctuations within an individual (**states**). If someone is 'trait-anxious', they have a predisposition to be anxious in many situations. A 'state-anxious' person on the other hand is displaying a mood related to the particular situation, as with the previously mentioned novice player.

17.1.2 Eysenck's theory of the biological basis of personality

There are many theories of personality but one of the most useful is that devised by Hans Eysenck, who emphasized the importance of genetic factors in determining intelligence and personality. He put forward a theory about the biological basis of extroversion/introversion (E) and neuroticism/stability (N) (Eysenck, 1967). He delineated two major independent dimensions along which the personality of individuals can vary (Figure 17.1).

Extroverts are highly sociable, enjoy the limelight, enjoy taking risks, crave excitement, are easy-going, optimistic and impulsive. They much prefer doing things to planning them, tend to be aggressive and can be expedient and unreliable. Introverts, on the other hand, are quiet, reserved and withdrawn. They enjoy their own company, like a well-ordered life and are cautious and

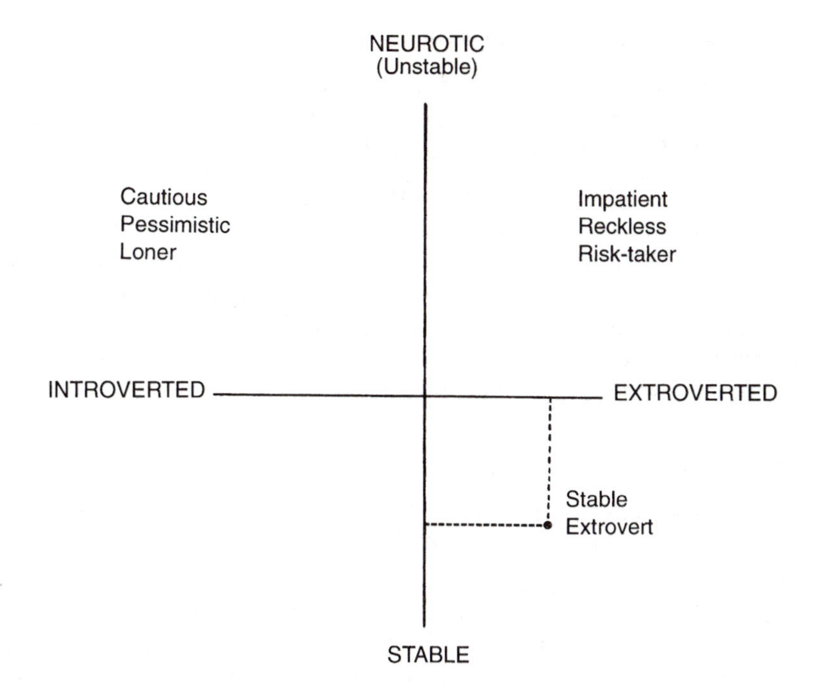

Figure 17.1 Dimensions of personality according to Eysenck (1967).

pessimistic, preferring to plan things ahead and leave nothing to chance. They are serious-minded, keeping feelings under control, reliable, and tend to have high ethical standards.

Those high on neuroticism are typically anxious, moody, frequently depressed and easily knocked out of their stride. They tend to be preoccupied with feelings of anxiety. The stable individual shows little emotional reaction to situations, being even-tempered and unworried. The level of neuroticism interacts with level of extroversion. Thus, neuroticism tends to reinforce the negative characteristics of those high on extroversion and introversion.

Eysenck postulated that an individual's level of E is a function of the nature of the ascending reticular activating system (ARAS). This system is believed to be responsible for non-specific arousal in the cerebral cortex. He argued that extroverts are chronically under-aroused and that introverts are chronically over-aroused. In other words, in extroverts, the ARAS tends to be inhibitory and, in introverts, excitatory in its effects on cortical arousal. Hence, introverts constantly attempt to reduce stimulation, whereas extroverts are 'stimulus hungry'.

Eysenck further postulated that the level of neuroticism is related to characteristics of the autonomic nervous system (ANS). Those high on neuroticism have a more volatile and responsive ANS.

17.1.3 Implications of Eysenck's theory

A great advantage of the theory is that it is testable, particularly in relation to the extroversion dimension. For example, evidence has been found to give support to the following predictions from the theory.

- Introverts are better at vigilance tasks than extroverts, i.e., they are less likely to suffer from lapses of concentration.
- Introverts have a relative dislike of strong stimuli. The more introverted players are the ones most likely to be negatively affected by a noisy crowd.
- Extroverts are more pain tolerant and have higher pain thresholds than introverts, e.g., they are more likely to 'play through' injuries, shrugging off injuries that would incapacitate the introvert.

17.1.4 Personality measurement

Given that personality is a large part of what makes a person unique, and that people frequently make value judgements regarding the personality of others, it is not surprising that psychologists have attempted and continue to attempt to measure 'personality'. Most psychologists would maintain that questionnaires provide the best means available of measuring personality.

Two of the best trait questionnaires are the Cattell 16PF and the Eysenck Personality Inventory (EPI). Over the past 20 years, the Cattell 16PF has featured in more research papers on the personality of athletes than any other instrument. The Profile of Mood States (POMS), which measures the negative scales of tension, depression, anger, confusion and fatigue together with the positive scale of vigour (energy), has also been extensively used with athletes. It differs from most other personality measures in being a state rather than a trait test.

The current popularity of the EPI can be attributable to its face validity, its lack of complexity and the fact that it is underpinned by a testable theory. It consists of 57 items to which the respondent must answer 'Yes' or 'No'. Twenty-four items relate to the Extroversion–Introversion dimension (E scale), 24 to the Neuroticism–Stability dimension (N scale) and nine items constitute the Lie (L) scale.

The value of information gained about a player from questionnaire responses should not be overestimated. Nor should it be underestimated. For example, a player's 'personality profile' may provide useful 'thumb nail sketch' information to the team coach, particularly where the player is new to the squad.

17.1.5 Personality and performance

Statements such as 'team sport athletes are different from individual sport athletes' are supported by the research evidence (e.g. Schurr *et al.*, 1977).

Nevertheless, it can also be that there are no identified personality profile patterns which are a prerequisite for successful participation in sport in general and soccer in particular, even at the highest levels.

It is quite apparent that elite players differ significantly in personality profiles. For example, Gary Lineker and Paul Gascoigne, both talented players, differ markedly in personality. The latter's behaviour suggests that he is an unstable extrovert in that he appears to be emotional, outgoing and impulsive.

Recent research reflects a growing emphasis on 'cognitive interactionism', with more concern for the player's perceptions of the environment and the way these perceptions affect behaviour. Thus, when attempting to predict a player's behaviour, the investigator might wish to know how anxious that player feels, how confident he or she feels, their perception of ability *vis-à-vis* their opponent(s) and their interpretation of success and failure.

17.2 MOTIVATION

Motivation relates to why the individual selects a particular activity in the first place, to the intensity and direction which is brought to performance, and to the extent that the individual maintains a commitment to the activity over time. Knowledge about what motivates the player is obviously of great importance to managers and training staff. Preparation for competition involves fine-tuning of motivational intensity. The way a player reacts to rehabilitation will in part be determined by his or her motivation to succeed. It will be influenced by the player's levels of general and sport-specific self-confidence.

Carron (1988) stated that 'motivation is the term used to represent the reasons why certain actions are chosen over others, carried out with energy and enthusiasm and adhered to with a high level of commitment' (p. 177). Psychologists have attempted to indicate its importance by suggesting that **performance = skill × motivation**. But, clearly, motivation is not just a matter of a player trying as hard as possible during a match. First, it needs to be controlled. There is little doubt that when Paul Gascoigne damaged his cruciate ligament in a lunging tackle at the beginning of the 1991 FA Cup Final, he was over-motivated and trying too hard. Secondly, motivation can be long-term as well as short-term, and it is certain that there are far more players who show impressive motivation during matches than are able to sustain motivation for long-term objectives. Getting psyched up for a match is easy compared with finding the motivation to endure the years of rigorous training and self-sacrifice demanded of the top-class player. A manifestation of the truth of this is the unfit player who sustains injury during a demanding game as a consequence of being highly motivated.

17.2.1 Participation motives of players

Anyone advising a player about training, rehabilitative exercise or about specific psychological problems is likely to be greatly assisted by a knowledge of participation motives. Research over the last quarter of a century has identified a variety of motives for participation in sport. Kenyon (1968) concentrated on six broad motives in investigating attitudes towards physical activity.

- Social experience.
- Health and fitness.
- Risk-taking.
- Aesthetic experience.
- Tension reduction.
- Ascetic experience (hard training).

Mathes and Battista (1985) used Kenyon's dimensions together with the following additional elements.

- The pursuit of victory: participating to experience winning.
- The demonstration of ability: participating in order to demonstrate skill or capability.
- Competition: participating to experience competition.

Their study demonstrated the predominant importance of the health and fitness motive. Whilst the professional soccer player may have health and fitness motives for participating, it must be remembered that soccer is a job, which provides a living that may be extremely lucrative.

17.2.2 Exercise adoption and maintenance

There has been much research in recent years into adherence to exercise and fitness programmes (see Chapter 12 regarding adherence to injury rehabilitation programmes). Initial research was in relation to the ability of patients to comply with prescribed medical treatment regimes – so-called secondary prevention. Researchers have also studied primary prevention adherence, i.e. the extent to which individuals comply with risk-reduction regimes. The coach and the specialist training staff will often be concerned with secondary prevention, such as would be the case when prescribing rehabilitative treatment following injury to a player. However, they will inevitably also be very much involved with players on primary prevention and achievement of match fitness, i.e. training regimes designed to maximize functional fitness and minimize injury susceptibility.

Dishman (1986) provided a useful review of the mainly American research on exercise compliance and his main conclusions are outlined in Chapter 11 on the psychology of injury. For the player wishing to achieve and sustain match fitness, research emphasizes the importance of self-motivation and the active support of the coaching staff.

17.2.3 Arousal and performance

Motivational intensity and physiological arousal are so highly correlated as to make them virtually synonymous. The more the backroom staff know about arousal (motivation) and performance, the more effective they will be in advising the healthy player and in treating the injured player. Less than optimum performance is often associated with inappropriate physiological arousal. Similarly, there are links between arousal level and injury. The underlying principle is that the player's physiological arousal can be manipulated in ways which can enhance performance and reduce the risk of injury.

The **inverted-U theory** maintains that, as arousal increases, performance increases up to a point, beyond which further increases in arousal will lead to a decrease in performance. Substandard performances could result from, on the one hand, a player being complacent or inattentive (under-aroused) and on the other, from a player 'trying too hard' (over-aroused).

A version of this theory, the Yerkes Dodson Law, maintains that optimal arousal for simple tasks is higher than for complex tasks (Yerkes and Dodson, 1908). Thus, the optimal arousal for chipping the ball over the advancing goalkeeper (complex task requiring sensitivity and fine motor control) will be lower than that required for booting the ball upfield to relieve pressure on the defence (relatively simple task, requiring gross motor effort). Because task complexity is related to the player's skill level, the optimal level of arousal for

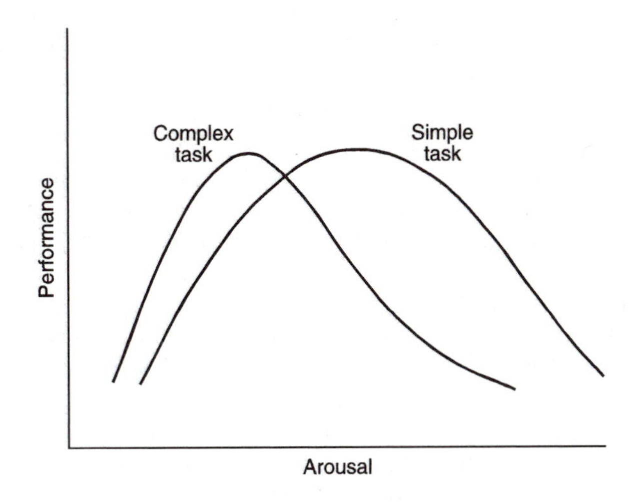

Figure 17.2 The performance–arousal relationship.

a novice performing a task will be lower than that of an expert performing the same task.

Although the inverted-U hypothesis is intuitively appealing, it has limited predictive capability and has been more often discussed than observed. Recently, more sophisticated theories have been developed, notably L. Hardy's (1990) Catastrophe Theory (see section 17.4.3 below).

17.2.4 Achievement motivation

The McClelland–Atkinson model of **achievement motivation** proposes that two factors determine an individual's need to achieve: the motive to achieve success and the motive to avoid failure (McClelland *et al.*, 1953). The motive to achieve success is a positive motive and is very strong in outstanding players. Their dedication is high. They work harder than others, and they have an intense desire to win. The motive to avoid failure is associated with high trait and/or state anxiety. It may mean that the individual avoids competition altogether, or adopts a cautious defensive strategy in competition with the main emphasis on avoiding failure. As Australian rugby coach Alan Jones noted after a poor performance by the England team in 1986, 'England pick a man for his first cap but he runs out to play in mortal fear that it will be his only cap. There must be a licence to fail, to make a mistake'.

Extrinsic motivation in the form of praise, rewards, trophies can be helpful in motivating those individuals whose intrinsic motivation is low and whose motive to avoid failure is high.

Cox (1994) described an inverse relationship between the probability of success and the incentive value of success. In other words, the incentive value increases as the probability of success decreases, as demonstrated by many FA Cup 'giant-killers' over the years. More important however is the relationship between probability of success and performance. Although the incentive value of success will be high when there is no chance of winning, the actual achievement need will be low. An apparent example of this occurred in Newcastle United's first European game for 17 years in September 1994. The away leg against Royal Antwerp was expected to be difficult, but Newcastle scored within 47 seconds and were two goals up after 10 minutes. Thereafter, Antwerp appeared only to be attempting to limit the embarrassment, and were eventually beaten 5–0.

The underlying principle is that the player with a high achievement need will be most highly motivated when the probability of success is about 50%, i.e. aiming for challenging, yet realistic, goals. Those players with a significant fear of failure, perhaps resulting from high trait anxiety or as a consequence of traumatic injury, are likely to benefit from training or competitive situations in which there is appreciably greater than a 50% chance of success.

The notion of perceived contingency recognizes that many players have long-term goals which are facilitated by progressive success at less elevated levels, e.g., progressive success through the rounds of the FA Cup is necessary for the realization of the longer-term goal of obtaining a Cup Winner's medal. In the non-anxious player who has a high motive to achieve success, a contingent achievement situation will be highly motivating. However, for those anxiety-prone players with a high motive to avoid failure, a performance situation with high perceived contingency will be demotivating. Advisers should seek to persuade such players to focus on short-term rather than long-term goals – 'a one step at a time' strategy, where the Cup match is treated like any other game.

Self-confidence

A key element in achievement motivation is self-confidence. It is a major factor discriminating between those who are high or low in achievement motivation. Bandura's self-efficacy theory (Bandura, 1977) suggests that 'success breeds success' in that successful performance leads to enhanced expectations of future success. Failure stimulates a vicious circle effect of declining self-confidence. Consequently, players, particularly inexperienced or traumatized individuals, should have their training and competitive programmes carefully managed to ensure success. Cox (1994) noted that 'once strong feelings of self-efficacy develop through repeated success, occasional failure will be of small consequence' (p. 220).

Vealey (1986) defined sport confidence as 'the belief or degree of certainty individuals possess about their ability to be successful in sport' (p. 222). She has developed a situation-specific model of sport confidence in which both trait and state sport confidence are identified. State sport confidence is influenced not only by trait sport confidence but by the individual's perception of what constitutes success.

The differentiation between trait and state sport confidence is useful, as is the identification of the mediating role of the competitive orientation (how success is construed). It suggests that state sport confidence can be enhanced not only by experiencing success, but by redefining success, e.g. in terms of process (mastery) rather than product (outcome). In other words, a useful strategy for the player low on self-confidence would be to focus on personal improvement rather than on winning *per se*.

17.3 CAUSAL ATTRIBUTION

Causal attribution is based on a 'common sense' theory and is concerned with the individual's perception of the causes of outcomes such as success or failure. It can refer not only to the individual's self-perceptions but to perceptions concerning other people's actions.

When a team wins a tournament, there are many causal factors to which the success might be attributed. For example, it may be attributed by a player to ability ('We had the best players'), effort ('We really wanted to win this one'), or even luck ('We got the lucky breaks'). By the same token, defeat has its perceived causes. Often, and quite logically, ability is the acknowledged major factor ('We were beaten by a better team'). Potentially more interesting are the causes which sound more like excuses: over the years in League soccer, defeat has been attributed to: poor refereeing, biased refereeing, the weather, the state of the ball, the state of the pitch (muddy, sloping, artificial, etc.), illness, injury, lack of sleep, lack of practice, lack of fitness, opponent's unfair play, and so on. After Swindon Town were beaten 7–1 by Newcastle United in a Premiership match in March 1994, Manager John Gorman observed, 'We have 14 players ill or injured at the moment, but that's not an excuse for defeat'. If it was not an excuse, why was it mentioned? Sometimes defeat attributed to injury might be plausible. At other times, it would be seen as an attempt to preserve self-esteem, as when Jack Charlton attributed Ireland's 1994 World Cup defeat by Mexico in Orlando to the weather. The kind of explanations given for success and failure, whether plausible or otherwise, reveals something interesting about players' and managers' motivational beliefs.

By examining the subjective causal attributions of a player, an adviser can gain an insight into the person's personality, self-confidence and motivation. By exploring the player's perceptions about the causes of outcomes, the support staff will gain a greater insight into 'what makes the player tick' and will thereby enhance the chances of successful intervention.

17.3.1 The attribution model

Fritz Heider, acknowledged as the founder of causal attribution theory, made the fundamental distinction between factors within the person and factors within the environment, either or both of which determine a particular behavioural outcome (Heider, 1958). He was thus making the distinction between internal and external sources of attribution. For example, whether or not a soccer player taking a penalty is successful depends on factors within the person (ability, effort, reaction to stress) and factors within the environment (goalkeeper's ability, weather conditions, crowd noise).

Weiner (1972) placed Heider's four perceived causes of success or failure within two causal dimensions: locus of causality (internal or external) and stability (stable or unstable). Ability and effort are internal determinants of action (cf. Heider's personal forces), and task difficulty and luck are external determinants (cf. Heider's environmental forces).

Weiner's main contribution was the stability dimension. He argued that the individual's perception of his or her specific and general ability is relatively stable over time. Likewise, task difficulty is categorized as relatively unchanging. Effort and luck are perceived to be factors which vary considerably from time to time. Weiner's model, however, does not readily accommodate the soccer player's frequent attribution for defeat: 'the referee was biased!'.

17.3.2 Causal attribution and need achievement

The internal–external locus of control construct, developed by Rotter (1966), relates to the extent to which individuals perceive that they have power over what happens to them. Those with an internal locus of control perceive themselves to be responsible for their own lives, whereas those with an external locus perceive themselves to be at the mercy of outside forces such as fate, luck and powerful others. The practical implications of this idea are immediately apparent. An internal locus of control would generally seem to be more functional and mature.

- Achievement motivation will be higher in those players who believe they are in control of their own destiny. Those acting in support of players would generally wish to encourage the internal disposition where players accept appropriate responsibility for their actions.
- Internals would be more self-reliant, tending not to act like hapless victims of circumstance.
- Internals would tend to be more flexible in their attributions, only making internal attributions when appropriate.
- An internal locus can be promoted by emphasizing the relationship between player behaviour and performance outcome. Blame should be avoided.
- Internal locus players undergoing treatment would tend to be more aware of the importance of their own efforts in achieving full recovery.
- Externals tend to have high levels of trait anxiety (Watson, 1967), which is often counterproductive in competitive situations.

17.3.3 Attribution and expectancy

Research evidence, as well as logic, suggests that when an expected outcome occurs in achievement situations, attributions tend to be stable. Conversely, when the unexpected occurs, unstable attributions are made. Expected outcomes, of course, are largely dependent on past experience. Cox (1994) pointed out the danger in making stable attributions for failure in that it implies no change for the future. If failure is linked with, say, 'lack of effort', which is an internal, unstable and controllable attribution, then the player can hope for success based on greater effort in future. However, this argument needs qualification.

- A coach or other adviser should only encourage attributions which have some credibility.
- The gaining of ability with practice should be acknowledged where possible,

i.e. ability can be an unstable factor, as indicated by Weiner (1972) in his critique of the original model.

- Attributing failure to the unstable factor of 'bad luck' on a systematic basis could mean that motivational or technical problems are not sufficiently addressed. Consequently, it is recommended that, wherever possible, failure should be attributed to unstable and internal factors.

Learned helplessness is evident when an individual, following experience of failure coupled with maladaptive internal attributions, feels that he or she does not control events and that failure is inevitable. Such persons tend to make persistent attributional errors focusing on personal inadequacy. In a motivational context, Dweck (1986) referred to learned helplessness as a maladaptive achievement pattern, contrasting it with the adaptive achievement pattern in which the individual is mastery-oriented. It can be characterized as either a short-term phenomenon, perhaps experienced by a player for only part of a competitive event, or a chronic condition persisting over time. The defeatist mentality can occasionally beset the best players, and, in some, can persist for long periods. There are various strategies which can be adopted to encourage the development of more adaptive achievement patterns.

1. Attribution retraining: 'learned helpless' individuals have been retrained to attribute their failure to effort or strategy instead of ability – unstable internal rather than stable internal (Dweck, 1975). Cox (1994) emphasized the link between failure attributions and the player's self-esteem.
2. Goal-orientation training: focusing on the development of process or skill mastery goals rather than on product or outcome goals, e.g. concentrating on the development of one-touch control rather than on, say, number of goals scored.
3. Teaching coping skills, e.g. increasing the player's focus on factors which can be controlled with the aid of self-coping statements. These statements (e.g. 'keep trying', 'maintain concentration') can be invoked at critical moments in a competition when the learned helpless player would normally be caving in.

17.3.4 Attributions and superstition

Weiner (1972) described an experimental situation in which an animal, in making a response which inadvertently precedes an externally controlled reward, makes the inference that the reward was caused by the response. He referred to this as a 'superstitious belief'. The analogy with the superstitious player or manager is hard to resist, e.g. managers who have 'lucky' suits. Although superstitious behaviour could be seen as attributionally maladaptive, it may conceivably serve a useful ritualistic purpose in providing reassurance at times of stress.

17.3.5 Egocentrism in attribution

Most participants, even at recreational level, care deeply about their success in sport, which inevitably suggests ego involvement. Such ego involvement can influence the links between attribution and outcome. The **self-serving hypothesis** states that individuals take credit for positive outcomes by attributing success to internal factors such as personal ability and effort. Conversely, the player who is ego-threatened by defeat tends to deny responsibility for the outcome by attributing it to factors beyond his or her control, such as luck, incompetent refereeing, injury and so on. Hence, the player adopts a *self-serving bias*, prepared to accept responsibility for success but less likely to explain defeat in terms of lack of ability.

The research literature shows more evidence of the ego-enhancing strategy than the ego-protecting strategy (Cox, 1994). In other words, players are less inclined to make external attributions following failure than to make internal attributions following success. For example, Sanderson and Gilchrist (1982), in a study of male squash players, found that although winners were more internal than losers, the latter still assessed internal attributes to be the most important determinants of match outcome.

There are various reasons which could explain the lack of research support for an ego-protecting strategy amongst losers.

1. It is probable that many studies, using contrived laboratory-based tasks, have not been sufficiently ego-involving, a necessary condition for testing the self-serving hypothesis.
2. In achievement situations, the internal factors of ability and effort are emphasized by significant others as important determinants of outcome (Reiss and Taylor, 1984).
3. Social norms and constraints effectively prevent many players from articulating ego-protecting attributions (Scanlan and Passer, 1980). The sport culture disapproves of 'whingeing', excuses and poor losers.
4. Illogical attributions are most likely immediately after defeat, often being replaced by more rational attributions within a short time (Sanderson, 1989).
5. 'Losers', as defined by researchers, may perceive themselves to be successful by their own criteria.
6. Attributions may be wrongly classified. For example, although a player might cite 'lack of effort' (internal) as a causative factor in defeat, the lack of effort may be triggered by an external factor, such as poor officiating (Iso-Ahola, 1977).

Some evidence indicating a self-protecting strategy has been found in the context of team sports. Cox (1990) reviewed research into self versus team attributions. Whereas most studies agree that members of winning teams are, in effect, modestly saying 'I'm okay but the team's great', as suggested by Gill

(1980), other studies contradict Gill's finding of reverse ego-centrism in losers. Bird and Brame (1978) and Sanderson and Kneale (1988) found clear evidence of a self-protecting strategy amongst losing players – an 'I'm okay but the team's so-so' response.

17.3.6 Attributions and motivation

Deci (1971) argued that extrinsic rewards can have either a controlling aspect or an informational aspect, either of which have consequences for intrinsic motivation. The controlling aspect is evident when the individual perceives his or her behaviour to be controlled by extrinsic factors – an external locus of control – which decreases intrinsic motivation. The statement, 'make it worth my while' or the question, 'what's in it for me?' illustrate the attitude of the player who perceives not intrinsic but extrinsic motivation as a controlling influence. The informational aspect is apparent when the player perceives extrinsic rewards as primarily providing feedback or information. Positive feedback about accomplishments or progress, e.g. during rehabilitation from injury, will tend to increase self-esteem and intrinsic motivation.

17.4 STRESS AND ANXIETY

Former England captain Bryan Robson, in responding to a question about stress in soccer, observed:

> It is a cut-throat business. It always has been and it will be. You don't play well, you are not in the team. It's as simple as that and there are no other considerations. So there is a lot of pressure.

Stress, and one of its emotional consequences, anxiety, are constant companions of the player. Some thrive on stress. They seek out stressful situations and appear to enjoy being 'on the brink of catastrophe'. Others appear indifferent to stress, with no amount of 'pressure' seeming to affect their performances. Yet others regularly experience the deadly effects of 'nerves' and in the process are transformed from their normal competence to a self-destructive incompetence.

Anxiety, a frequent consequence of stress, is an important and necessary aspect of the human condition. Without anxiety, and the consequent need to reduce it, there would be little human progress or achievement. Our ancestors benefited considerably from anxiety in that it mobilizes the individual for 'flight' or 'fight'. The contemporary person typically has little need for physical flight or fight and yet is exposed to relatively high levels of stress and anxiety. Consequently as Levitt (1971) pointed out, the anxiety which once protected us now threatens to destroy us.

This modern predicament provides a strong rationale for sport and exercise as tension-reducing or cathartic. The stresses of modern living can be dissipated through the opportunities for 'flight and fight' offered by competitive sport in general and soccer in particular. The cathartic benefits of sport are evident, for example, in the feelings of well-being often experienced as a consequence of vigorous exercise.

Competitive sport, however, cannot only dissipate tension, but can actually generate stress and anxiety, not all of which is necessarily beneficial to the player. The fact that competitive sport is goal-directed, perceived to be serious, and ego-involving means that the player experiences significant stress. It is necessary therefore for soccer players and their advisers, including the sports medicine specialist, to understand the meaning of stress and anxiety, and to be familiar with their effects on performance. With understanding comes the possibility of harnessing and controlling anxiety reactions, making it more likely that the individual is able to achieve his or her goals.

17.4.1 Defining stress and anxiety

Stress and anxiety are not synonymous. Stress is experienced as a consequence of any demand placed upon the body by a stressor, which could be psychological and/or physical. Spielberger (1989) noted that:

> If a stressor is perceived as dangerous or threatening, irrespective of the presence of an objective danger, an emotional reaction (anxiety) is evoked. Thoughts or memories that are perceived as threatening can also evoke anxiety reactions as readily as real dangers in the external world. (p. 4)

Thus anxiety is an emotional manifestation of stress in the particular circumstances where there is a *perception* of danger or threat. Stress, like anxiety, is best defined in terms of the individual's *reaction* to a stressor. The stress response is thus determined by the individual's perception of the situation rather than the inherent potency of the stressor.

Eustress and distress

Selye (1975) made a useful distinction between pleasant stress (eustress) and unpleasant stress (distress), thereby indicating that the stress reaction is not necessarily counterproductive. Eustress is sought after by the sensation-seekers and risk-takers. Distress is associated with suffering of some kind and although it may well involve anxiety reactions, is not the same as anxiety.

Stress and anxiety are mediated by complex psychobiological processes, which means, for example, that it is quite possible to experience enjoyment and anxiety simultaneously. Harris (1980), in a paper entitled 'On the brink of catastrophe', described how some players are able to interpret the anxiety stimulated by threatening and dangerous situations as enjoyable (eustressful) rather than distressful.

Managers and coaches should be able to distinguish between 'sensation seekers' and those more susceptible to distress. It is important that they appreciate that anxiety is not necessarily something that the individual would wish or need to avoid. Zuckerman's (1979) Sensation Seeking Scale enables this trait to be measured.

State and trait anxiety

This important distinction is referred to in section 17.1 above. State anxiety fluctuates in intensity over time depending on the degree of perceived threat. The perceived threat is determined not only by the objective demand existing, but also by the person's level of trait anxiety or anxiety proneness. The anxiety-prone player will be predisposed to perceive threat, and thereby experience state anxiety in a wide variety of situations, many of which would be perceived as harmless by those with more 'laid back' dispositions.

The lesson for the support staff is that each player is unique in his or her reaction to stressors. The potency of stressors is dependent on the perceptions of the player as determined by objective demand, the meaning of the situation, past experience and level of trait anxiety. The coach and others should make every effort to be sensitive to these perceptions.

17.4.2 Measuring anxiety

The player's anxiety reactions and disposition can be quantified by using one or more of the validated measures of anxiety. Nevertheless, the value of the in-depth interview in providing detailed information about the player's psychological profile should not be overlooked.

Electrophysiological measures of, for example, heart rate and electrical activity in the brain, or blood pressure and skin conductance, have an objectivity which has encouraged their widespread use as indicants of state anxiety, most notably in the lie-detector test (polygraph). However, there are questions concerning the validity of such measures. For example, it is obvious that increases in heart rate can be induced by many factors other than anxiety.

There are also fundamental practical difficulties in using electrophysiological measurement in competitive situations. The treatment room, however, does offer a suitable environment for those specialists wishing to experiment with these measures.

Direct observation as a measure of state anxiety depends greatly on the experience and acuteness of observation of the observer. Knowing whether the player is dejected or just lost in concentration, whether the frown represents grim-faced determination or is the sign of impending defeat can only come with particular experience. Getting the players themselves to do the monitoring via a checklist offers potential benefits in that the self-perception of the player is sharpened, an awareness of the links between psychological state and performance is encouraged, and those advising the player can share in the insights provided with consequent improvements in the quality of support offered.

The questionnaire is easily the most common method of measuring state anxiety. Given the real limitations of electrophysiological measurement, if you wish to know how anxious a player is, the simplest and best approach is to ask. The assumption being made in these circumstances is that the player is responding insightfully and honestly. However, not all players are insightful about themselves, particularly the younger ones whose personalities have not fully matured. Additionally not all players are honest, all of the time, for a variety of reasons. Even so, the paper-and-pencil test is arguably the best method available.

The Competitive State Anxiety Inventory (**CSAI**) developed by Martens (1982) has proved popular over the past decade. Given that state anxiety has been found to be multidimensional, CSAI-2 (Burton and Vealey, 1989), which measures cognitive and somatic state anxiety, and self-confidence state anxiety, is proving particularly useful for measuring state anxiety in the competitive arena. Although CSAI-2 measures the intensity of state anxiety, high scores do not necessarily indicate that performance will be disrupted. This has led Jones (1991) to supplement CSAI-2 with a 'facilitative' dimension. For each of the items, Jones has suggested an additional response on a 7-point scale, ranging from 'very debilitative' through to 'very facilitative'. High scores on this dimension, say over 50, would indicate that the player's particular state is beneficial to performance, irrespective of the actual intensity of the state.

17.4.3 Research on stress and performance

On the basis of the research evidence, it can be hypothesized that soccer players would tend to experience lower levels of anxiety than individual sport participants. The competitor in individual sports events has an undiluted sense of success and failure and will tend to experience ego-threat more intensely and more frequently than the team sport player. This is not always the case, as demonstrated by the high levels of state anxiety experienced by soccer players during penalty 'shoot-outs'.

The importance of treating state anxiety as multidimensional is demonstrated by the evidence that cognitive and somatic state anxiety respectively have differential effects on performance. Worry tends to be damaging, whereas performance is best when somatic state anxiety is at an optimum level. The *interaction* between the two is also important. Hardy and colleagues, in their Catastrophe Theory of competitive state anxiety, argued that high somatic anxiety without worry is manageable, but with worry, it can be dramatically destructive (Hardy and Parfitt, 1991). Consequently Hardy urged players to be equipped with 'multiple relaxation strategies' and to be somewhat wary of 'psyching up' strategies prior to competition (L. Hardy, 1990).

Some of the most valuable research concerning competitive anxiety has been conducted by Mahoney and his associates. They have examined the differences in anxiety reactions of successful and less successful athletes, and some of the key finding are as follows.

- More successful athletes regulate, tolerate and use their anxiety better than less successful athletes (Mahoney and Meyers, 1989).
- Skilled athletes tend to be less anxiety focused and more task focused during competition.
- The more successful athletes tend to view anxiety as energizing and helpful.
- Less successful players appear to 'arouse themselves to near panic states' and to 'either "choke" and/or progressively deteriorate in their performances during the competition' (Mahoney and Avener, 1977).

Although experienced and skilled players learn to regulate and utilize their anxiety more effectively, veteran players, i.e. those who continue competing into their 30s and 40s, may suffer from increased levels of anxiety, perhaps as a consequence of declining physical prowess and associated lower self-confidence. Relevant in this context is Little's (1969) research concerning the susceptibility of some players to neurotic breakdown when sustaining injury towards the end of their careers.

17.4.4 Attentional narrowing and stress

Nideffer (1989) argued that stress interferes with the individual's ability to control attention. His Theory of Attentional and Interpersonal Style argues that for optimum performance, the individual must be focused on the most task-relevant cues. Stress, manifested as cognitive anxiety, leads to attentional narrowing, a kind of 'tunnel vision' effect, whereby important task-related cues may be ignored. Consequently, the errors in performance increase. With very high stress, panic may ensue, with the individual losing all control of attention (Nideffer, 1989), as evidenced by experienced players missing open goals or penalty kicks.

17.4.5 Life events and stress

There is considerable evidence linking the onset of illness to the accumulated stress of life: the greater the degree of life change, the greater the likelihood of illness (Holmes and Holmes, 1970). Life change, quantified in terms of positive and negative 'life events', such as 'marital separation', 'death of close friend', and 'change in financial state', has also been linked with incidence of injuries to players. The finding of Bramwell and colleagues (1975) that those players with greatest accumulated life change were more likely to experience injury can be explained in terms of stress-induced attentional narrowing. More recent researchers have found that the frequency of injury is related to negative rather than total or positive life stress (Passer and Seese, 1983).

Exercise is a useful buffer against life stress. Brown and Siegel (1988) reported that the negative impact on health of stressful life events diminished as the amount of time spent on vigorous exercise increased. They suggested that benefits may accrue from increases in self-efficacy following regular exercise. Alternatively, they argued that exercise may help shift the focus of attention away from the problems of life. This argument has some face validity in the context of amateur soccer. The professional player, however, who is working for a living and playing perhaps in excess of 50 games in the season, is more likely to experience the stressful effects of exercise.

Kobasa (1979) argued that **psychological hardiness** is an effective buffer against life stress. Psychologically hardy individuals tend to see change as challenging and stimulating, feel themselves to be in control of their lives and are happy to enter into commitments.

Other research has focused on hassles (irritating or unpleasant events experienced on a daily basis) and uplifts (pleasant and satisfying experiences) in relation to psychological well-being. Evidence suggests that the degree of experienced hassle is a better predictor of psychological well-being than either uplifts or major life events (Wolf *et al.*, 1988). This has obvious implications for the coaching staff whose responsibility it is to monitor the well-being of the players, many of whom are exposed to significant hassle, particularly at away matches or during tournaments. Hassle reduction could usefully be a major objective, as in the case of management of individual teams/professional clubs today.

17.4.6 Stress and burnout

Competitive soccer generates considerable stress, exposing many players and coaches to a chronic level of stress which can lead to burnout. Burnout is evident when an individual's physical, mental and emotional resources are severely depleted. For the player, it may mean physical withdrawal from competitive sport which had previously been fulfilling. It can be characterized by emotional exhaustion, impersonal attitudes to associates and decreased performance capability (Fender, 1989).

Most research has focused on individuals within the helping professions on the basis that those who help others (e.g. managers, coaches, physiotherapists) are at a greater risk of exhausting their physical and emotional resources. There is little evidence suggesting that burnout is a significant problem in sport, although Dale and Weinberg (1989) found that coaches who had a 'considerate' leadership style were relatively susceptible to burnout. It is possible of course that many players and managers who drop out of the game do so as a consequence of developing some or all of the symptoms of burnout.

Suggested interventions include developing self-awareness, taking time out, changing activity and learning to say 'no' (Fender, 1989).

17.5 PSYCHOLOGICAL PREPARATION OF THE PLAYER

Traditionally, getting into shape has meant working on cardiorespiratory fitness, strength, flexibility and technical aspects of performance. An increasing minority of players are now paying attention to psychological preparation. Many players can describe what it is like to be under pressure, psyched up or psyched out, but few of them can control their psychological states. Although there are a fortunate few who possess an inherent ability to remain calm under pressure, sustaining concentration in situations where most people would 'go to pieces', there is no doubt that psychological skills training offers hope of controlling emotions and maximizing performance in a more certain way than if the player is merely physically fit and technically excellent.

The intervention strategies available are many and varied. On the one hand there are stress management techniques, emphasizing the importance of relaxation. On the other hand, there are psyching-up strategies which increase arousal and motivation. Implicit in all the strategies is the concern for the optimization of concentration and self-confidence. The player and the coaching staff, faced with the bewildering array of strategies, should remember the following points at the outset.

1. Each player is unique, physically and psychologically. This implies that there is no exact formula for all players in terms of mental preparation strategies, e.g. using a particular relaxation strategy could lead to one player being finely tuned and another falling asleep.
2. Self-management of mental preparation is the goal. Particularly where younger players are concerned, there is always the danger of over-reliance on a manager, coach or other mentor, with a consequent inhibition of the development of self-awareness, and a disintegration of performance when the mentor is not there to offer support. There are frequent problems for players at the World Cup when they may be denied contact with their own long-standing mentors.
3. Being able to relax is a foundation skill underlying all stress management and performance preparation techniques.

17.5.1 Relaxation procedures

The importance of sensitivity and perceptiveness on the part of those giving advice to players cannot be overemphasized. The team captain should be able to understand how nervous a player is, whether he needs a calming influence rather than strong words. Relaxation procedures rather than exhortations to greater effort would often be more appropriate.

There are many specific techniques for inducing relaxation, some of which have been outlined by Cox (1994), e.g. hypnosis, progressive relaxation. Players themselves have their own idiosyncratic techniques, e.g. listening to music. Whether the player uses well-known relaxation techniques or those of his or her own making, it must be appreciated that the player needs to learn how to relax, and the learning process can take weeks or even months of effort. As outlined in section 17.3, the player needs to induce optimum arousal, i.e. the player must not only be able to relax, but to relax optimally. This emphasizes the importance of control. The player must also be able to induce relaxation quickly, perhaps during a match. The penalty-taker does not have the time to go through a complex and time-consuming ritual in order to maximize the chances of success.

There are commercially available relaxation packages, e.g. the National Coaching Foundation's guide for sports. This consists of a booklet and an audio-tape.

17.5.2 Cognitive strategies

These are psychological procedures which can be used by players to prepare themselves for competition.

Imagery

Imagery is the visualization of situations, which can be used to practise skills, to instil confidence and to motivate the player whether for competition or for rehabilitation. Smith (1987) identified the five following points with regard to the application of imagery to sport.

1. Imagery skills can be developed. Although some individuals have 'photographic memories' or 'vivid imaginations' and can conjure up images with ease, those without such natural facility can learn to visualize situations effectively. It should also be remembered that some individuals have strong negative imaginations and tend to concentrate on 'worst possible' scenarios: their positive imagery skills need to be developed.
2. The player should have a positive attitude about imagery: a belief in the efficacy of the technique is necessary.
3. Skilled players appear to make more effective use of imagery than less skilled players. This is because skilled players are experienced in the situations they may wish to visualize. Also, the more familiar a situation, the more vivid and precise the imagery.
4. Relaxation is a prerequisite for effective imagery: this emphasizes the importance of relaxation techniques in underpinning performance enhancement strategies.
5. Two distinct forms of imagery – internal and external – can be identified.

Internal imagery is kinaesthetic in nature, the players imagining themselves to be, say, scoring a goal. External imagery is visual in nature: the players imagine, for example, that they are watching themselves (Mahoney and Avener, 1977).

Some research suggests that internal imagery is more effective than external imagery (e.g. Mahoney and Avener, 1977). This may be because the technique induces micro-contractions in the relevant muscles.

Effectiveness of imagery

There has been a considerable amount of research into the effectiveness of imagery/mental practice, although much of this has been in laboratory-contrived situations and has little ecological validity. However, several generalizations have emerged.

- The larger the cognitive component in an activity (the more thought that is required), the greater will be the effectiveness of imagery (Ryan and Simons, 1981).
- The use of imagery for skill rehearsal is better than no practice at all (Feltz and Landers, 1983). This suggests that in circumstances where a player is prevented from practising physically, by injury for example, mental practice could be particularly helpful.
- Woolfolk *et al.* (1985) found that visualizing being beaten (negative outcome imagery) is more potent in causing decrement in performance than imagining success (positive outcome imagery) is in facilitating performance. Because of the disproportionate power of negative outcome imagery, the importance of cognitive strategies to counteract it is stressed.
- Imagery can be learned, and Hickman (1979) provided some elementary exercises to help players develop their abilities to use imagery effectively.
- Orlick and Partington (1988) surveyed 235 Olympic athletes and found that 99% used mental imagery as a preparation strategy.

17.5.3 Goal-setting

Goal-setting encompasses both the long and short-term aspects of motivation and every player and coach should be aware of the benefits to performance of setting and achieving goals. On one level, we all set goals and aim to achieve them. Although a player aims to play well, aims to recover from injury and so on, the principles of goal-setting are more subtle than that. In general, goals can enhance performance by focusing and directing activity, mobilizing and regulating effort, increasing persistence and encouraging the use of optimum strategies (see Chapter 12 on goal-setting in the context of rehabilitation from injury).

Research into goal-setting, well summarized by Locke and Latham (1985), suggests the following guidelines for players and their advisers.

1. Giving the player a specific goal is much more effective than a general goal, e.g. telling a player, 'do your best' is too vague to be useful and does not specify what the player should actually do. On the other hand, telling a striker during shooting practice to hit, say, five consecutive volleys on target informs the player exactly what the requirements are.
2. Goals should be challenging but not so severe as to be unrealistic. Theoretically the player should be capable of meeting the challenge, but it should not be so easy as to excite little interest or effort.
3. Short-term goals should be used as an incremental means of achieving long-term goals. A talented junior's aspiration to be a world class senior player will be less overwhelming if progress is made via a series of itemized shorter-term goals, e.g. national junior squad, Premier League first team, Cup Winner's medal, national team squad, World Cup squad, and so on.
4. Goals need to be quantifiable so as to allow comparison between aspiration and achievements. In practice sessions and rehabilitation programmes, goals are easily quantified, but it is a little more difficult in matches. Computerized match analysis is promising in this respect.

   The effectiveness of objective information is maximized if various kinds of feedback charts are devised. Detailed records over time of performance in training and competition against stated goals will reveal the level of progress and the extent to which goals need modification to remain challenging. Players should be encouraged to keep a daily log relating to performance. Feedback charts can also generate healthy competition, leading to further improvements in performance.
5. Goal-setting is an on-going activity involving a constant evaluation of goals and setting of new goals. For example, progress during a season could be different from that expected at the outset, and goals need to be modified accordingly in order to avoid complacency on the one hand and despondency on the other. Goals for matches need to be personally challenging but not so much that they become demotivating.
6. Goals are most likely to be effective if they are accepted by the player rather than merely imposed by the coach. Ideally goals for a particular player should emerge as a result of dialogue with the coach so that the player feels involved. With this approach, it is also likely that the goals will be specifically tailored for the individual.

17.5.4 Intervention programmes

Having outlined general information regarding relaxation procedures and cognitive strategies for performance enhancement, it is appropriate to focus on more individualized and packaged intervention programmes, which in many cases derive from the foundation techniques described earlier.

Visual-Motor Behaviour Rehearsal (VMBR) was developed by Suinn (1972). A desensitization technique, it consists of relaxing the player by using progressive relaxation, training the player to use relevant imagery, and using imagery to simulate participation in a stressful situation. The emphasis is on the use of internal kinaesthetic imagery, but with performance errors visualized through external imagery, i.e. the visualization of oneself from outside one's body. Research suggests that VMBR is effective in reducing state and trait anxiety (Seabourne *et al.*, 1982).

Stress Inoculation/Management Training (Meichenbaum, 1977) is a desensitization technique designed to increase the player's ability to cope with stress. It does so by exposing the player progressively, through imagery, films, role-playing and real-life situations, to greater and greater stress.

17.5.5 Psyching-up strategies

Relaxation procedures and many of the cognitive intervention strategies are designed to help the player *reduce* arousal to an optimum level. Arousal reduction is a common requirement, not because players are particularly anxiety prone but because competitive sport creates negative stress for the participants. There are, however, occasions in some sports and with some players when arousal needs to be *increased* to an optimum level, thus requiring psyching-up strategies. There are various circumstances in which psyching-up strategies would be employed.

- The player may be known to be a slow-starter in a match.
- The player may be excessively 'laid-back' by nature, which could be a problem both in competition and during rehabilitation.
- When a player is feeling a little jaded, psyching-up may be appropriate. However, if there are signs of burnout, withdrawing the player from competition should be considered.
- The player or team may be over-confident and consequently complacent.
- The player or the team may have never been beaten by the particular opposition.
- The team might be safe from relegation and feel that 'there's nothing to play for'.
- Psyching-up is more likely to be appropriate in sports involving strength, explosive power, endurance and speed (Weinberg and Jackson, 1985). It is therefore a particularly appropriate technique for soccer players.

The dramatic psyching-up strategies of weight-lifters can readily be appreciated, but soccer players generally demand more measured and controlled effort. Consequently, it is difficult to imagine a weight-lifter who is too psyched-up, but easier to imagine this problem in soccer. The following psyching-up strategies are suggested.

1. **Goal-setting**. The value of goal-setting for preventing complacency against unrated opponents is emphasized, e.g. setting personal performance goals for the game.

2. **Team talks**. Used with discretion and acknowledging the varying needs of individuals, team talks can be effective. A combination of a reasoned, calm, analytical approach, reminding the players of their individual responsibilities and, when necessary, a more emotive approach is recommended. Former manager of the Northern Ireland soccer team Billy Bingham had the experience to recognize that team talks are not always for psyching-up. After the draw with England at Wembley in 1985 which enabled them to qualify for the World Cup in Mexico, Bingham commented, 'Before the match I deliberately kept the team talk low key because I sensed my players were on edge. Like me, they were terribly nervous.'
3. **Fan support**. Typically beneficial to the player, this is one reason why teams generally have better home than away records. Also important is social support in the form of positive peer modelling (e.g. established players acting as mentors of junior players).
4. **Self-activation**. Examples of positive self-statements are: 'You are going to play well', 'You are a good player', 'Don't give in' and 'Keep the pressure on'.
5. **Pre-competition work-out**. Husak and Hemingway (1986) found that a pre-competition work-out reduced feelings of tension and anxiety. More generally, pre-competition preparation is important: players should ensure that they are physically well prepared and rested, that their kit is in order, that they know about the opponent(s), and that they arrive at the venue in plenty of time.
6. **Identification**. Gaining inspiration and motivation by identifying with an admired player or an admired philosophy can be helpful.
7. **Videos**. Watching videos of excellent performances prior to competition can also be motivational, particularly if the player is able to identify with the style of performance.
8. **Audio-tapes**. Many competitors use audio-tapes as a means of psyching-up, with material taking on a variety of forms, e.g. exhortations from a coach, inspirational music, positive self-statements.
9. **Positive mental imagery**. The player imagines playing really well, recreating images from successful matches and mentally rehearsing strategies in a variety of match situations. With practice, the imagery will become more vivid and, providing the player concentrates on creating positive mental images, it could be very beneficial just prior to, or even during, a match. It should not, however, be seen as a substitute for physical practice.

17.5.6 Attentional focus training

Good concentration, which implies full attention being given to the task, is necessary for success in soccer. Losing players will often attribute their defeats to a lack or loss of concentration. For others, concentrating well is second nature, even to the extent of having 'peak experiences'. Effortless concentration is what every dedicated player seeks to achieve.

Orlick and Partington (1988), after surveying the mental preparation strategies of a sample of Olympic athletes, commented:

> The best players had taken the time to discover what kind of focus worked best for them in competition. They had developed a refined plan to draw upon this focus in the competition. In almost all cases the best focus was one that kept the player connected to what he or she was doing (his or her job). In contrast, the worst focus was one in which the player was dwelling on factors over which he or she had no direct control, such as other competitors, final outcome, or other distractions.

Most players, for a variety of reasons, experience problems with concentration. They appreciate neither how important good concentration is as a part of general competition strategy nor how much they can control it.

Nideffer (1986) identified two independent dimensions of concentrating: **width** and **direction**. Width varies from broad to narrow and direction from internal (focusing on internal cognitions) to external (focusing on the external environment). This produces four kinds of attentional focus.

- **Broad-internal**: concentrating on a broad range of cognitions – a player rehearsing various strategies for the game.
- **Narrow-internal**: concentrating on a narrow range of internal cues/cognitions – a soccer player mentally rehearsing the personal role in a set-piece play.
- **Broad-external**: focusing on a large number of external stimuli – a soccer player scanning the field to pick out a team-mate for a pass.
- **Narrow-external**: focusing narrowly on an external stimulus – a soccer player taking a penalty kick.

The player will need to adopt each of the four types of attentional focus depending on the requirements at any time. Because individuals vary in their attentional style, they will vary in their ability to be appropriately flexible in their attentional focus, e.g. some might have a predisposition to be broadly focused, finding it extremely difficult to adopt a narrow focus when it is needed. A further complicating factor is that stress/arousal leads to attentional narrowing. This means, for example, that the midfield player in soccer who often needs a broad-external focus may develop tunnel vision during a stressful game.

17.5.7 Thought stopping and centring

These are techniques which the player can use to improve attention control. When the player is experiencing stress, it is common for negative thoughts to prevail – self-doubt, defeatism, feelings of injustice – and concentration to be disrupted.

Thought stopping involves learning to recognize when you are dwelling on negative thoughts and losing control ('We're going to lose, there's no point in trying') and *stopping* such thoughts. Centring entails the induction of a relaxed state with the aid of deep breathing and slow exhalation.

- Stand comfortably, consciously relaxing neck, arm and shoulder muscles.
- Take a slow deep breath, noticing that you are extending your stomach.
- Consciously maintain the relaxation in your chest and shoulders.
- Exhale slowly, feeling yourself getting heavier as your muscles relax.

Thought stopping and centring can be used in real-life situations, but can be practised in contrived circumstances. Displaced negative thought should be immediately replaced with positive thoughts and task oriented ideas.

17.5.8 Refocusing

The player, whether competing or recovering from injury, is subject to distractions that can prevent the achievement of the goal. There are dozens of situations in competition which present a natural threat to concentration, e.g. making an unforced error, poor officiating, overly aggressive opponents and the various reactions to the ebb and flow of a match. What often happens is that the player focuses on the error, on the bad refereeing decision, when he should refocus on the task. In soccer, teams are often distracted by their own success, losing concentration as a consequence of scoring a goal. It is remarkable how often the opposing team takes advantage of this collective lapse in concentration by scoring soon afterwards.

Refocusing involves recognizing that the distraction is occurring, concentrating on the current objectives, ignoring what cannot be changed, concentrating on the *process* rather than the outcome (Jackson and Roberts, 1992) and concentrating on what happens *next*.

Players often adopt a preparatory routine or ritual which helps to settle them down prior to action, e.g. the goalkeeper bouncing the ball. Routines like these can be employed by the player as a means of facilitating refocusing.

17.5.9 Self-defeating thoughts

The stress experienced by a player can elicit a variety of self-defeating and irrational thoughts which must be challenged if performance is to be maximized or injury is to be overcome. The training staff have an important role to play in helping the player quickly to identify and reject faulty, unhelpful and counterproductive thoughts. Examples of such thoughts include:

- belief that the chances of winning are so remote that there is no point in trying;
- belief that a career is over, perhaps because an apparently similar injury ended someone else's career;

- negative reactions to the therapy, perhaps because of unrealistic expectations of immediate improvements.

17.6 SOCIAL PSYCHOLOGY

We have learned that there is an important psychological dimension to being successful in soccer. Top-class players have psychological as well as physical and physiological characteristics which contribute to their success. The players' personality determines how dedicated and how psychologically robust they are in the pursuit of excellence. Much stress has been placed on the uniqueness of the individual player. It is important, however, to remember that the soccer player operates within a social context and is much influenced in his or her behaviour by others, whether they be managers, coaches, doctors, team-mates, opponents or fans. The interaction between the player and these others will have significant consequences concerning, for example, the coach–player relationship, aggressive conduct, team cohesion and team performance.

17.6.1 Aggression in soccer

Aggression can be defined operationally in terms of its primary motives.

- Hostile retaliation or **reactive aggression** where the primary motive is to inflict physical harm on another person. It involves anger and is often impulsive. By definition, it involves some form of real or imagined provocation.
- **Instrumental aggression**, where the intention is to injure or risk injuring another person but with the primary motive of, for example, winning a contest. If the primary motive is achieved, the instrumental aggression is reinforced and is likely to occur again. With instrumental aggression, the intention is not necessarily to injure someone in pursuit of victory but to risk injuring an opponent. Thus in competing for a high ball the flailing elbow may 'create space' for the defender and provide an advantage, but can also inflict serious injury on an opponent. Similarly, the 'professional foul' may prevent a certain goal, but may also cause injury.

Theories of aggression

Some researchers (e.g. Freud) have maintained that aggression is an innate biological drive, that individuals need to find expression for it and that it is best controlled by positive 'outlets' being provided. Sport is one such positive, socially acceptable and cathartic outlet. Although intuitively appealing, there is not much evidence in support of this argument.

The **Frustration–Aggression Hypothesis** (Dollard *et al.*, 1939) states that aggression is a consequence of frustration. But others have noted that frustration does not always lead to aggression and aggression is not always preceded by frustration. Berkowitz's reformulation (Berkowitz, 1969) maintains that frustration increases the predisposition to aggression, that the individual can *learn* not to respond to frustration with aggression and that, although the individual may experience transient satisfaction upon completion of an aggressive act (cf. instinct theory), such satisfaction (reinforcement) makes aggression more likely in the future.

The **social learning** theory of aggression (e.g. Bandura, 1973) maintains that aggression is not instinctual but learned. For example, if a soccer player believes that his instrumental aggression enhances his team's chances of success, the aggression will be positively reinforced (learned) and will recur. In effect, the player conducts a cost–benefit analysis of aggressive behaviour. If the benefits (rewards) outweigh the costs (sanctions) then aggression will be positively reinforced.

It is possible that the aggressive player is using a Freudian defence mechanism known as **projection**: the aggressor, feeling guilty at hating his opponent who represents a threat to goal achievement, projects these feelings ('he hates me'), thereby legitimizing his aggressive actions. This psycho-dynamic interpretation appears to underlie the maxim, 'get your retaliation in first'.

17.6.2 Team cohesion

The term 'cohesion' is used by social psychologists to describe the extent to which groups become united in a common purpose. It has become a major focus of interest in sports psychology because of the wish to understand why a team is successful or otherwise. Is success a consequence of the sum of the individual talents? Clearly not necessarily so, as evidenced by the failure of many expensively assembled and talented professional soccer teams. Is success then anything to do with how well the players get on with each other? Perhaps, although there are examples of successful teams whose members are not in harmony. Although many coaches and managers attempt to buy success in the transfer market, others, perhaps through necessity, focus on 'added value', i.e. establishing a level of team spirit and cohesion which produces a standard of performance better than that which would be predicted from the abilities of the individual team members. Understanding the dynamics of team cohesion should facilitate the achievement of added value.

Carron (1988), a leading researcher into team cohesion, defined it as 'a dynamic process which is reflected in the tendency for a group to stick together and remain united in the pursuit of its goals and objectives'.

Social cohesion refers to the extent to which team members are attracted to each other and enjoy each other's company. **Task cohesion** refers to the degree to which team members work together to achieve a common objective.

The distinction between social and task cohesion is a useful one in that they can operate independently (Lenk, 1969). Thus the apparent contradiction of a discordant team (low social cohesion) doing well (high task cohesion) can be explained.

Measuring team cohesion

Sociometry is a technique developed for the measurement of attraction and repulsion amongst group members. Each team member is asked the same question concerning how he or she feels about other team members. On the basis of the responses, a *sociogram* is created to illustrate the attractions. It will be appreciated that this long-established technique has a great deal of flexibility, depending on the nature of the question asked. For example:

- With which member of the team do you enjoy the greatest friendship?
- Name the three people in the team with whom you are most friendly, in order of preference.
- Who is the best player/who are the three best players in the team?
- Who is the best *team* player?
- In a pressure situation, who would you have most/least confidence in?

Sociograms can be used not only for charting popularity, but also for the illustration of task cohesion, e.g. how often each player passes to various other team members. This information on task interactions can then be compared with standard sociometric data (Cratty, 1989).

In the German Democratic Republic, where sociometric analysis of team structure was popular, a more sophisticated procedure was developed. The Group Analysis Procedure (GAP) allows all members of the team to be included in the assessment. All members are asked to make an assessment of their relationship with every other team member. The typical question is: 'How willingly would you (e.g. train) with each of your team-mates?' Answers are given on a seven-point scale ranging from 'very willingly' to 'very unwillingly'. Results provide a detailed picture of the individual's standing within the group. Former GDR sports psychologist Hans Schellenberger has used the GAP extensively and considered it a 'highly valuable complement to the social-psychological methods for assessing social relationships within the team' (Schellenberger, 1990).

Determinants of team cohesion

According to Cartwright and Zander (1968), the degree of team cohesion is determined by cooperation, team stability, homogeneity and size. Cooperation is particularly important in groups where there is significant interaction amongst members, as in soccer (Johnson *et al.*, 1981). Stability has been measured in terms of the **half-life** of a team, i.e. the time it takes for a particular team to reach the point where only half its original members remain. Research suggests that there is an optimal half-life of five years (Donnelly *et al.*, 1978). Regarding homogeneity, Eitzen (1975) has noted:

> The more alike the members of a group, the more positive the bond among the members. This is usually explained by the assumption that internal differentiation on some salient characteristic such as religion, race and socioeconomic status leads to greater likelihood of clique formation. (p. 41)

For obvious reasons, team cohesion tends to be greater in smaller groups (Widmeyer *et al.*, 1988). The phenomenon of **social loafing** – a decrease in individual effort within a team context – is more prevalent in large groups, possibly as a consequence of a dilution of responsibility for the outcome and a feeling that it is possible to 'hide in a crowd' (C.T. Hardy, 1990). No research has been undertaken in soccer but the phenomenon of social loafing will be well recognized by coaches.

Carron (1988) has listed the following additional antecedents of cohesiveness.

- Contractual responsibilities: professional and amateur soccer teams are very different in this respect. Social cohesion for the amateur team could be sufficient in itself.
- Normative pressures: when an individual joins a team, there are pressures to support the team ethos. Opting out or selfish behaviour is likely to be strongly discouraged.
- Physical and functional proximity of team members: friendships tend to develop as a consequence of close interaction between team players.
- The way the leader behaves has an important effect on cohesiveness: the way he or she communicates, clarifies roles, provides feedback, and so on.
- Long-term team success: there is more evidence to suggest that success leads to cohesion than that cohesion leads to successful performance.
- Shared negative experience: there are situations where setbacks or threats draw a team closer together. There are many examples of soccer teams reduced to ten men playing outstandingly well. In the 1994 FA Cup quarter-final between Manchester United and Charlton Athletic, United's goalkeeper Schmeichel was sent off shortly before half-time – at that time, the score was 0–0. United scored three goals in the second half with ten men. Manager Alex Ferguson commented:

  > The determination in the dressing room at half-time was brilliant. They were spitting in the tunnel to get back out on to the pitch. If you've got players with personal motivation it makes your job a lot easier. They said to themselves, this is a challenge to us, the chips are down. And they were brilliant.

- Group permeability: the more self-contained the group the more it is reliant on its own resources and the more cohesive it tends to be.

Consequences of team cohesion

Carron observed that 'cohesive groups work harder toward the achievement of the group's goal; they are more successful' (Carron, 1988, p. 170). Cohesive teams are more stable – fewer drop-outs, less absenteeism, greater punctuality (Carron, 1988) and greater tolerance of adversity (Brawley *et al.*, 1988). They also experience more satisfaction (Williams and Hacker, 1982). However, cohesiveness can distort the perceptions of a team by making them look inwards, overvaluing themselves and undervaluing outsiders. This can cause difficulties for new recruits to the team.

Developing team cohesion

The following general principles for facilitating cohesion are suggested.

- Clear role definition for players, along with the recognition of the importance of the role. Getting substitutes or substituted players to appreciate the importance of their roles is a particular challenge to coaches.
- Each player should have a clear understanding of the roles of other team members and the important contribution they make to the team.
- The coach or other mentor (e.g. team doctor) should get to know each player and demonstrate that they care about the individual.
- Each player should feel a sense of being involved, and this can be achieved by providing them with the opportunity to contribute to decision-making and generally air their views.
- Watch out for cliques, which can form, according to Cox (1990), as a result of (1) constant losing, (2) players' needs not being met, (3) players not being given enough opportunities to play and (4) the effects of the coach's favouritism and scapegoating.
- Be positive, even in defeat.

17.6.3 Crowd effects

There can be no doubt that supporters play a major part in professional soccer. Their vociferous support at matches adds to the atmosphere and would seem to be a major contributory factor to the phenomenon of the home advantage. Here the evidence is reviewed concerning both the positive and negative effects on performance of the audience, and the way in which the audience contributes to home advantage.

Social facilitation

The suggestion that the mere presence of others facilitates performance is intuitively appealing, but research in recent years has revealed a more complex set of relationships. For the soccer player, 'presence of others' means not only spectators but also co-actors, i.e. team-mates and opponents. Most research in this area has focused on Zajonc's theory that the sheer presence of an audience increases the arousal of the individual who, as a consequence, performs better on well-learned tasks but displays a performance decrement on poorly learned tasks (Zajonc, 1965). In effect, this theory suggests that skilled players will have their performance facilitated by an audience, whereas less skilled players will tend to perform poorly.

However, the player is not faced with the 'mere presence' of others. As Cox (1994) pointed out, audiences in sport tend not to be merely present but to make their presence felt by their noise, their applause, their disapproval and their hostility. And team-mates and opponents interact with the player. This is certainly true in soccer. This makes most of the 'effects of an audience' research inapplicable to sport in general and soccer in particular.

Home advantage

The existence of a home advantage in a wide variety of sports is well established (e.g. Schwartz and Barsky, 1977; Agnew and Carron 1994). Using archival data, Schwartz and Barsky found that in baseball, the home team won 53% of the games, in American football, 60%, and in basketball and hockey, 64%. In soccer, Pollard (1986) found that the home team gained about 64% of all the points in the English Football League and that this pattern has been consistent over a number of decades.

Whereas there is no doubt that home advantage does exist, there has been considerable debate as to why it exists. The research evidence regarding the hypothesized causes is summarized below, using Pollard's list of hypothesized causative factors.

1. **Crowd support**. Despite the widespread belief that crowd support is a crucial factor in home advantage, the evidence is equivocal. On the one hand, Schwartz and Barsky (1977) found that in Major League baseball, the home advantage varied as a function of the crowd density: the greater the density, the better the home advantage. But in soccer, Pollard's (1986) exhaustive research failed to find such an effect, leading him to conclude that, 'If local crowd support does contribute to home advantage then the effect appears to be independent of its intensity and to operate at low levels of support' (p. 245). Mizruchi's (1985) findings in NBA basketball that home advantage is greatest for teams with a strong tradition, with intense local pride and with a city centre location would seem worthy of testing within the context of soccer.
2. **Travel fatigue**. This is based on the assumption that the visiting team will have experienced disruption of routine and fatigue as the consequence of travelling to the away game. The evidence for this effect is largely anecdotal, leading Pollard (1986) to conclude that if there is an effect it is likely to be small. If the effect is small, it would be very difficult to isolate.

3. **Familiarity with conditions**. Here the assumption is that the home team is advantaged by familiarity with the local conditions, particularly the playing surface. However, Pollard reports research which fails to demonstrate support for this assumption. Teams with pitches and/or stadia that are atypical in size did not significantly differ in home advantage from others. Moreover, Queens Park Rangers demonstrated an average home advantage at a time when the team was playing home matches on an artificial surface. Pollard suggested that 'familiarity' may be more subtle in its effects in that the home player, for example, will have a more exact and instantaneous awareness of the location of himself and other players in a fast-moving game. He concluded that 'familiarity remains a plausible but unsubstantiated contributing factor to home advantage and one in need of further research' (p. 246).
4. **Referee bias.** There is some evidence of at least subconscious bias amongst 'home-based' umpires in cricket (Sumner and Mobley, 1981), and it is widely assumed, for example, that soccer referees are more reluctant *per se* to give penalty decisions against the home team. Nevill *et al.* (1995) provided evidence the visiting team in English and Scottish Football League matches are penalized more than the home team and that the imbalance increases the greater the size of the crowd.
5. **Special tactics.** Typically, home teams play an attacking role whereas away teams tend to be more defensive, and it may be that this is generally of benefit to the home team. Pollard (1986) speculated that there may be a connection between the particularly marked home team advantage in European Cup competition and the fact that teams on the initial away leg are often ultra-defensive, even regarding a 1–0 defeat as a success.
6. **Psychological factors**. The factors listed above are all 'psychological' to some extent. Under this heading, Pollard was thinking in particular of the 'vicious circle effect' of the belief amongst the participants that there is a home advantage. Such an effect may be present, thus contributing to home team confidence (not to mention away team anxiety), but proving the existence of such an effect would be problematic.

In summary, various factors have been hypothesized as contributing to home advantage in sport, but much of the evidence concerning their effects is equivocal. It is likely that the factors interact in complex ways to produce home advantage, which suggests that multiple regression research designs should be employed when investigating the phenomenon. Pace & Carron (1992) used a multiple regression design in investigating the effects of various travel factors on home advantage in the National Hockey League and found that travel factors accounted for less than 2% of the variance in game outcome. Multiple regression analysis of a comprehensive range of potential contributory factors is needed.

17.6.4 Leadership in sport

When we examine who has managed very successful soccer teams in recent years in the English League, we are faced with leaders as diverse in personal qualities as Matt Busby and Brian Clough, Bill Shankly and Bob Paisley and Kenny Dalglish and Alex Ferguson. What qualities do they have in common which mark them out as exceptional leaders? Cox (1994) mentioned the 'unquenchable desire to succeed, to excel and to win' as characteristic of great American coaches and team captains (p. 321). Vince Lombardi's view that 'winning isn't everything, it's the only thing!' has its British counterpart in Bill Shankly's 'football isn't a matter of life and death, it's much more important than that!'.

Leadership involves more than single-mindedness. Stogdill (1950) stated that 'leadership is the process of *influencing* the activities of an organized group in its efforts toward goal-setting and goal achievement'. Carron (1980) has characterized leaders as operating an *influence system* or a *power system* of leadership, with the latter tending to isolate themselves from the players.

The distinction is made between those leaders who emerge from the group as a consequence of their competence and assertiveness (*emergent leaders*), and those who occupy formal leadership roles (*appointed or prescribed* leaders). In soccer, leaders such as team captains, player–managers and managers are most often emergent, and research suggests that they are often the most effective leaders (Cratty and Sage, 1964).

Carron (1988) suggested that every prescribed leader in every profession has two common responsibilities: first, to ensure that the team is effective in meeting the goals of the organization, and secondly, to ensure that the legitimate needs and aspirations of the team members are met. If these two responsibilities are fulfilled, then the leadership is effective.

Leadership style

Fiedler (1974) argued that leadership effectiveness depends on the leader's style of interacting and on the 'situation favourableness'. Leadership style varies along a continuum from relationship-motivated (person-oriented) to task-motivated (task-oriented). A person-oriented leader effectively says, 'If we get on well together, then we'll do the job', whereas the task-oriented leader says, 'If we get the job done successfully, then we'll feel good about one another'.

Fiedler argues that whether the situation is favourable or not depends on three factors: leader–member relations, task structure and the power position of the leader, with leader–member relations being most important. The truth of this latter point is evidenced by manager Brian Clough's failure at Leeds United in the 1970s. Although he had unquestionable leadership qualities, as evidenced previously at Derby County and subsequently at Nottingham Forest, his leadership style was not appropriate for the well-established and mature Leeds team. His appointment lasted 43 days.

Using the multidimensional model of leadership, Chelladurai and Carron (1978) argued that player satisfaction and performance are a function of the interaction of leader behaviours prescribed by the situation, leader behaviours preferred by the particular group of players and actual leader behaviour. Team performance and satisfaction will be maximized when there is a congruence amongst the three kinds of behaviour, i.e. when the actual leader behaviour is preferred by the players and matches the needs of the situation.

Summary

In discussing the ways in which psychology relates to soccer, the concepts of personality, motivation, causal attribution, stress and anxiety, performance preparation and social psychology have been examined. A portrait emerges of the soccer player (or a manager) as a unique individual with a unique view of the world, motivated by diverse extrinsic and intrinsic factors, stimulated or threatened by stress and operating within a social context which may facilitate or inhibit his or her performance. The key point is that a greater understanding by players, managers and coaches of the psychological factors underlying performance will help facilitate the realization of potential for all those involved.

REFERENCES

Agnew, G. and Carron, A.V. (1994) Crowd effects and the home advantage. *International Journal of Sport Psychology*, **25,** 53–62.

Bandura, A. (1973) *Aggression: a Social Learning Analysis*, Prentice-Hall, Englewood Cliffs, NJ.

Bandura, A. (1977) Self-efficacy: toward a unifying theory of behavioral change. *Psychological Review*, **84,** 191–215.

Berkowitz, L. (1969) *Roots of Aggression*, Atherton Press, New York.

Bird, A.M. and Brame, J.M. (1978) Self versus team attributions: a test of the 'I'm OK but the team's so-so' phenomenon. *Research Quarterly*, **49,** 260–8.

Bramwell, S.T., Masuda, M., Wagner, N.N. and Holmes, T.H. (1975) Psychosocial factors in athletic injuries: development and application of the social and athletic readjustment rating scale (SARRS). *Journal of Human Stress*, **1,** 6–20.

Brawley, L.R., Carron, A.V. and Widmeyer, W.N. (1988) Exploring the relationship between cohesion and group resistance to disruption. *Journal of Sport and Exercise Psychology*, **10,** 199–213.

Brown, J.D. and Siegel, J.M. (1988) Exercise as a buffer of life stress; a prospective study of adolescent health. *Health Psychology*, **7,** 341–53.

Burton, D. and Vealey, R. (1989) *Competitive Anxiety*, Human Kinetics, Champaign, IL.

Carron, A.V. (1980) *Social Psychology of Sport*, Mouvement Publications, Ithaca, NY.

Carron, A.V. (1988) *Group Dynamics in Sport*, Spodym Publishers, London/Ontario.

Cartwright, D. and Zander, A. (eds) (1968) *Group Dynamics: Research and Theory*, 3rd edn, Harper & Row, New York.

Chelladurai, P. and Carron, A.V. (1978) *Leadership*. Canadian Association for Health, Physical Education and Recreation, Sociology of Sport Monograph Series, CAHPER, Ottawa.

Cox, R.H. (1990) *Sport Psychology: Concepts and Applications*, W.C. Brown, Dubuque, IA.

Cox, R.H. (1994) *Sport Psychology: Concepts and Applications*, rev. edn, Brown & Benchmark, Dubuque, IA.

Cratty, B.J. (1989) *Psychology in Contemporary Sport*, Prentice-Hall, Englewood Cliffs, NJ.

Cratty, B.J. and Sage, J.N. (1964) The effects of primary and secondary group interaction upon improvement in a complex movement task. *Research Quarterly*, **35,** 164–75.

Dale, J. and Weinberg, R.S. (1989) The relationship between coaches' leadership style and burnout. *The Sports Psychologist*, **3,** 1–23.

Deci, E.L. (1971) Effects of externally mediated rewards on intrinsic motivation. *Journal of Personality and Social Psychology*, **18,** 105–15.

Dishman, R.K. (1986) Exercise compliance: a new view for public health. *Physician and Sportsmedicine*, **14(5),** 127–45.

Dollard, J., Miller, N., Doob, L. *et al.* (1939) *Frustration and Aggression*, Yale University Press, New Haven, CT.

Donnelly, P., Carron, A.V. and Chelladurai, P. (1978) *Group Cohesion and Sport*. Canadian Association of Health, Physical Education Recreation, Ottawa.

Dweck, C.S. (1975) The role of expectations and attributions in the alleviation of learned helplessness. *Journal of Personality and Social Psychology*, **31,** 674–85.

Dweck, C.S. (1986) Motivational processes affecting learning. *American Psychologist*, **41,** 1040–8.

Eitzen, D.S. (1975) Group structure and group performance, in *Psychology of Sport and Motor Behavior* (eds D.M. Landers, D.V. Harris and R.W. Christina), College of HPER, Pennsylvania State University, University Park, PA.

Eysenck, H.J. (1967) *The Biological Basis of Personality*, C.C. Thomas, Springfield, IL.

Feltz, D.L. and Landers, D.M. (1983) The effects of mental practice on motor skill learning and performance. *Journal of Sport Psychology*, **5,** 25–57.

Fender, L.K. (1989) Athlete burnout: potential for research and intervention strategies. *The Sports Psychologist*, **3,** 63–71.

Fiedler, F. (1974) The contingency model – new directions for leadership utilization. *Journal of Contemporary Business*, **4,** 65–79.

Gill, D.L. (1980) Success–failure attributions in competitive groups: an exception to egocentrism. *Journal of Sport Psychology*, **2,** 106–14.

Hardy, C.T. (1990) Social loafing: motivational losses in collective performance. *International Journal of Sport Psychology*, **21,** 305–27.

Hardy, L. (1990) A catastrophe model of performance in sport, in *Stress and Performance in Sport* (eds J.G. Jones and L. Hardy), Wiley, London.

Hardy, L. and Parfitt, C.G. (1991) A catastrophe model of anxiety and performance. *British Journal of Psychology*, **62,** 163–78.

Harris, D.V. (1980) On the brink of catastrophe, in *Psychology in Sports: Methods and Applications* (ed. R.M. Suinn), Burgess Publishing Company, Minneapolis.

Heider, F. (1958) *The Psychology of Interpersonal Relations*, Wiley, New York.

Hickman, J.L. (1979) How to elicit supernormal capabilities in athletes, in *The Coach, Athlete and the Sports Psychologist* (eds P. Klavora and J.V. Daniel), Human Kinetics, Champaign, IL.

Holmes, T.S. and Holmes, T.H. (1970) Short-term intrusions into the lifestyle routine. *Journal of Psychosomatic Research*, **14,** 121–32.

Husak, W.S. and Hemingway, D.P. (1986) The influence of competition day practice on the activation and performance of collegiate swimmers. *Journal of Sport Behavior*, **9,** 95–100.

Iso-Ahola, S.E. (1977) Immediate attributional effects of success and failure in the field: testing some laboratory hypotheses. *European Journal of Social Psychology*, **7,** 275–96.

Jackson, S.A. and Roberts, G.C. (1992) Positive performance states of players: towards a conceptual understanding of peak performance. *The Sport Psychologist*, **6,** 156–71.

Johnson, D.W., Maruyama, G., Johnson, R.T. *et al.* (1981) Effects of cooperative, competitive and individualistic goal structures on achievement: a meta-analysis. *Psychological Bulletin*, **89,** 47–52.

Jones, G. (1991) Stress and anxiety, in *Sport Psychology: a Self-Help Guide* (ed. S.J. Bull), The Crowood Press, Marlborough.

Kenyon, G.S. (1968) Six scales for assessing attitude toward physical activity. *Research Quarterly*, **39,** 566–74.

Kobasa, S.C. (1979) Stressful life events, personality and health: an inquiry into hardiness. *Journal of Personality and Social Psychology*, **37,** 1–11.

Lenk, H. (1969) Top performance despite internal conflict, in *Sport, Culture and Society* (eds J.W. Loy and G.S. Kenyon), Macmillan, New York.

Levitt, E.E. (1971) *The Psychology of Anxiety*, Paladin, London.

Little, J.C. (1969) The athlete's neurosis – deprivation crisis. *Acta Psychiatrica Scandinavica*, **45,** 187–97.

Locke, E.A. and Latham, G.P. (1985) The application of goal-setting to sports. *Journal of Sport Psychology*, **7,** 205–22.

Mahoney, M.J. and Avener, M. (1977) Psychology of the elite player: an exploratory study. *Cognitive Therapy and Research*, **1,** 135–41.

Mahoney, M.J. and Meyers, A.W. (1989) Anxiety and athletic performance: traditional and cognitive-developmental perspectives, in *Anxiety in Sports: An International Perspective* (eds D. Hackfort and C.D. Spielberger), Hemisphere, New York.

Martens, R. (1982) *Sport Competition Anxiety Test*, Human Kinetics, Champaign, IL.

Mathes, S.A. and Battista, R. (1985) College men's and women's motives for participation in physical activity. *Perceptual and Motor Skills*, **69,** 719–26.

McClelland, D.C., Atkinson, J.W., Clark, R.W. and Lowell, E.L. (1953) *The Achievement Motive*, Appleton-Century-Crofts, New York.

Meichenbaum, D. (1977) *Cognitive Behavior Modification*, Plenum Press, New York.

Mizruchi, M.S. (1985) Local sports teams and celebration of community: a comparative analysis of the home advantage. *Sociological Quarterly*, **26,** 507–18.

Nevill, A.M., Newell, S. and Gale, S. (1995) Can the crowd influence the result of soccer matches? *Journal of Sports Sciences*, **13,** 69.

Nideffer, R.M. (1986) Concentration and attention control training, in *Applied Sports Psychology* (ed. J.M. Williams), Mayfield Publishing, Palo Alto, CA.

Nideffer, R.M. (1989) Anxiety, attention and performance in sports: theoretical and practical considerations, in *Anxiety in Sports: an International Perspective* (eds D. Hackfort and C.D. Spielberger), Hemisphere, New York, pp. 117–86.
Orlick, T. and Partington, J. (1988) Mental links to excellence. *The Sports Psychologist*, **2,** 105–30.
Pace, A.D. and Carron A.V. (1992) Travel and the home advantage in the National Hockey League. *Canadian Journal of Sports Sciences*, **51,** 60–4.
Passer, M.W. and Seese, M.D. (1983) Life stress and athletic injury: examination of positive versus negative events and three moderator variables. *Journal of Human Stress*, **9,** 11–16.
Pollard, R. (1986) Home advantage in soccer: a retrospective analysis. *Journal of Sports Sciences*, **4,** 237–48.
Reiss, M. and Taylor, J. (1984) Ego-involvement and attributions for success and failure in a field setting. *Personality and Social Psychology Bulletin*, **10,** 536–43.
Rotter, J.B. (1966) Generalized expectancies for internal versus external control of reinforcement. *Psychological Monographs*, **80(1),** 1–28.
Ryan, E.D. and Simons, J. (1981) Cognitive demands, imagery, and frequency of mental rehearsal as factors influencing the acquisition of motor skills. *Journal of Sport Psychology*, **3,** 35–45.
Sanderson, F.H. (1989) Analysis of anxiety levels in sport, in *Anxiety in Sports: an International Perspective* (eds D. Hackfort and C.D. Spielberger), Hemisphere, New York.
Sanderson, F.H. and Gilchrist, J. (1982) Anxiety and attributional responses of competitive squash players. *British Journal of Sports Medicine*, **16,** 115.
Sanderson, F.H. and Kneale, S.K. (1988) Attributional responses of 4-a-side soccer players, in *Science and Football*, (eds T. Reilly, A. Lees, K. Davids and W. Murphy), E. & F.N. Spon, London, pp. 545–51.
Scanlan, T.K. and Passer, M.W. (1980) Self-serving biases in the competitive sport setting: an attributional dilemma. *Journal of Sport Psychology*, **2,** 124–36.
Schellenberger, H. (1990) *Psychology of Team Sport*, Sport Books, Ontario.
Schurr, K.T., Ashley, M.A. and Joy, K.L. (1977) A multivariate analysis of male athlete characteristics: sport type and success. *Multivariate Experimental Clinical Research*, **3,** 53–68.
Schwartz, B. and Barsky, S.F. (1977) The home advantage. *Social Forces*, **55,** 641–61.
Seabourne, T.G., Weinberg, R.S. and Jackson, A. (1982) Effect of visuo-motor behavior rehearsal in enhancing karate performance. Unpublished manuscript, North Texas State University, Denton, TX.
Selye, H. (1975) *Stress Without Distress*, New American Library, New York.
Smith, D. (1987) Conditions that facilitate the development of sport imagery training. *The Sports Psychologist*, **1,** 237–47.
Spielberger, C.D. (1989) Stress and anxiety in sports, in *Anxiety in Sports: an International Perspective* (eds D. Hackfort and C.D. Spielberger), Hemisphere, New York.
Stogdill, R.M. (1950) Leadership, membership and organisation. *Psychological Bulletin*, **47,** 1–14.
Suinn, R. (1972) Removing emotional obstacles to learning and performance by visuo-motor behavior rehearsal. *Behavioral Therapy*, **31,** 308–10.
Sumner, J. and Mobley, M. (1981) Are cricket umpires biased? *New Scientist*, **91,** 29–31.

Vealey, R. (1986) Conceptualization of sport-confidence and competitive orientation: preliminary investigation and instrument development. *Journal of Sport Psychology*, **8,** 222–46.

Watson, D. (1967) Relationship between locus of control and anxiety. *Journal of Personality and Social Psychology*, **6,** 91–3.

Weinberg, R.S. and Jackson, A. (1985) The effects of specific vs non-specific mental preparation strategies on strength and endurance performance. *International Journal of Sport Psychology*, **8,** 175–80.

Weiner, B. (1972) *Theories of Motivation: from Mechanism to Cognition*, Markham, Chicago.

Widmeyer, W.N., Brawley, L.R. and Carron, A.V. (1988) How many should I carry on my team? Consequences of group size. *Psychology of Motor Behavior and Sport: Abstracts*, North American Society for the Psychology of Sport and Physical Activity, Knoxville, TN.

Williams, J.M. and Hacker, C.M. (1982) Causal relationships among cohesion, satisfaction and performance in women's intercollegiate field hockey teams. *Journal of Sport Psychology*, **4,** 324–37.

Wolf, T.M., Elston, R.C. and Kissling, G.E. (1988) Relationship of hassles, uplifts and life events to psychological well-being of freshmen medical students. *Behavioral Medicine*, Spring, 37–45.

Woolfolk, R.L., Murphy, S.M., Gottesfeld, D. and Aitken, D. (1985) Effects of mental rehearsal of task motor activity and mental depiction of task outcome on motor skill performance. *Journal of Sport Psychology*, **7,** 191–7.

Yerkes, R.M. and Dodson, J.D. (1908) The relationship of strength of stimulus to rapidity of habit formation. *Journal of Comparative Neurology and Psychology*, **18,** 459–82.

Zajonc, R.B. (1965) Social facilitation. *Science*, **149,** 269–74.

Zuckerman, M. (1979) *Sensation Seeking: Beyond the Optimal Arousal*, Erlbaum, Hillsdale, NJ.

Football and society: the case of soccer hooliganism

18

John Minten

Introduction

The other chapters in this book have addressed the application of a wide range of scientific disciplines to soccer. As with sport in general, football operates within the wider context of the society in which it is located; the development of the differing codes of football found throughout the world reflects the interaction between the early forms of football and the host culture. In this chapter an attempt is made to illustrate aspects of this relationship by focusing on a specific code of football, namely soccer, and the particular issue of soccer hooliganism in Great Britain. There are a number of interesting issues in other codes of football, and within soccer itself, that are deserving of serious study. Examples are the history and development of women's soccer (Williams and Woodhouse, 1991) and the wider relationships between soccer and its fans (Taylor, 1992).

Throughout the past three decades there has been much research and academic work published on soccer hooliganism, as well as many government reports and much media coverage. It is hoped that introducing the main issues and theoretical concepts will enable students to read more deeply into the topic.

Initially, a brief overview of the history and development of soccer in Britain will be covered as this is crucial in understanding more recent

Science and Soccer. Edited by Thomas Reilly. Published in 1996 by E & FN Spon, London. ISBN 0 419 18880 0.

issues. The nature of hooliganism will then be considered, and how concerns have changed over the past three decades. At this stage the major theoretical, predominantly sociological, explanations of soccer hooliganism will be discussed. Finally government reports and attempts to solve soccer hooliganism will be discussed, focusing on the Taylor Report (1990) into the Hillsborough disaster, which already has had a major impact on the game in Britain.

18.1 THE ORIGINS OF SOCCER IN GREAT BRITAIN

The complex social changes that accompanied the industrial revolution in Great Britain in the eighteenth and nineteenth centuries were responsible for the development of sport, and specifically soccer, as it is known today. Many forms of 'football' existed before this time, based around holidays and festivals, some of which have survived to the present day. An example is the Shrovetide game in Ashbourne, Derbyshire. These games were very different from modern soccer, with no set numbers of participants, playing area or rules of conduct. In fact there were different rules for each different locality in which it was played (Mason, 1980). The decline in the 'rural sports' in the latter part of the eighteenth century was due to many factors but resulted in a reduction in opportunities in terms of both time and space for active recreation. It was not until the later half of the nineteenth century that these folk sports were reconstructed under different social conditions. Again there were many factors involved in this reconstruction, not least the influence of the public schools, the specific details of which are well documented elsewhere (for example, Bailey, 1978; Mason, 1980).

The governing body of soccer, the Football Association (FA), was founded in 1863 and based the rules of the game on those agreed by graduates at Cambridge University in an attempt to standardize the various public school games. The game was initially dominated by the upper and middle classes, though following the establishment of the FA Cup in 1871, the social base of those involved in soccer widened. Literally thousands of clubs were formed in the period up to the end of the century with the game being carried with 'missionary zeal' to the masses by ex-public school boys and 'Muscular Christians'. The aim was for the working classes to play the sports to gain the benefits of healthy, 'rational' recreation; even so, spectatorship grew very quickly in this period.

As with rugby football, along with spectatorship and commercialization came pressures for professionalization, particularly from the clubs in the North of England. Unlike rugby which split into the Union and League codes, professionalism in soccer was accepted as legitimate by the FA in 1885 in order to avoid a similar division. The growth of soccer as a professional, spectator sport continued with the formation of the Football League in England in 1888 to control the professional game, and the similar leagues in Scotland in 1889 and Ireland in 1890. By the 1913–14 season the average attendance in the First Division of the English League was 23 100.

It would be wrong to assume that the composition of the early crowds was exclusively working class and male. There is evidence that initially at least the crowds had significant proportions of middle class, female and, particularly, skilled working class spectators (Walvin, 1975). Pressures in the later part of the century were to change this to fit more with the current perception of predominantly male, lower working class soccer crowds. Dunning *et al.* (1988) suggested that:

> the mixing of classes in non-work situations became increasingly problematic . . . At the same time, the frequent reports of crowd disorderliness are likely to have had a deterrent effect on the attendance, not only of the higher social classes but probably of women as well.

It can be seen therefore that concerns regarding crowd behaviour were raised from the early days of the game of soccer. This, and developments in the twentieth century, will be revisited later in the chapter and when considering the work of Dunning and his co-workers at the Norman Chester Centre for Football Research at Leicester University.

18.2 THE NATURE AND DEVELOPMENT OF SOCCER HOOLIGANISM

There is no catch-all offence of 'soccer hooliganism' in British law, although the Popplewell Report (1986) suggested that this should be considered, and the Taylor Report (1990) proposed the separate offences of throwing a missile, chanting obscene or racist abuse, and going on the pitch without reasonable excuse. In popular terms soccer hooliganism consists of not only the swearing and unruly behaviour which often occurs in other sporting situations, but also the more serious pitch invasions and fighting in and around soccer grounds or in transit to games. It is usually assumed that this is carried out by lower working class males. As will be discussed in more detail later, forms of hooligan behaviour have been present at games of soccer since its very early days. In fact in the 'folk' games of football this behaviour often came from the participants, though at times there was little distinction between the spectators and the players.

As would be expected over a period of more than a century, the type of hooligan behaviour has changed. In the pre-war period it often took the form of attacks on officials, attacks on players and vandalism. Dunning *et al.* (1988) claimed that all modern manifestations of hooliganism were to be found occurring in this period, including fights between rival fans, and trouble at away matches. Many people assume that soccer hooliganism began in the 1960s. Though this is clearly not true, the perception may be due, in part, to the fact that at this time television began to show both the games and the hooligan acts for the first time. This increased the exposure of hooliganism and shaped public attitudes towards it, though at this time there was, in fact, an increase in the prevalence and a change in the nature of the hooliganism.

In the past three decades soccer hooliganism has been at the top of the list of problems for soccer and sport, and has come to be perceived as a major social problem. Throughout this time expression of the phenomenon has changed, not least because of the attempts to solve the problem. In the 1960s the hooliganism was largely confined within the soccer grounds themselves, with concerns centring on pitch invasions, fights between fans and the 'taking of ends'. As the strategies of the police became more sophisticated and grounds became segregated, the hooliganism was much more likely to occur in the surrounding neighbourhoods, and in transit to games. Some academics would suggest that the attempts to solve the problem had, in fact, made it more difficult to contain and caused more inconvenience for the surrounding populace.

Against this background of major domestic problems there was a growing concern regarding the behaviour of British soccer fans abroad. This initially was focused on English club sides with trouble reported in a variety of European cities in 1974, 1977, 1980 and 1981 (Williams *et al.*, 1984). This reached its nadir in May 1985 at the Heysel Stadium in Brussels when 39, mostly Italian, fans were killed under a collapsed wall following disturbance by Liverpool fans. This occurred against a background of poor segregation and poor stadium safety. The disaster placed hooliganism high on the political agenda of a British government elected on a 'law and order' ticket. There was little protest then when as a result of the Heysel incident English clubs were banned from European competitions indefinitely. Also at this time more legislation was proposed to try to combat the problem, including compulsory club membership schemes, although this was never implemented.

There was also concern in the 1980s about the rise of so-called super-hooligan groups. These were much more sophisticated in their organization, such as the 'Inter City Firm' of West Ham United and the 'Headhunters' of Chelsea. This led to the formation of specialist police sections and undercover operations, which had limited success in combating these groups. It appeared at this time that there was a rise in middle class hooligans, wearing smart designer clothes. The vast majority of arrests, though, were still of working class youths, the clothes more of a statement of 'hardness' than an indication of class position.

After the ban by the Union of European Football Associations (UEFA) on English clubs competing in Europe, attention was focused on the behaviour of English fans supporting their national team abroad. Incidents were intensely covered by the media at the European Championships in Germany in 1988 and Sweden in 1992, and at the World Cup in Italy in 1990. There is no doubt that other countries in Europe had equivalent, or at times worse, problems of soccer hooliganism. Despite this English fans were *perceived*, at least, to be the worst in Europe and hooliganism became known as the 'English Disease'.

18.3 THEORETICAL EXPLANATIONS OF SOCCER HOOLIGANISM

The kinds of explanations that are put forward by many politicians, those involved in football and the ordinary citizen do not stand up to close examination as deep causes of soccer hooliganism. The most common of these are alcohol, violent acts by the players and the effect of unemployment and the 'permissive society'. Not all hooligans drink, there is no close correlation between incidents on the field and hooligan acts and there have been periods of high levels of trouble when there were neither high levels of unemployment nor a 'permissive society'. These may very well be related factors which may initiate hooligan acts or which may make incidents worse; none, however, is the cause of the problem.

There has been much written in the academic literature, and many explanations of soccer hooliganism have been put forward over the past two decades. In this section the three most adequate theories that have been put forward are discussed, those of Taylor (1971a, 1971b, 1982) and Clarke (1978), Marsh *et al.* (1978) and Dunning *et al.* (1988). Students who have no background in the social sciences may have difficulty in reconciling the fact that differing theories may co-exist to explain the same phenomenon. In the physical and life sciences theories are usually rejected and superseded by new, more accurate explanations of phenomena. However, the social sciences deal with the actions and interactions of humans as a subject matter, which can be very complex: they have consciousness, feelings and may often act irrationally. Therefore these human actions and interactions are much more open to different interpretations than in other areas of study.

18.3.1 Marxist theories of soccer hooliganism

The theories of Ian Taylor (1971a, 1971b, 1982) and John Clarke (1978) are usually classified together, and referred to as either Marxist or sub-culturalist theories. This is because they both explained the phenomenon as resulting from the relationships between soccer, modern capitalist society, the working classes and the existence of a 'soccer sub-culture'.

These authors argued that soccer in the past fitted in with the traditional working class culture; it matched their patterns of leisure and was expressive of the working class values of masculinity and active and collective participation. Taylor suggested that there existed a 'participatory democracy': fans thought that they exerted at least some control over the club and the players, though he recognized that this may only ever have been illusory.

Since the Second World War three things have undermined this close fit between soccer and working class culture. First, the game has become 'professionalized' and 'bourgeoisified'. Management became increasingly important, with the adoption of more complex playing tactics and the emergence of highly paid players and inflated transfer fees. Players became distant from the fans with a 'jet setting' lifestyle. There was also more emphasis on international competition, both for clubs and the national team. Secondly, Clarke (1978) referred to the 'spectacularization' of the game, with an attempt to appeal to the upper working class and middle classes, to go upmarket. Thirdly, the fan started to be treated as a consumer, rather than a supporter. Taylor (1971a) summed this up when he suggested that:

> We are presented with a soccer that is dominated by contractual relationships between the club and the player, and between the player and the supporter, a soccer in which the clubs are increasingly concerned to provide a passive form of spectacle, and a soccer dominated by financial rather than sub-cultural relationships.

The football hooligans from the lower working class therefore were 'doubly alienated', both from society due to their social position, and from soccer because of the changes in the game. The fans no longer believed that their opinion could influence the clubs or that the clubs represented their local interests. The hooligans from the soccer sub-culture were therefore attempting to reclaim *symbolically* the class values that are specific to soccer. The violent actions were an attempt to recover the forms of masculinity lost in the modern players, and through their collective, participatory actions to challenge the middle class values of the passive consumer. The pitch invasions were an attempt to reclaim control of the game, which the hooligans at least thought that they had previously. Clarke (1978) also suggested that changes in the social situation of working class youth in the 1960s meant that effective controls over young people at matches were removed, such as the influence of older people on the terraces.

The hooligans, therefore, could be described as constituting a working class resistance movement, part of a working class youth sub-culture attempting to reclaim the lost values of the game. This is not to suggest that these were conscious, deliberate attempts, but rather symbolic actions in response to the fact that they possessed little real power to control other aspects of their lives.

Since this theoretical position explains soccer hooliganism in terms of the relationships between the working classes, soccer and society, it is not surprising that any solutions put forward would also be framed in these terms. Taylor (1971a) proposed that:

> It is likely that the conflict in and around soccer grounds will only be fully resolved by changes in the club's relationship to its supporters and also (very importantly) by changes in the situation of the rough working class in wider society.

The latter of these solutions appears to be unlikely; in fact since the election of the Conservative government in Britain in 1979 the poorer sections of British society have become poorer in material terms at least. Attempts have been made by soccer clubs to improve their relationship with fans, though with the new, more commercial era of the 1990s, some clubs may further alienate supporters.

As with all social theory, this approach is subject to criticism, though it must be remembered that it was formulated in the 1970s. It was also one of the first attempts to explain the causes of hooliganism, rather than considering it as random and meaningless, and to explain why it was carried out by certain members of soccer crowds. One central criticism is the lack of an empirical base: for example, there is little hard evidence that this 'participatory democracy' actually existed. It also presumes that hooliganism, at least in a particular form, began in the 1960s, a point which Dunning *et al.* (1988) would strongly reject.

Soccer hooliganism may also be studied adopting a wider Marxist analysis of crime and 'law and order'. For example Hall *et al.* (1978) discussed crime and the concept of 'moral panic', linked to a crisis in capitalist 'hegemony', the term 'hegemony' being used to describe the ways in which dominant groups in society ensure that their interests and values are accepted by other groups in society. The political and media interest in creating this moral panic is to frighten people into adopting a right-wing position, and to accept increasingly coercive measures. Soccer hooliganism is exaggerated by the media and hooligans are portrayed as 'folk devils', manifesting violence that is meaningless and incomprehensible. This supports the imposition of repressive state policies such as increased policing. This identification of 'folk devils' is not restricted to soccer hooligans and various groups in society have been targeted by right-wing politicians to act as scapegoats for social problems or to legitimize harsh 'law and order' rhetoric.

18.3.2 The theory of ritualized aggression

Marsh *et al.* (1978) put forward a theory with social anthropological and social psychological roots to explain soccer hooliganism. It was formulated around the same time as the sub-culturalist theory. The field work on which the theory was based was carried out primarily at Oxford United Football Club and involved interviews, observation, participant observation and psychological testing.

Marsh *et al.* (1978) challenged the 'common sense' view of hooliganism that was reflected in the media, which suggested that it was a widespread, serious problem of violent disorder that often involved innocent bystanders. In contrast they suggested that arrests at soccer matches were in line with any large crowds, that over half of injuries in soccer crowds were the result of accidents and that incidents rarely involved bystanders. They argued against the assumption that hooliganism was meaningless and disorderly violence but was what they termed 'aggro'.

The basis of the theory is that soccer hooliganism is a form of 'ritualized aggression'. In non-human species animals often fight in a ritualistic manner, for example over mates, to avoid serious injury or death, the aim being to establish dominance. Marsh *et al.* (1978) argued that, although aggression in humans is biologically based, its nature and form are culturally controlled:

> We suggest that aggro is one contemporary means available to man for coping with aggression. In other words, the function of aggro is precisely that of its analogues described by animal ethologists.

They proposed that hooliganism is a form of 'aggro' in a number of ways, centred around the suggestion that there is a social structure, and therefore order, present on the terraces. They provided evidence of group hierarchies and careers on the terraces, with what they termed 'novices' (the young pre-teenagers), 'rowdies' (older boys involved in mock fights and noise) and 'tomboys' (youths that took part in the more serious trouble). They also suggested that there were roles held by members of the hierarchy: for example, the 'chant leader' who would initiate and lead the singing on the terraces, the 'organizer' of the activities surrounding the match, and the 'nutter' who failed to conform to the norms of the hooligan group and was considered crazy even by the hooligans themselves.

The researchers also found evidence of ritualistic behaviour containing symbols and rules of behaviour. The wearing of emblems, to distinguish between groups, and the roles that the person held provide examples. They found recognized patterns of behaviour regarding match days, ritual insults and the denial of the opposition's masculinity and, importantly, rituals and rules regarding fighting. These determined who the opponents were, the way in which the fights started and the way in which the fight was conducted: on most occasions the real violence would be limited by the 'victors' if it was causing too serious harm to the opposition.

Marsh *et al.* (1978) argued that because this 'aggro' was functional there is a contradiction in trying to suppress it, and if no other outlet is provided real violence may ensue. They suggested that the contemporary attempts to solve the problem were in fact making the problem worse. They proposed that:

> If we accept that there are, from one significant standpoint at least, *rules* of disorder, we might be able to develop management strategies which have far more purpose and effect than those which have currently emerged from the atmosphere of moral outrage and collective hysteria.

The research at Oxford was based on empirical work, though there are criticisms of the theoretical stance. The concept of aggro may underestimate the violence that occurs at matches, perhaps reflective of the fact that the researchers failed to study events outside the soccer ground itself. The forms of hooliganism that occur at soccer that do not conform to the concept of aggro may also have been overlooked. It is hard to see how missile throwing may be considered as a controlled, symbolic act of violence, and acts of aggro may easily escalate into serious violence. There is also a key problem for sociologists in accepting the biological determinism inherent in this ritualized aggression theory.

18.3.3 The historical/developmental theory

Dunning and his co-workers developed their theory in the 1980s, although they had begun to address soccer hooliganism before this time. The work was laid out in the comprehensive text *The Roots of Football Hooliganism: an Historical and Sociological Study* (1988), although much other research has also been published from the Norman Chester Centre for Football Research.

The work is based on a historical analysis of the development of football and associated acts of disorder. The researchers provided strong evidence that soccer hooliganism is not a new phenomenon and that the kinds of actions that are now labelled as such were present from the first matches. Therefore explanations that rely solely on factors in recent history are liable to be flawed. They also disproved the myth that soccer hooliganism is only an English problem, citing much evidence of disturbances in a variety of European and South American countries. What has, in fact, happened is that hooliganism has gone through cycles, with many examples before the First World War, less between the wars, and a particular escalation since the 1960s. What accounts for this variation is the involvement in soccer of the lower working classes.

The theory has two basic strands to the argument: the relationship between sport and violence, and the involvement of the lower, or 'rough' working classes. It is argued that sport, as with the rest of society, has gone through a 'civilizing process', and examples of this can be seen in the way in which many forms of violence have been outlawed from sport, and in the development of referees and governing bodies to enforce tight sets of rules. Recently two problems have undermined this civilizing process in sport. First is the growing seriousness of sport, with massive increases in prizes and rewards and the greater involvement of non-participants such as coaches, managers and agents. Secondly, sport has become much more important in our culture. Sport has come to fulfil much of the role previously performed by religion; it has become an enclave for the expression of outmoded forms of masculine identity and a source of group identity for many.

This civilizing process in sport has therefore been restricted by its seriousness and cultural centrality, especially for the lower working class. There are aspects of the attributes of this sector of society which account for this, particularly what Dunning *et al.* (1988) referred to as their 'violent masculine style'. In the 1960s there was a split, or polarization, of the working classes. The upper sections of the working classes, with a growing affluence, aspired to the norms and values of the middle class, particularly with respect to acts of violence. In contrast, for the lower working class violence was still part of their lives and a means of solving problems and gaining status.

Soccer hooliganism is, therefore, the ritualistic and real violence that comes from 'rough' working class masculinity. The lower working class began to perceive soccer games as somewhere that violence could be expressed, which was reinforced by the media, starting an 'amplification spiral'. The more the media highlighted the problem the more it became defined to the hooligans as a place to express this behaviour. Particularly around the time of the World Cup in England in 1966, concern emerged about the behaviour of fans at matches, with the 'respectable' supporters either not attending or disapproving of the activity of the fans. Part of the problem though was the attitude of society towards the actions of fans, with previously tolerated activities now labelled as hooligan. Old films of matches in the 1920s and 1930s often show men running onto the pitch and celebrating goals, actions which at one time were deemed to be 'high jinks'. Today these supporters would be ejected from the ground and labelled as hooligans.

Dunning *et al.* (1988) summed up their approach when they stated:

> We have argued that an adequate understanding of football hooliganism requires not only a short-term analysis of developments since the Second World War but, more crucially, a longer term, developmental account, firstly of the manner in and degree to which heavily masculine values have been produced and reproduced in the working class over time and, secondly, of the varying extent to which the football context has formed an arena for the expression of such values.

In common with other academic explanations of hooliganism, Dunning *et al.* (1988) were critical of the attempts to solve the problem by the government and football authorities, suggesting that they were at best ineffectual, and at worst added to the problem. The more sophisticated the strategies by the police the greater the organization of the hooligans, locked into what they called a process of mutual reinforcement. The key to reducing the problem as perceived by the researchers at Leicester is to soften this violent masculine style at soccer matches, perhaps by the greater involvement of women in all aspects of the game.

Study of soccer hooliganism had been dominated in the past decade by the workers at the University of Leicester. This has to some degree restricted alternative theoretical exploration and criticism, though some limitations of the theory are apparent. The theory is based only on English fans and neglects cultural factors such as religion which underlies violence in, for example,

Scotland. There is also doubt cast on the reliability of official statistics used by Dunning *et al.* (1988), and the researchers' ability to explain adequately the more recent phenomenon of the football 'firms' and their particular forms of confrontation.

18.4 GOVERNMENT REPORTS AND ATTEMPTS TO ERADICATE HOOLIGANISM

Each of the theories discussed has been critical of attempts to solve the phenomenon of hooliganism. If the deeprooted causes of the problem were never taken on board by government and the soccer authorities, then this is not really surprising. It is necessary though, since these attempts have changed the face of soccer in Britain, to consider these attempts further and to discover their origins. There is not scope in this chapter to cover in detail the numerous reports over the past three decades: Table 18.1 outlines the major reports into soccer and soccer hooliganism and the main points of interest and recommendations. The Taylor Report (1990) deserves more detailed attention.

Table 18.1 Summary of government reports into soccer and soccer violence

Report	*Date*	*Main points*
Chester	1966	Wide-ranging report on state of the game including soccer hooliganism
Harrington	1968	Recommended many, now common, methods of deterrence, e.g. segregation, all-ticket matches, stricter penalties
Lang	1969	Stronger and more sophisticated policing. Improvements in grounds and tighter operation by clubs
McElhone	1977	Similar recommendations to previous reports with emphasis on restrictions on alcohol
Smith	1978	First real report from a social science perspective. Noted failure of past measures
Teasdale	1984	Better implementation of existing strategies and careful timing of fixtures, introduction of closed circuit TV, membership schemes, better international liaison
Popplewell	1986	Considered Bradford fire and hooliganism. Proposed searches before entry, additional powers for police and specific offences
Taylor	1990	Report into Hillsborough disaster (see main text)

On 15 April 1989, 95 Liverpool fans were crushed to death at the Hillsborough Stadium in Sheffield. Although the incident was not caused by hooliganism, major factors in the tragedy were the attitude and the actions of the police who had been conditioned into treating all fans as potential hooligans, and the erection of perimeter fencing designed to stop pitch invasions. The Taylor Report (1990) into the disaster was viewed by many as a turning point in the attitude of the establishment to football fans. The report was seen as escaping the law and order rhetoric common in the earlier reports, and Ian Taylor (1991) stated that the Hillsborough disaster represented the end point of discourse around soccer solely in terms of hooliganism. The report was seen as displaying a more sensitive understanding of issues relating to soccer, and an attempt to understand hooliganism.

The specific measures proposed were numerous and mostly related to safety. The most relevant to the hooliganism debate were the recommendations for the suspension of the proposed membership schemes and the introduction of all-seater stadia, initially in the top two divisions of the league. The former was generally welcomed as most people in the game had considered such schemes unworkable, though doubt has been cast as to the benefits of all-seater stadia when set against the enormous cost, and whether they would have any effect on hooliganism. There is little doubt that many stadia, mostly built in the early part of the century, needed upgrading with better facilities for fans. This has been one of the motivations for the subsequent sale of television rights for the Premier League to a satellite company which would reach a relatively tiny audience, but who paid large amounts for the privilege.

Summary

Soccer is not the only sport where there is crowd trouble, though it has undoubtedly had the most. The question that remains is why, with a similar class composition and large crowds, have there been relatively few incidents at professional Rugby League matches? Whatever the merits of other theories it appears that the researchers at Leicester University offer the best answer. It is because, unlike Rugby League, soccer has come to be associated with a violent masculine style, a location for young, predominantly working class males to demonstrate behaviour to some degree outmoded in the rest of society. After the Hillsborough disaster, however, it appears that there has been a 'deamplification spiral' with soccer hooliganism becoming unacceptable even to many fans themselves who, although not actually hooligans, were in the past ambivalent in their attitudes toward the troublemakers. For example, pitch invasions after the removal of perimeter fences were condemned by fans, seen almost as disrespectful to those who perished at Hillsborough. In some ways the new attitude of the authorities was seen in the Taylor Report (1990), even though many of the ideas had been put forward many years before by academics.

Recently the situation regarding soccer spectators seemed to have improved with a reduction in arrests and ejections at soccer matches from 7.4 per 10 000 attendances in 1988–9 to 4.2 in 1991–2 (Football Trust, 1993). Attendances, which had declined steadily in the early 1980s, have risen since the low point of the 1985–6 season (see Figure 18.1) and there has developed a range of constructive literature around soccer such as fanzines, and associations with other, non-hooligan, youth subcultures. It remains to be seen how long this situation will last and there have been problems since Hillsborough, specifically with travelling England fans. Is the reduction in the hooliganism problem simply a downturn in a cycle with more serious problems around the corner as influences change, or political focus shifts, or have the major manifestations permanently died out? Time will tell; if there are problems in the future perhaps those with power to influence policy will learn from recent history and adopt sensitive and well-thought-out strategies.

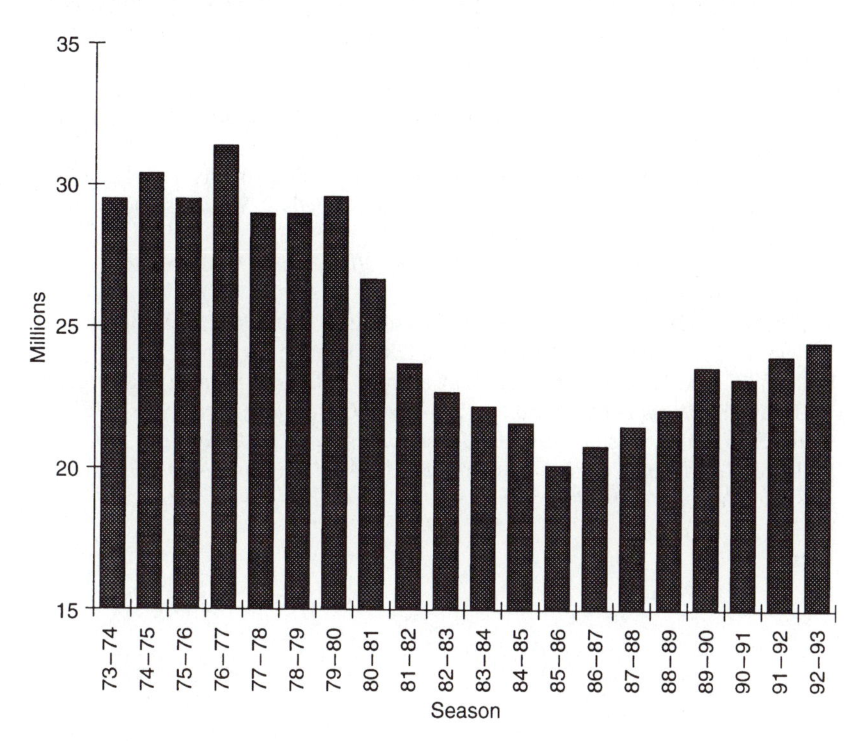

Figure 18.1 Total attendances per session at Football League, League Cup and FA Cup Matches, 1973–4 to 1992–3. (Based on data from the Football Trust.)

REFERENCES

Bailey, P. (1978) *Leisure and Class in Victorian England*, Routledge & Kegan Paul, London.

Clarke, J. (1978) Football and the working class: tradition and change, in *Football Hooliganism: the Wider Context* (ed. R. Ingham), Inter-Action Imprint, London, pp. 37–60.

Dunning, E., Murphy, P. and Williams, J. (1988) *The Roots of Football Hooliganism: an Historical and Sociological Study*, Leicester University Press, Leicester.

Football Trust (1993) *Digest of Football Statistics 1991–1992*, Sir Norman Chester Centre for Football Research, Leicester.

Hall, S., Critcher, C., Jefferson, T. and Roberts, B. (1978) *Policing the Crisis. Mugging, The State, and Law and Order*, Macmillan, London.

Marsh, P., Rosser, E. and Harre, R. (1978) *The Rules of Disorder*, Routledge & Kegan Paul, London.

Mason, T. (1980) *Association Football and English Society*, Harvester, London.

Popplewell, O. (1986) *Crowd Safety and Control at Sports Grounds: Final Report*, HMSO, London.

Taylor, I. (1971a) Soccer consciousness and soccer hooliganism, in *Images of Deviance* (ed. S. Cohen), Penguin, Harmondsworth, pp. 134–64.

Taylor, I. (1971b) Football mad: a speculative sociology of football hooliganism, in *Sociology of Sport: a Collection of Readings* (ed. E. Dunning), Frank Cass, London, pp. 352–77.

Taylor, I. (1982) The soccer violence question. Football hooliganism revisited, in *Sport Culture and Ideology* (ed. J. Hargreaves), Routledge & Kegan Paul, London, pp. 152–96.

Taylor, I. (1991) English football in the 1990s: taking Hillsborough seriously? in *British Football and Social Change: Getting into Europe* (eds J. Williams and S. Wagg), Leicester University Press, Leicester, pp. 3–24.

Taylor, The Rt Hon. Lord Justice (1990) *The Hillsborough Stadium Disaster: Final Report*, HMSO, London.

Taylor, R. (1992) *Football and its Fans*, Leicester University Press, Leicester.

Walvin, J. (1975) *The People's Game*, Allen Lane, London.

Williams, J. and Woodhouse, J. (1991) Can play, will play? Women and football in Britain, in *British Football and Social Change: Getting into Europe* (eds J. Williams and S. Wagg), Leicester University Press, Leicester, pp. 85–108.

Williams, J., Murphy, P. and Dunning, E. (1984) *Hooligans Abroad*, Routledge & Kegan Paul, London.

Football violence: an interdisciplinary perspective

19

Benny Josef Peiser

While there is now considerable public awareness of the existence and extent of football violence, we doubt whether the public is aware of the catastrophic result from such criminal acts. The Board frequently deals with cases of people scarred for life, sometimes with cases of people seriously and permanently maimed and occasionally with people who are killed. We welcome the efforts which the courts, the police and many sporting organisations are making to *attempt* to lessen the number of such crimes.

Criminal Injuries Compensation Board, 1980

Introduction

Football in its varying forms is, without doubt, the most popular sport in the world. Whilst rugby or American football are almost exclusively played in western and Commonwealth countries, soccer is by far the biggest global sport. In 1993, the Féderation Internationale de Football Associations (FIFA) represented 179 national soccer associations which in total represent around 200 million active members (both male and female). According to FIFA, some 150 million are active players. Altogether, 1.3 million referees officiate weekly at soccer matches for some 600 000 soccer clubs which represent 4.1 million soccer teams world-

Science and Soccer. Edited by Thomas Reilly. Published in 1996 by E & FN Spon, London. ISBN 0 419 18880 0.

wide. Together with an estimated 200 to 300 million additional soccer players who are not organized in clubs or associations affiliated to FIFA, in particular pupils and students involved in school sport, there are an estimated 400 to 500 million active soccer players worldwide. This impressive number indicates that some 10% of the world's population are – in one way or another – participating in soccer.

In December 1863 some 20 British representatives of 11 football clubs, or teams, founded the Football Association (FA). From the onset, association football (soccer) succeeded in conquering and transforming the world and its cultures. Association football is the only sport which attracts almost the entire world population. For example, the 1994 World Cup final was watched by well over 2 billion people around the globe. It has become a multi-trillion-US$ business, thereby transforming a nineteenth-century elite sport into the single biggest leisure industry of the world. The universal impact of football means that the ever increasing problem of football-related violence is of major social, political and scientific concern.

19.1 THE HISTORICAL PERSPECTIVE

The origins of modern football are rooted in pagan blood rituals. Ancient ball games such as the Persian *buzkashi*, the Mesoamerican *peloya*, or the Roman *harpastum* were extremely violent and cruel contests which involved killings, blood sacrifices and serious injuries. When, during the fourth century AD, Christianity became the official state religion of the Roman Empire, all pagan rituals, in particular Greek and Roman athletic and gladiatorial contests, were radically suppressed. The majority of these traditional games were associated with pagan and sacrificial cults which Christianity tried to wipe out. The violent nature also contradicted the very basis of Judeo-Christian ethics. Jews and Christians were uncompromising in their attitude toward pagan spectacles. They attacked the very nature of these cruel combat sports, finding them incompatible with the idea of the holiness of life. The belief that God made man according to His image and likeness contradicted the participation in physical activities which involved the infliction of pain and injuries upon other living beings. Despite continuous efforts of the church, kings and magistrates throughout the last 1500 years of history to ban these games, neither the various types of mob football nor ancient blood sports could be entirely erased. The pagan pastimes proved to be so popular that even the most powerful authorities finally gave up their battle against unlawful sports.

In particular, the repeated attempts by English lawmakers during the Middle Ages to outlaw football once and for all failed. Between the fourteenth and seventeenth century, football was banned on more than 30 occasions in

Table 19.1 Selected list of prohibitions of football by the state and local governments

1314	Edward II	1477	Edward IV
1331	Edward III	1478	(London)
1349	Edward III	1488	(Leicester)
1365	Edward III	1491	James III of Scotland
1388	Richard II	1496	Henry VII
1401	Henry IV	1570	(Peebles)
1409	Henry IV	1572	(London)
1410	Henry IV	1581	(London)
1414	Henry V	1594	(Shrewbury)
1424	James I of Scotland	1608	(Manchester)
1450	(Halifax)	1609	(Manchester)
1454	(Halifax)	1615	(London)
1457	James II of Scotland	1655	(Manchester)
1467	(Leicester)	1666	(Manchester)
1471	James III of Scotland	1667	(Manchester)
1474	Edward IV		

Source: Dunning and Sheard (1979, p. 23).

England. However, England's most popular folk sport was always revived (cf. Table 19.1).

In most of these cases, football was outlawed because of its inherent violence. Medieval football was a wild and brutal game played according to oral rules which allowed a high level of tolerated physical violence. During the nineteenth century, British pedagogues finally abandoned their struggle against the brutal sport. Since they could not deny its popularity, they now joined in. As violence and rebellion in England's public schools rose steadily, school masters discovered that football could be functionalized according to their own interests as a safety valve and means of controlling extremely violent pupils. Instead of banning the sport outright, new regulations were introduced in order to give it a more organized and less brutal character. These developments finally constituted the basis of modern football.

The codification and reformation of the rules of the game during the nineteenth century led to a reduction of the most blatant brutality of traditional football. Yet, assaults and grievous harm were still legitimate as long as physical violence occurred within the rules of the game. Despite attempts of nineteenth century reformers to 'civilize' football and its rules, there has been a steady increase of football violence ever since. Given the unacceptable levels of football violence, it is doubtful whether nineteenth century football rules are still compatible with the demands of a civilized and law-abiding society. The objective and legal criticism of today's football codes, together with a call for a reform of out-moded regulations, constitute a central component of the science of football.

19.2 WHAT IS FOOTBALL VIOLENCE?

Violence can be defined as any form of behaviour which inflicts pain, harm or injury on another living being or oneself, thereby violently disturbing the homeostasis of the victim. Participants in all codes of football normally accept a certain degree of painful and injurious attacks on their bodies as legitimate. Bodily contact is acknowledged in the laws of the game. All significant body contacts that occur in tackles, collisions and so on are deemed legal according to the rules of the game. Physical violence in football is frequently the result of reckless or intentional behaviour. Because of a positive relationship with victory and success, 'tough play' is demanded by fans, coaches and managers. Each football player learns from an early age how to use the various parts of the body in order to tackle, to block and to foul. Thus, both minor and extreme forms of violence are still inherent in today's football and resulting assaults and injuries are regarded as normal by-products of the game.

There is little doubt that a barbarizing process has occurred during this century. Two world wars, more than ten genocides and uncountable regional and civil wars have documented only too dramatically that civilizations and highly developed cultures can collapse and regress into barbaric states. In recent decades, almost all Western democracies have seen an increase of crime and social violence (Figure 19.1). Against this background, the number of violent incidents in sports, in particular in football, seem to have equally mushroomed. The modern codes of football, considered by many sociologists as outstanding examples of an alleged 'civilizing process', are simultaneously accused of regressing into new forms of blood sports (Atyeo, 1979). Has the so-called civilizing process, perhaps, merely suppressed violent collective rituals and pastimes without wiping them out altogether? Will they probably come to the surface again in critical times?

Violence in football is of utmost interest in the social and sports sciences. The growing violence in modern sports and its social consequences are not only major fields of academic interest. Football-related violence has been the focus of political, sociological, moral and judicial debate for many years. Contact sports, particularly American football, rugby and soccer, have all tolerated dangerous behaviour by players contrary to the spirit and letter of the rules. Despite the fact that research in the field of football violence has quickened its pace in the course of the past decade, very little has been done to combat the violence itself. Since 1980, the US Congress has made several attempts to outlaw player violence. Recently, the British Law Commission has recommended that players who commit serious injuries, intentionally or recklessly, should be brought before the courts (Law Commission, 1994). If their proposals are accepted by government and Parliament, reckless play and intentional injuries could soon become a criminal offence. Yet, without a thorough understanding of the sociological and psychological dynamics of

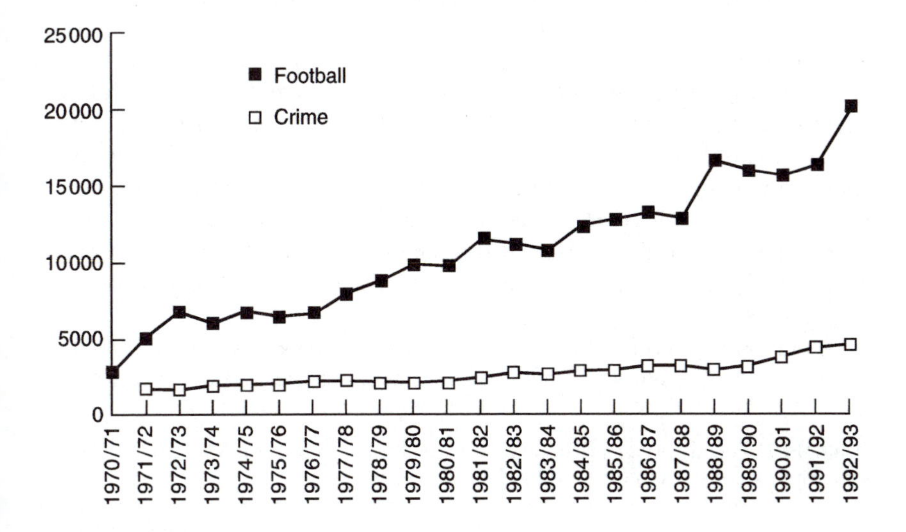

Figure 19.1 Development between 1970 and 1993 of field offences (cautions, sendings off) dealt with by the English Football Association, and crime rate for England and Wales (figures in thousands). (Sources: figures for field offences from *The History of Field Offences for Clubs dealt with by the Football Association from 1970/1 to date*, FA Disciplinary Committee, 1994; crime figures from *Criminal Statistics England and Wales, 1982* and *1992*, HMSO, London.)

football violence, there is unlikely to be an adequate legal solution to the pressing problem.

19.3 THE PRESENT STATE OF FOOTBALL VIOLENCE

Association football is undoubtedly one of the most violent of today's sports. Hooliganism or spectator violence has long been an international phenomenon causing riots and destruction throughout the world. Consequently, research has focused on football-related mass violence (see Chapter 18). However, it is not only spectator violence which is an increasing worldwide problem; according to many observers, brutality, violence and criminal assaults are also increasing on the football field. When, in January 1995, Eric Cantona of Manchester United attacked a spectator, studs first, his assault highlighted this alarming development. Violence in football receives publicity through the spectators and the media. Because various forms of football violence are legitimate or tolerated, they act as reinforcers.

Since the 1970s, there has been a steady increase in scientific research on football violence (Bakker *et al.*, 1993, pp. 81ff). The growing concerns about intolerable levels of football violence demonstrate the relevance of these kind of investigations and the need for significant changes. The past two decades have also brought an apparent upsurge in the level of brutality, violence and death in many areas of contact sports. Since 1945, more than 640 football players have been killed in North America alone as a consequence of injuries (Mueller and Blyth, 1988). Statistics show that there has also been a dramatic increase in cautions and dismissals from play in the FA leagues (cf. Figure 19.1 and Table 19.2). However, these data have not been analysed yet. Consequently, it is still unclear if this significant change reflects a dramatic increase of player violence or, alternatively, if it is the result of new rules and regulations.

As research conducted over the past two decades has shown, the average number of fouls committed in a professional soccer match ranges between 40

Table 19.2 The history of soccer field offences dealt with by the FA compared with the development of the crime rate in England and Wales

	Cautions	*Sendings off*	*Total*	*Notifiable offences recorded by police (× 1000)*
1970/71	2450	246	2696	
1971/72	4673	324	4997	1690
1972/73	6223	496	6719	1658
1973/74	5640	415	6055	1963
1974/75	6397	483	6880	2106
1975/76	6072	474	6546	2136
1976/77	6350	453	6803	2463
1977/78	7620	496	8116	2396
1978/79	8411	572	8984	2377
1979/80	9428	600	10 028	2521
1980/81	9492	658	10 150	2794
1981/82	11 059	723	11 782	3088
1982/83	10 683	904	11 587	3071
1983/84	10 455	682	11 137	3314
1984/85	12 050	779	12 829	3426
1985/86	12 296	900	13 196	3660
1986/87	12 740	971	13 711	3716
1987/88	12 321	1055	13 376	3550
1988/89	16 070	1147	17 217	3706
1989/90	15 625	969	16 594	4364
1990/91	15 194	1047	16 241	5075
1991/92	15 881	1182	17 063	5383
1992/93	**19 785**	**1175**	**20 960**	**5230**

Sources: Cautions and sendings off reported for clubs dealt with by the FA: full members to season 1987/88 and from 1988/89 all English clubs in the 'Six' FA Leagues. Figures of field offences: *The History of Field Offences for Clubs dealt with by the Football Association from 1970/71 to Date*, FA Disciplinary Committee, 1994. Crime figures in: *Criminal Statistics England and Wales 1982* and *1992*, HMSO, London.

and 75, of which only 13% are penalized (Begerau, 1981). Provided that the average number of fouls per match has not significantly changed during the past 25 years, we can observe a significant change in the number of dismissals and cautions.

The average number of illegitimate physical assaults in one professional soccer match is about 30 fouls per game. If we multiply the average foul rate by 16 000 (i.e. the 1993 number of professional FA matches), we can estimate a total number of almost 480 000 incidents per year of actual bodily harm.

19.4 THE MEDICAL PERSPECTIVE

The risks of injury to individuals in general and of football injuries in particular have been neglected for a long time. As a result of violence, indifference or foul play in sports, millions of people each year require medical care after sports accidents or acts of physical violence. At a time when economic crises are jeopardizing efforts to improve the health of mankind, violence-related injuries cost the world community almost 500 billion US$ in medical care, sick pay and lost productivity every year (World Health Organization, 1993). In 1977, the average claim cost for high school soccer injuries in the United States was $127 (Pritchett, 1981). Although this is an outdated figure, it can be calculated that an estimated 250 million football injuries may cost the world community almost $100 billion every year. A substantial reduction of sports and football injuries is consequently of major importance to the World Health Organization.

A football injury can be defined to be the presence of pain, discomfort or disability arising during and as a result of playing. However, many sports physicians do not accept this definition. Accordingly, most researchers only register incidents which lead to serious injuries causing players to miss at least one game or practice session (Nielsen and Yde, 1989). As a result, there has not been any agreement about a general definition of football injuries, resulting in conflicting data on football injury rates (Schmidt-Olsen *et al.*, 1991).

Yet, the level of football violence can best be demonstrated by analysing injury rates. Football at both the professional and amateur level is a relatively dangerous sport resulting in a high rate of injuries. Whilst in soccer about 75% of all players sustain one or more injuries per year, the injury rate in rugby and American football is nearly 100% (Atyeo, 1979).

It is estimated that the risk of sustaining a serious injury in American football is five times higher than in soccer. Between 1945 and 1984, 643 fatal injuries occurred in American football. Minor changes in the rules, which were made in 1976, helped to reduce the number of head and cervical spine fatalities but did not produce a reduction in the number of overall injuries (Mueller and Blyth, 1988). In the USA, it is estimated that over 300 000 American football-related injuries are treated annually in hospital emergency rooms. American football is so physically debilitating that a professional football player's life span is approximately five years shorter than that of the average male (Atyeo, 1979).

Football injuries have become an increasing interest in sports medicine. It is estimated that 50–60% of all sports injuries and up to 10% of all hospital-treated injuries in Europe are due to soccer (Franke, 1980). In soccer, there are fewer injuries inflicted during training sessions than there are during real matches. The more intense and competitive football becomes, the greater the likelihood of foul and injurious play. Although the likelihood of sustaining an injury in soccer is relatively high, it is considerably lower compared to the risk factor in American football and rugby. The vast majority of football injuries result from collisions, direct kicks, tackles, hits by a kicked ball or from falling. On average, 30–40% of traumatic soccer injuries are caused by foul play. In many cases these injuries are self-inflicted because the player committing the foul may sustain the more serious injury.

Most football injuries would be preventable to a large extent if significant rule changes were to be made. Why, in that case, are these changes not enforced? This question has been raised by many medical researchers who increasingly demand preventive measures. The development of a preventive strategy in football has to take into account the compulsory use of protective equipment (Ekstrand *et al.*, 1983) and, most of all, significant changes in the rules of the game. These reforms are necessary not only to reduce the high levels of violence and injurious conduct but, more importantly, to minimize legitimate bodily contacts, which are responsible for the vast majority of serious injuries.

19.5 THE PSYCHOLOGICAL PERSPECTIVE

Does participation and watching football lead to a reduction or to a promotion of aggression and violence? This is one of the main research questions which sport psychologists have been discussing for many years.

The most popular misconception regarding football violence holds that aggression on and off the football field acts as a social and individual safety-valve. According to the so-called catharsis theory of aggression, expressions of violence in football reduce the overall level of social aggression among both players and spectators. Aggression not acted out in football or any other sports will otherwise explode in a much worse form of unsocial or even criminal behaviour (Russell, 1983; Bennett, 1991).

Consequently, participation in football events is widely regarded as a socially adequate means of discharging pent-up hostility and frustration, thereby improving the peaceful and lawful nature of society. Empirical research, however, shows unequivocally that participation in football has quite the opposite effect (Coakley, 1994). Far from reducing the level of aggression, playing or watching football reinforces aggressive impulses and leads to an increased likelihood of violent behaviour on and off the field. Due to its inherent agonal structure and dynamics as a mock battle, football creates its own aggression. Sports psychologists have found that watching the varying forms of football significantly increased the levels of aggression (Russell *et al.* 1988/89; Young and Smith, 1988/89). During the past 20 years, none of the major psychological studies on football and sport violence found any support for the catharsis theory (Bakker *et al.*, 1993; Coakley, 1994; Russell, 1993). Furthermore, since aggression is an essential prerequisite of victory in football, its catharsis is neither intended nor desired.

Psychological studies of the various forms of football have unambiguously shown that aggressive feelings and behaviour are increasing in both players and spectators. The observation of foul play or bodily harm on the field produces a high level of emotional arousal and thus leads to a significant increase in hostility among players and spectators. As a result, many social scientists consider spectator violence a direct response to the violence observed on the football field (Russell, 1993).

According to social learning theory, aggressive behaviour is mainly learned through observing and imitating aggressive conduct of others. Observing football violence often results in imitation of such behaviour (Young and Smith, 1988/89). Football undoubtedly provides legitimate and socially acceptable possibilities for learning aggressive and violent behaviour. Since many football players act as popular role models, their violent conduct is reproduced by other players and observers. Learning to play football, therefore, in most cases involves also imitating aggressive and foul play. This is part of the socialization process and is, in many cases, supported by coaches, parents and peers alike.

On average, 75% of the fouls in soccer matches are committed by defensive players. Yet, few fouls are committed within the penalty area – mostly by attacking players (Bakker *et al.*, 1993). These data underline the fact that particular football rules can significantly reduce violent behaviour, if rule violation is severely punished.

Violence in football is also related to the great insecurity associated with participating in a highly competitive team sport. Football players are often willing to use violent behaviour to 'prove' themselves. Violence can easily become a means for players to prove their worth and establish membership on their teams. Football violence can also become a way to reaffirm manhood. Accordingly, injuries may become symbols of courage if the injured player endures the pain and remains in the game.

Generally, football players and spectators are very concerned with victory. This attitude regularly results in frustration, and frustration often enhances aggression and violence. Foul play and violent conduct are frequently used as a strategy in football. A strategy of tough play is often used to win games. In most cases, this increases the likelihood of injurious conduct. Evidence shows that an ever-increasing level of violence is accepted by football players as a means to gain victory.

According to Coakley (1994), football is socially constructed in ways that not only ritualize aggression but tie the expression of aggression to certain forms of masculinity. Being able to be tough and play violently has become a part of gaining respect 'as a man' within many male groups. Football players often view aggression and violence as legitimate and 'natural'. In fact, many do not recognize injurious acts as violence, even when committed intentionally, as long as they are committed within the rules of the game.

Many boys and men participate in the various forms of football because they 'have learned to define masculinity in terms of being tough enough to participate in the give and take of violent confrontations' (Coakley, 1994, p. 176). And when the give and take of violence leads to injury and pain, they are expected to 'be a man' and stay in the game.

19.6 THE LEGAL PERSPECTIVE

What causes acts of violent play in football and what legal consequences have violent players to face? In the first place, the structure of football as a contact or collision sport makes harmful bodily contacts inevitable. In contrast to non-contact sports, the rules of football legitimize and endorse a whole range of actual bodily harm, so that the border to criminal violence is often unclear.

Law courts must deal with an increasing number of cases of football violence. This is true particularly in North America, where more and more football players sue one another for excessive unlawful bodily contacts. Players may be liable for any intentional or reckless violence that results in injury.

In 1984, the New Zealand Rugby Football Union tried to modify the laws of rugby in order to make the game safer. Significant changes of the rules subsequently led to a dramatic fall in the numbers of cervical spinal cord injuries, which had caused many fatal and near fatal injuries in the past. This example shows that adequate rule changes have positive effects. More importantly, it underlines the obligation of sporting and governmental bodies to adjust the rules according to medical advice and safety requirements. If these necessary adjustments are ignored, sporting bodies risk legal action. In consequence, they could face litigation and huge compensation bills. This was the case in 1987, when an Australian court found the government guilty of negligence and awarded a Rugby League player, who had become tetraplegic, A$2m compensation.

In the eyes of many health and medical advisers, sports administrations should be held legally responsible for damage claims if they fail to change certain rules and practices of football which are known to be extremely hazardous. According to legal experts, sporting bodies could face criminal persecution, if the continuous increase of violent and injurious conduct in football is not adequately challenged and reversed through rule changes and harsh measures of deterrence. Since the sporting bodies ignore calls for rule changes although they are well aware of the imminent dangers of certain injurious conduct in football, they could soon be found negligent in not acting accordingly.

Most dangerous assaults and battery, such as direct hits and kicks against an opponent, are strictly prohibited in soccer. Actual bodily harm and serious injury, however, constitute legitimate bodily contacts as long as they occur within the rules of the game.

According to criminal law, actual bodily harm is defined as any hurt or injury intended to interfere with the health or comfort of the victim. The pain or injury does not have to be serious or permanent but must be more than trifling. Pain or discomfort resulting from bruises or swellings can be sufficient evidence of an assault occasioning actual bodily harm.

Grayson (1990), a sports law specialist, has criticized the general tolerance towards sports violence. In his view, violent breaches of sporting laws and rules of play, condoned or inadequately disciplined by over-tolerant administrations, coaches and referees, have created a misconception that in sport the criminal law stops at the touchline of the boundary.

On average, 30% of football injuries are caused by foul play. Yet, only a quarter of these violence-related injuries are penalized by the referee (Begerau, 1981). In most cases, fouling, fighting and violent acts by football players result in advantages for the offender's team and disadvantages for the team of the victim. Only one out of seven fouls in soccer is penalized by a free kick. Moreover, fouls are not only inadequately penalized, but the free kick in itself is no adequate sanction of violent conduct. The advantages of foul play for the offending team heavily outweigh any benefits to be gained as a result of free kicks. As a result of this inherent deficiency, foul and aggressive play increases the overall chances of success (Begerau, 1981).

Although free kicks are generally an inadequate means of deterring foul play, research has shown that the degree of violent conduct in football significantly decreases after players have been cautioned or sent off by the referee (Begerau, 1981). Effective punishment of unlawful conduct could, therefore, serve as a deterrent, both at the individual and general level. Consistent punishment of violent behaviour dissuades the individual offender from re-offending during a particular match. Furthermore, it would discourage other potential offenders (both on and off the football field) if the examples were to be set by the referee or by additional sanctions laid down by sporting bodies and/or law courts.

Due to the public pressure of huge stadium crowds and fears of unpopular interruptions and the consequent slowing down of the game, existing means of deterrence (i.e. free kicks, penalty kicks, cautions, sendings off, suspensions, etc.) are not used adequately. By consequence, FIFA has introduced stricter regulations for referees which were inaugurated during the World Cup tournament in 1994. As a result, more players than usual were cautioned or dismissed. However, it is still uncertain if this reform will lead to a significant reduction of football violence.

Before FIFA's reform, by comparison, the average number of cautions in soccer was only 1.3 per match, whilst the average number of dismissals was less than 0.1 per match. In view of these diminutive numbers, it comes as no surprise that the traditional penalty system in football has failed to deter unlawful and even criminal conduct. The evidence of approximately 500 000 unlawful incidents each year in professional soccer alone in Great Britain clearly demonstrates that today's rules in football have lost their coercive force. The legitimacy of physical violence on the field has helped to create the general belief that violence constitutes an acceptable means of obtaining an objective.

The real danger lies in the effects of habitual football violence on the respect for the law. If one continues to deter and punish actual bodily harm and physical violence inadequately, the public could lose respect for the law. In conclusion it seems most reasonable to tighten significantly the out-dated penalty system in all codes of football. Evidence, however, suggests that football violence will increase as long as the courts are reluctant to curb actual bodily harm on the football field and the responsible sporting bodies fail to ban the most dangerous practices (Grayson, 1990).

19.7 IS TODAY'S FOOTBALL VIOLENCE IN THE PUBLIC INTEREST?

Since the mid-nineteenth century, football has served to teach and stimulate physical violence in Europe and North America alike. Many anthropologists have suggested that violent sports have been promoted by the authorities for the purpose of training young males in the skills and characters that are necessary in warfare. Contact sports are not only considered to generate strength, courage and competitiveness, but above all they 'may fit people for defence, public as well as personal, in times of need' (Sir Malcolm Foster, quoted in Law Commission, 1994, p. 21).

The Duke of Wellington is believed to have said that 'the Battle of Waterloo was won on the playing fields of Eton'. To this, more recent historians have added that Britain's war successes which followed were equally determined by its sport tradition. Many historians believe that the Anglo-American world has developed efficiency and courage in battle from its experience of competitive sports, the emphasis being on physical courage, instant decision and action. According to this pragmatic view of contact sports, Britain and America's supremacy as military world powers stems in large part from their various football traditions. Wherever political or military circumstances require quick and instinctive action, firm endurance of pain and hardship or coolness in the face of danger, they can rely on their experience on the football pitch. Accordingly, football has been generally approved of as serving the public interest.

More importantly, however, is the fact that football is an inexpensive sport which provides excellent physical exercise for many millions of people. Consequently, despite its inherent violence, football, in particular soccer, contributes greatly to public health and well-being. The participation in football also increases social involvement. Football contains social mechanisms for bringing people together and establishes cohesive and integrated social relationships.

Last but not least, football is a global industry which generates public wealth and creates tens of thousands of jobs. Compared to the wealth and public health generated by the football industry, the direct treatment costs resulting from football injuries are relatively low (Nicholl *et al.*, 1991). In view of these merits, the negative side-effects of football are generally minimized. The public interest in retaining today's structure of football is consequently overwhelming. In contrast, public pressure on sporting bodies to reform the game's regulations in order to reduce violent and injurious conduct in football is still lacking.

Summary

This chapter focuses on some fundamental problems of football violence. In particular, the psychological, medical and legal perspectives of player violence are discussed. In addition, this interdisciplinary presentation provides historical data and questions whether today's football violence is in the public interest.

REFERENCES

Atyeo, D. (1979) *Blood and Guts. Violence in Sports*, Paddington Press, New York.

Bakker, F.C., Whiting, H.T.A. and van der Brug, H. (1993) *Sport Psychology. Concepts and Applications*, John Wiley, New York.

Begerau, R. (1981) Aggressives Verhalten im Bundesliga-Fussball. *Sportwissenschaft*, **11(3),** 318–29.

Bennett, J.C. (1991) The irrationality of the catharsis theory of aggression as justification for educators' support of interscholastic football. *Perceptual and Motor Skills*, **72,** 415–18.

Coakley, J.J. (1994) *Sport in Society. Issues and Controversies*, 5th edn, Mosby, Chicago.

Dunning, E. and Sheard K. (1979). *Barbarians, Gentlemen and Players. A Sociological Study of the Development of Rugby Football.* Martin Robertson, New York.

Ekstrand, J., Gillquist, J. and Liljedahl, S.-O. (1983) Prevention of soccer injuries. *American Journal of Sports Medicine*, **11(3),** 116–20.

Franke, K. (1980) *Traumatologie des Sports*, Georg Thieme Verlag, Stuttgart.

Grayson, E. (1990) Sports medicine and the law, in *Medicine, Sport and the Law* (ed. S.D.W. Payne), Blackwell Scientific Publications, Oxford, pp. 29–76.

Law Commission (1994) *Consent and Offences Against the Person*, Consultation Paper No. 134, HMSO, London.

Mueller, F.O. and Blyth, C.S. (1988) Forty years of head and cervical spine fatalities in American football: 1945–1984, in *Science and Football* (eds T.L. Reilly, A. Lees, K. Davids and W.J. Murphy), E. & F.N. Spon, London, pp. 224–9.

Nicholl, J.P., Coleman, P. and Williams, B.T. (1991) *Injuries in Sport and Exercise. Main Report*, A Report to the Sports Council, Sheffield.

Nielsen, A.B. and Yde, J. (1989) Epidemiology and traumatology of injuries in soccer. *American Journal of Sports Medicine*, **17(6),** 803–7.

Pritchett, J.W. (1981) Cost of high school soccer injuries. *American Journal of Sports Medicine*, **9(1),** 64–6.

Russell, G.W. (1983) Psychological issues in sports aggression, in *Sports Violence* (ed. J.H. Goldstein), Springer, New York, pp. 157–81.

Russell, G.W. (1993) *The Social Psychology of Sport*, Springer, New York.

Russell, G.W., Di Lullo, S.L. and Di Lullo, D.D. (1988/89) Effects of observing competitive and violent versions of a sport. *Current Psychology*, Winter, 312–21.

Schmidt-Olsen, S., Jorgensen, U., Kaalund, S. and Sorensen, J. (1991) Injuries among young soccer players. *American Journal of Sports Medicine*, **19(3),** 273–5.

Young, K. and Smith, M.D. (1988/89) Mass media treatment of violence in sports and its effects. *Current Psychology*, Winter, 298–311.

World Health Organization (1993) *Handle Life with Care. Prevent Violence and Negligence*, WHO, Geneva.

PART FOUR

Match analysis

Notational analysis

20

Mike Hughes

Introduction

A considerable amount of research has been devoted to establishing the need for objective forms of analysis and their importance in the coaching process. There are clearly established difficulties facing any single individual attempting to analyse and remember objectively the events occurring in complex team games, such as soccer. One of the main solutions to these inherent problems has been the use of notational analysis systems. Consciously or unconsciously, coaches, scouts and managers have adopted, designed and developed systems for gathering information for as long as people can remember. Over the past three decades these have been improved by both coaches and sports science researchers, to the point where the design of the systems has become an end in itself. The aim of this chapter is to review not only the data that have been produced, but also assess the major innovations and developments in the systems used for notational analysis in soccer.

20.1 HISTORICAL PERSPECTIVE

General, rudimentary and unsophisticated forms of notation have existed for centuries. For at least five centuries attempts had been made to devise and develop a system for notating movement. Further, the Egyptians, thousands of years ago, made use of hieroglyphs to read dance, and the Romans employed a primitive method of notation for recording salutatory gestures.

Science and Soccer. Edited by Thomas Reilly. Published in 1996 by E & FN Spon, London. ISBN 0 419 18880 0.

The earliest recorded form of music notation was conceived in the eleventh century, although it did not become established as a uniform system until the eighteenth century. Historical texts give substantial evidence pointing to the emergence of a crude form of dance notation much later, in about the fifteenth century. The early attempts at movement notation may well have 'kept step' with the development of dance in society, and as a consequence the early systems were essentially designed to record particular movement patterns as opposed to movement in general.

It is apparent, then, that dance notation actually constituted the 'starting base' for the development of a general movement notation system. Arguably the greatest development in dance notation was the emergence of the system referred to as 'Labanotation' or 'Kinetography-Laban', so-called after its creator Rudolph Laban in 1948. Laban (1975) highlighted three fundamental problems encountered in the formulation of any movement notation system:

1. recording complicated movements accurately;
2. recording this movement in economical and legible form;
3. keeping abreast with continual innovations in movement.

It was these three fundamental problems that left dance in a state of flux, incapable of steady growth, for centuries. As already mentioned, the development of Labanotation represented a major factor in the evolution of notation, and Kinetography-Laban, or Labanotation is the most widely used of all movement notation systems. The next 'step' in the development of movement notation came in 1947 with the conception of another form of dance notation. Details of 'Choreology' were published by Jean and Rudolph Benesh in 1956. In this form of notation, five staves formed the base or matrix for the human figure.

i.e. Top of head
Top of shoulder
Waist
Knees
Floor

All notation was completed on a series of these five-line grids with a complex vocabulary of lines and symbols.

The major underlying disadvantage of both the Benesh and Laban methods of notation in terms of sport is that they are both utilized primarily for the recording of patterns of movement rather than its quantification. Later researchers attempted to develop a system of movement notation based entirely on the mathematical description of movement in terms of the degrees of a circle in a positive or negative direction. As with Labanotation and Choreology, this system did not allow the description of movement in terms familiar to sport or everyday life.

Movement notation systems, developed primarily in the field of expressive movement, gradually diversified into game analysis, specifically sports and games. Ensuing research proved severely limited both in variety and detail. As recently as the late 1970s, the majority of what little research there was published in game analysis was concerned with basketball and soccer – and this was at a fairly global and unsophisticated level. More recently work in this area has expanded to encompass most team and individual sports. With this increased interest in notation, an awareness of the potential of this form of analysis has been expressed in a number of ways.

The major purposes of notation are:

1. analysis of movement;
2. tactical evaluation;
3. technical evaluation;
4. statistical compilation.

Many of the traditional systems are concerned with the statistical analysis of events which previously had to be recorded by hand. The advent of on-line computer facilities overcame this problem, since the game could be digitally represented via data collection directly on to the computer, and then later documented via the response to queries pertaining to the game. The major advantage of this method of data collection is that the game is represented in its entirety and stored in ROM (read only memory) or some other method. A database is therefore initiated and this is a powerful tool of great potential. Team sports can benefit immensely from the development of computerized notation. The sophistication that data manipulation procedures make available would aid the coach in understanding and improving performance.

The information derived from this type of computerized system can be used for several purposes.

- Immediate feedback.
- Development of a database.
- Indication of areas requiring improvement.
- Evaluation.
- As a mechanism for selective searching through a video recording of the game.

All of the above functions are of importance to the coaching process, the initial *raison d'être* of notational analysis. The development of a database is a crucial element, since it is sometimes possible, if the database is large enough, to formulate predictive models as an aid to the analysis of different sports, subsequently enhancing training programmes and competitive performance.

Both types of notation, manual or computerized systems, have their advantages and disadvantages. Hand notation systems are cheap and are accurate, if fully defined operationally and used correctly. The disadvantages of these types of systems are that the time required for data processing can be very long. When the systems become more sophisticated, in order to gather data in more complex analyses of match-play, which is often the case with a team game such as soccer, then considerable learning and training time is necessary in order to ensure accuracy and reliability of the operator.

One advantage of using computers is that the data analysis, once the software has been designed, written and tested, takes only as long as the computer and/or printer takes to process the output. This can be done in a very short time, which is very important to coaches and athletes for immediacy of feedback of performance. Secondly, the learning time required to use notation systems can be reduced considerably, especially with the variety of advances in special keyboards, digitization pads, graphical user interfaces and voice interactive systems that have become available for computer hardware. Finally, the rapid developments of computer graphics, word-processing, database and multi-media packages to present the analyses in clear and appealing ways have enabled athletes and coaches to understand complex data. This potential has only just begun to be tapped. The disadvantages of computerized systems are that they are expensive and can be less accurate than hand notation systems, unless very carefully designed and validated.

Skilful management of data input and output can make the use of the systems, and the understanding and assimilation of the data, easier for the coach or player. The advances made in tackling these problems in computerized notation analysis provide a useful structure with which to explore the developments in this field.

20.1.1 The development of sport-specific notation systems (hand notation)

The first attempt to devise notation systems specifically for sport analysis was due to the work of Messersmith, together with a number of other workers. Most of the work was motion and work-rate analysis in basketball and American football, comparisons of men and women in basketball and the effects of rule changes on work-rate in basketball. The first and last of these pieces of work have almost the same titles: they trace the development of a subject area through its growth and development by Messersmith over 15 years – a formative period for notational analysis.

Notation systems were commercially available for analysis of American football as early as 1966, and the Washington Redskins were using one of the first in 1968. Interestingly, American football is the only sport that has, as part of its rules, a ban on the use of computerized notation systems in the stadium. How this could be enforced is not clear. All clubs do claim to use similar hand notation systems, the results of which are transferred to computer after the match. Clubs exchange data, just as they exchange videos, on opponents. Because of the competitive nature of this, and other 'big money' sports, actual detailed information on results of these analyses is seldom available.

The first publication of a comprehensive sport notation system in Britain was that by Downey (1973), who developed a detailed system which allowed the comprehensive notation of lawn tennis matches. Detail in this particular system was so intricate that not only did it permit notation of such variables as shots used, positions and so on, but it also catered for type of spin used in a particular shot. Downey's notation method has served as a useful base for the development of systems for use in other racket sports, specifically badminton and squash.

Because of its simplicity as a game, squash has attracted the attention of a number of sports analysts. Several systems have been developed for the notation of squash, the most prominent being that by Sanderson and Way (1977). The different squash notation systems possess many basic similarities. Sanderson and Way's method made use of illustrative symbols to notate 17 different strokes, as well as incorporating court plans for recording accurate positional information. The system took an estimated 5–8 hours of use and practice before an operator was sufficiently skilful to record a full match whilst play was in progress. Processing the data could take as long as 40 hours of further work. This system was used to gather a database and show that competitors in squash play in the same patterns, winning or losing. It would seem that the majority of players are unable to change the patterns in which they play. The major pitfall inherent in this system, as with all long-hand systems, was the time taken to learn how to use it. A second problem was that the amount of raw data generated required so much time to process.

20.2 HAND NOTATION SYSTEMS IN SOCCER

The definitive motion analysis of soccer, using hand notation, was by Reilly and Thomas (1976), who recorded and analysed the intensity and extent of discrete activities during match-play. They combined hand notation with the use of an audio tape recorder to analyse in detail the movements of English First Division soccer players. They were able to specify work-rates of the players in different positions, distances covered in a game and the percentage time of each position in each of the different ambulatory classifications (Figure 20.1). They also found that typically a player carries the ball for less than 2% of the game. Reilly (1990) has continually added to this database enabling him to define clearly the specific physiological demands in soccer, as well as all the football codes. The work by Reilly and Thomas has become a standard against which other similar research projects can compare their results and procedures. A detailed analysis of the movement patterns of the outfield positions of Australian professional soccer players was completed in a similar study to that above (Withers *et al.*, 1982). The data produced agreed to a great extent with that of Reilly and Thomas (1976); both studies emphasized that players cover 98% of the total distance in a match without the ball, and were in agreement in most of the inferences made from the work-rate profiles.

An alternative approach towards match analysis was exemplified by Reep

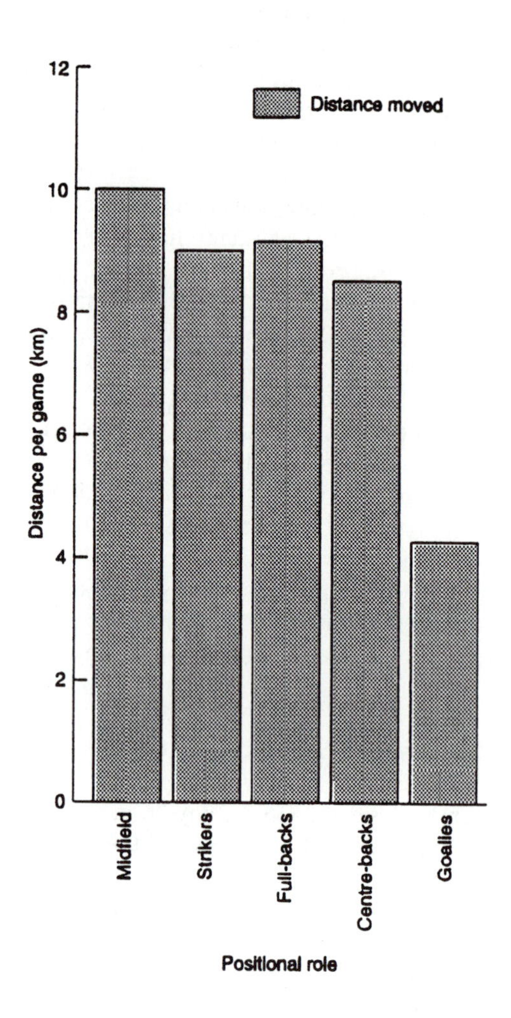

Figure 20.1 Distance travelled per game of soccer with respect to the position of the player. (Data from Reilly and Thomas, 1976.)

and Benjamin (1968) who collected data from 3213 matches between 1953 and 1968. They were concerned with actions such as passing and shooting rather than work-rates of individual players. They reported that 80% of goals resulted from a sequence of three passes or less. Fifty per cent of all goals came from possession gained in the final attacking quarter of the pitch.

More recently, Bate (1988) found that 94% of goals scored at all levels of international soccer were scored from movements involving four or less passes, and that 50–60% of all movements leading to shots on goal originated in the attacking third of the field. Bate explored aspects of chance in soccer and its relation to tactics and strategy in the light of the results presented by Reep and Benjamin (1968). It was claimed that goals are not scored unless the attacking team gets the ball and one, or more, attacker(s) into the attacking third of the field. The greater the number of possessions a team has, the greater chance it has of entering the attacking third of the field, therefore creating more opportunities to score. The higher the number of passes per possession, the lower the total number of match possessions, the total number of entries into the attacking third, and the total chances of shooting at goal. Thus Bate rejected the concept of possession football and favoured a more direct strategy. He concluded that to increase the number of scoring opportunities a team should:

1. play the ball forward as often as possible;
2. reduce square and back passes to a minimum;
3. increase the number of long passes forward and forward runs with the ball;
4. play the ball into space as often as possible.

These recommendations are in line with what is known as the 'direct method' or 'long-ball game'. The approach has proved successful with some teams in the lower divisions of the English League. It is questionable whether it provides a recipe for success at higher levels of play.

Harris and Reilly (1988) considered attacking success in relation to team strategy and the configuration of players around the point of action, by concentrating mainly upon the position of attackers in relation to the defence and overall the success of each attacking sequence. This was a considerable departure from many of the systems previously mentioned which have tended to disseminate each sequence into discrete actions. Harris and Reilly provided an index describing the ratio of attackers to defenders in particular instances, while simultaneously assessing the space between a defender and an attacker in possession of the ball. These were analysed in relation to attacking success, whereby a successful attack resulted in a goal, an intermediate attack resulted in a non-scoring shot on goal, and an unsuccessful attack resulted in an attack ending without a shot. Successful attacks tended to involve a positive creation of space, where an attacker passes a defender – an unsuccessful attack involved a failure to use space effectively due to good organization of defensive lines.

It is evident that hand notation systems provide a detailed record of behaviour during soccer match-play. It is possible also to derive theories of play from such analysis. A complete theory must take the opposition into consideration, and this requires highly sophisticated modelling which is beyond the scope of this chapter.

20.3 COMPUTERIZED NOTATION SYSTEMS

Using computers does introduce extra problems in notational analysis, of which the system users and programmers must be aware. Increases in possibilities of error are enhanced by either operator errors, or hardware and software errors. Any system is subject to limitations of human perception where the observer misunderstands an event, or incorrectly fixes a position but the computer–operator interface can result in the operator thinking the correct data are being entered when this is not the case. This is particularly so in real-time analysis when the data must be entered quickly. Hardware and software errors are introduced by the machinery itself, or the programs of instructions controlling the operation of the computer. Careful programming can eradicate this latter problem. To minimize both of these types of problems, careful validation of computerized notation systems must be carried out. Results from both the computerized system and a manual system should be compared and the accuracy of the computerized system quantitatively assessed.

Although computers have only recently impinged on the concept of notational analysis, this form of technology is likely to enhance manipulation and presentation of data due to the ability of computers to process large amounts of data easily and the improved efficiency of the graphics software.

20.3.1 Data entry

A fundamental difficulty in using a computer is entering information. The traditional method employs the QWERTY keyboard, but studies using this method to enter data about soccer into the computer have shown that this can be a lengthy and boring task.

An example of this type of application of computerized systems calculated the time spent by three professional soccer players in different match-play activities. The analysis was completed post-event from video tapes, using a specially designed computer program. The results indicated that soccer is predominantly an aerobic activity, with only 12% of game time spent in activities that would primarily stress the anaerobic energy pathways. The mean time of 4.4 s for such high-intensity work indicated the alactic acid energy supply system was the anaerobic system of primary importance (Mayhew and Wenger, 1985). The interval nature of soccer was partly described, and suggestions for the design of soccer-specific training programmes were offered. Considerable keyboard skills, and time, were required to enter the data into the computer. However, the work did not really extend in any way the previous efforts of Reilly and Thomas (1976).

An alternative to the approach using the QWERTY keyboard is to use a specifically designed keyboard. One of the major developments in computerized notation was the mini-system devised by researchers at the University of British Columbia. They configured a keyboard on a minicomputer to resemble the layout of a soccer field and designed a program which yielded frequency tallies of various features of play (Franks and Nagelkerke, 1988). The path of the ball during the game was followed, so off-ball incidents were considered extraneous. A video was time-locked into the system so that relevant sections of the match could be replayed visually alongside the computer analysis.

Minimal consideration had been given to the number of games to be notated prior to the establishment of a recognized system of play. This is important, since any fluctuation in the patterns and profile will affect the inferences made, particularly with reference to the match outcome. Teams may also vary their system and pattern of play according to their opposition, although these factors were not considered by any of the researchers mentioned so far. Furthermore, the existence of patterns of play peculiar to individual players was not illustrated.

An alternative to the specially designed keyboard is the use of digitization pads. In England researchers worked with the 'Concept Keyboard', in Canada they used the 'Playpad', but both systems were similar. These are touch-sensitive pads that can be programmed to accept input to the computer, via the pad, over which specially designed 'overlays' can be placed. This enabled a pitch representation, as well as action and player keys to be specific and labelled (Figure 20.2). The digitization pads considerably reduced the time to learn the system, and made the data input quicker and more accurate. The system enabled an analysis to be performed of patterns of play, both at team and player levels, and with respect to match outcome.

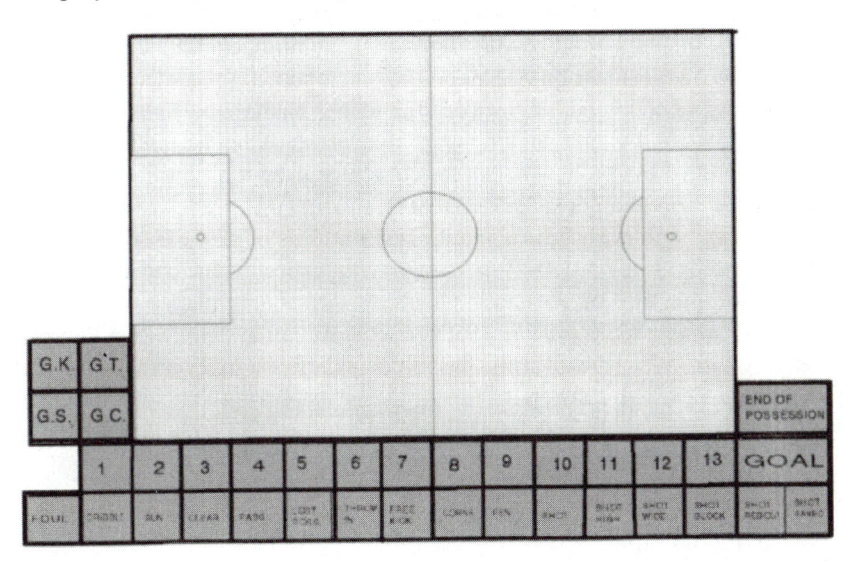

Figure 20.2 The overlay for the Concept Keyboard for data input for notational analysis of soccer. (Reproduced with permission from Hughes *et al.*, 1988.)

Hughes, Robertson and Nicholson (1988) used a concept keyboard and hardware system, with dedicated software, to analyse the 1986 World Cup finals. Patterns of play of successful teams, those teams that reached the semi-finals, were compared with those of unsuccessful teams, i.e. teams that were eliminated at the end of the first rounds. The main observations were as follows.

1. Successful teams played significantly more touches of the ball per possession than unsuccessful teams.
2. The unsuccessful teams ran with the ball and dribbled the ball in their own defensive area in different patterns compared with the successful teams. The latter played up the middle in their own half, the former used the wings more.
3. This pattern was also reflected in the passing of the ball. The successful teams approached the final sixth of the pitch by playing predominantly in the central areas while the unsuccessful teams played significantly more to the wings.
4. Unsuccessful teams lost possession of the ball significantly more in the final sixth of the playing area, both in attack and defence.

This work was further extended, analysing attacking plays only, to examine whether such unsuccessful teams use different attacking patterns from successful teams. An attack was defined as any move or sequence of moves that culminated, successfully or otherwise, in an attempt on goal. A total of 37 individual action variables and 18 different pitch divisions were employed in the data collection program. The data analysis program employed chi-square test of independence to compare the frequency counts of each action available, with respect to position on the pitch, between successful and unsuccessful teams. It was concluded that successful teams passed the ball more than unsuccessful teams when attacking, particularly out of defence and in the final attacking end of the pitch. The successful teams used the centre of the pitch significantly more than unsuccessful teams. Further differences demonstrated that successful and unsuccessful teams used patterns of play that vary significantly in attack. Implications were drawn with respect to the optimization of training and preparation for success in elite soccer match-play.

Partridge and Franks (1989a, 1989b) used a similar system, with a Playpad to enter their data (the overlay for which is shown in Figure 20.3), and produced a detailed analysis of the crossing opportunities from the 1986 World Cup. They carefully defined how they interpreted a cross, and gathered data on the following aspects of crosses:

1. build-up of attacks;
2. area of build-up;
3. area from which the cross was taken;
4. type of cross;

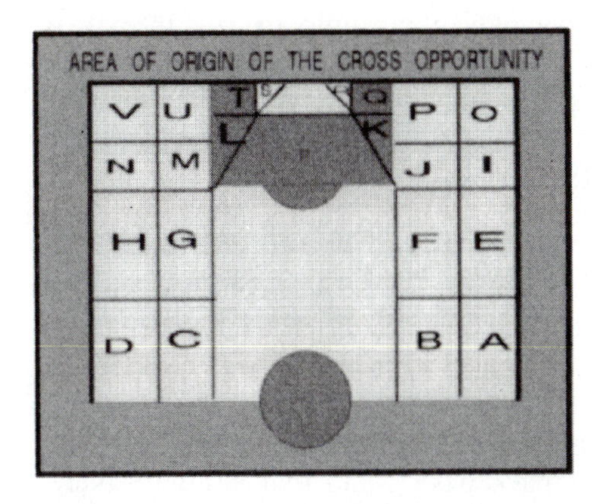

Figure 20.3 The overlays for the Playpad for data input for notational analysis of crosses by players in soccer. (Reproduced with permission from Partridge and Franks, 1989a.)

5. player positions and movements;
6. specific result of the cross;
7. general result, if the opportunity to cross was not taken.

Fifty of the 52 games of the competition were analysed from video tape, using specifically designed software on an IBM XT Microcomputer that enabled each piece of information relating to crossing opportunities to be recorded and stored. The program recorded the time at which all actions took place during the match, for extracting visual examples post-analysis, in addition to the usual descriptive detail about the matches, i.e. venue, teams and so on. A second program was written to transform and download these data into dBASE III +. After this, the database was queried to reveal selected results. The authors summarized their results by considering what they termed 'key factors', these being the opportunity to cross the ball, the nature of the cross, the actions of the target players, the supporting players, the position from which the cross is made and the aims of the crossing player. Partridge and Franks related their results to the design of practices to aid players understand their roles in the successful performance of crossing in soccer.

These systems are now updated to enable a graphical user interface to input the data, thus eliminating the need for the concept keyboard or digitization pad. These systems use on-screen graphics that are controlled and interfaced by means of the computer mouse. They have been used recently to investigate the effect of the changes in the rules of soccer in Premier Division matches in England, in the 1992/93 season. It was found that the time-wasting of goal-keepers was significantly reduced, that the midfield was more congested as a result of fewer back passes, and that although there were more errors in defence, there were not any more goals or goal-scoring possibilities. It was felt that more research on the changes in patterns should be continued through the following seasons, to assess how the players adjust to the new rules.

The use of digitization pads has considerably eased some of the problems of data entry, but one of the more recent innovations in input is the introduction of voice entry of data into the computer. It has been demonstrated that this type of system can be used by the computer 'non-expert' for notation systems, and that, as prices come down and hardware efficiency and flexibility improve, this method of data entry will be the future of interfacing sports data with computers.

Data input has moved from the use of difficult QWERTY keyboards and complex learning skill problems, through special keyboards specifically task-designed, concept keyboards, digitization pads, voice interactive systems and graphical user interfaces in the attempt to simplify the task and make it more accurate. As the technology improves, the easier it will become to access the computer and all the vast potential it offers to the athlete, coach and sports scientist.

20.3.2 Data output

In practical applications to sport, it is imperative that the output from notation systems is immediate and, perhaps more important, clear, concise and to the point. The first systems produced tables of data, often incorporated with statistical significance tests, that were difficult for the non-scientist to understand. Some researchers attempted to tackle the problem before the advent of computerized graphics, but it was debatable whether the type of presentation was easier to understand than the tables of data. Frequency distributions across graphical representations of the playing area offer a compact form, which coaches of different nationalities have found easy to assimilate.

Yamanaka, Hughes and Lott (1993) took this a step further in presenting their data (see Figure 20.4). They demonstrated the ethnic differences in

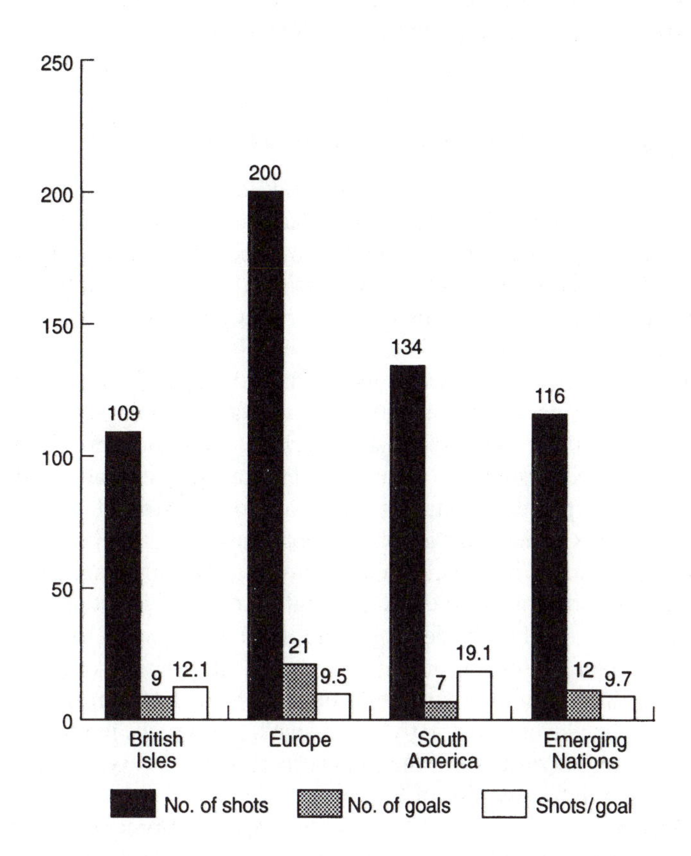

Figure 20.4 Shots per goal in the 1990 World Cup: an example of data output in graphical form, demonstrating the facilitation of assimilation of the data. (Reproduced with permission from Yamanaka *et al.*, 1993.)

international soccer by analysing the 1990 World Cup. By defining four groups, British Isles, European, South American and Emerging Nations, and by analysing the respective patterns of play in matches with respect to pitch position, they were able to differentiate between the playing styles of these international groups. They also presented data in a case study of Cameroon, who had had such a successful World Cup. The data for Cameroon were compared to those of the other groups to examine the way in which the former had developed as a footballing nation.

The recent development and growth of integrated software packages incorporating programming languages, graphics packages, databases and word processors have enabled the development of whole systems that immediately access graphics in both the data input stages, as well as the data processing. These systems have great potential to provide quantitative and qualitative feedback.

20.3.3 Applications of computerized notation systems

At the Second World Congress of Science and Football a considerable amount of work on computerized analysis of football was presented; this is all collated in the proceedings (Reilly *et al.*, 1993). A number of these have already been considered because of their contribution to the development of hardware and software as well as their contributions to research. The more significant applications will be reviewed.

Using computer-assisted video feedback and a specific algorithm for the statistics, Dufour presented an analysis of an evaluation of players' and team performance in three aspects: physical, technical and tactical. The ability of the computerized systems to provide accurate analysis and feedback for coaches on their players and teams was clearly demonstrated.

Gerisch and Reichelt used graphical representation of their data to enable easier understanding by the coach and players. Their analyses concentrated on the one-on-one confrontations in a match, representing them in a graph with a time-base, so that the development of the match could be traced. Their system can also present a similar time-based analysis of other variables, interlinking them with video so that the need for providing simple and accurate feedback to the players is attractively achieved. Despite the limited amount of data presented, the results and their interpretation were very exciting in terms of their potential for analysis of the sport.

Winkler presented a comprehensive, objective and precise diagnosis of a player's performance in training and match-play using a computer-controlled dual video system. This was employed to assess physical fitness factors employed in training contexts. In addition, he used two video cameras, interlinked by computer, to enable a total view of the playing surface area. This, in turn, permitted analysis of all the players in a team throughout the whole match, on and off the ball – something that not many systems have been able to produce.

Data from the 1990 World Cup have been analysed with a view to the interpretation of goal scoring and the importance of possession. Previous data may have been misinterpreted because they had not been 'normalized', i.e. the number of goals scored should be divided by the frequency of number of passes per possession. There were 88% of possessions with four passes or less, and so there should be more goals scored, proportionately, from these possessions, if they were truly the best 'chance' (Figures 20.5–20.7). Observations showed that the passing tactics of a team are dependent upon the skill levels of the players in each team, and the tactics best suited to a particular team can be gauged by match analysis (Hughes, 1993).

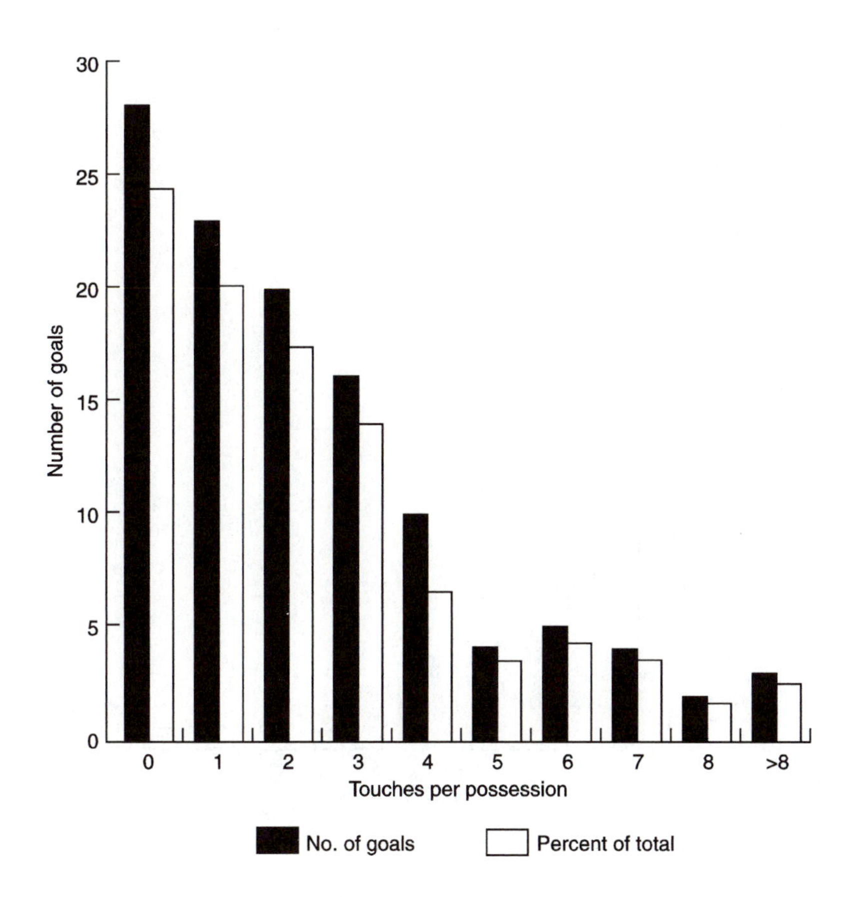

Figure 20.5 Goals scored in the 1990 World Cup with respect to the number of touches per possession.

20.4 COMPUTERS AND VIDEO

The ability of computers to control the video image has introduced exciting possibilities for enhancing feedback. An inexpensive IBM-based system was developed for the team sport of field hockey; this system has been modified to analyse and provide feedback for ice-hockey and soccer (Franks and Nagelkerke, 1988). Following the game, a menu-driven analysis program allowed the analyst to query the sequentially stored time-data pairs. Because of the historical nature of these game-related sequences of action, it was possible to perform both post-event and pre-event analysis on the data. That is to say, questions relating to what led up to a particular event or what

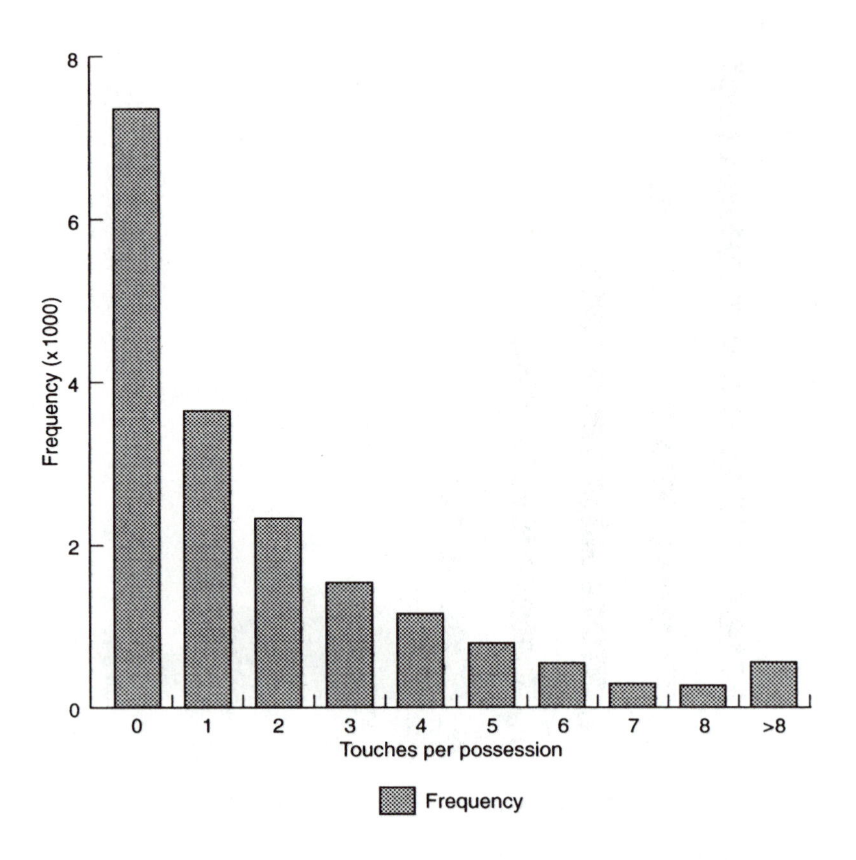

Figure 20.6 The frequency of the different numbers of touches per possession in the 1990 World Cup finals.

followed a particular event could now be asked. In addition to presenting the sports analyst with digital and graphical data of team performance, the computer can also be programmed to control and edit the video tape record of the game action.

The interactive video computer program could access, from the stored database, the times of all specific events such as goals, shots, set plays and so on. Then, from a menu of these events, the analyst could choose to view any or all of these events within one specific category. The computer may be programmed to control the video such that it finds the time of the event on the video and then plays back that excerpt of game action. It is also possible to review the same excerpt with an extended 'lead in' or 'trail' time around that chosen event.

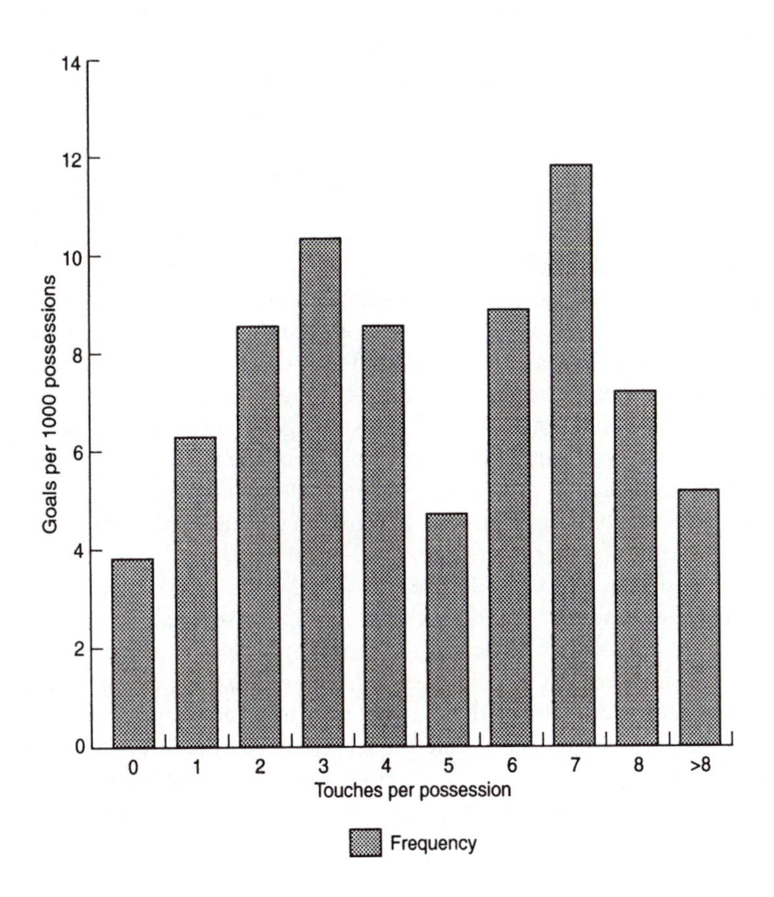

Figure 20.7 The normalized frequency of goals scored (goals per 1000 possessions) with respect to the number of touches per possession.

20.5 FUTURE DEVELOPMENTS

There are several developments that will extend notational analysis over the next few years. The first will be the development of 'all-purpose', generic software. Work in some centres has almost reached this point now. Another technological advance that will make computerized notation more easily handled by the non-specialist will be the introduction of 'voice-over' methods of data entry. This is possible, but relatively expensive, at present. The integration of both these technological developments with computerized video feedback will promote detailed objective analysis of competition and the immediate presentation of the most important elements of play. In addition, as the use of CD-ROM (read only memory) becomes easier and less expensive, the accessibility of databases, both visual and numerical, will become truly integrated and simple. Computerized systems now commercially available enable the analysis, selection, compilation and re-presentation of any game on video to be processed in a matter of seconds. The coach can then use this facility as a visual aid to support the detailed analysis.

As these systems are used more and more, and larger databases are created, a clearer understanding of game processes in each sport will follow. The mathematical approach will make these systems more and more accurate in their predictions of behaviour in competition. At the moment the main functions of the systems are analysis, diagnosis and feedback – few sports have gathered enough data to allow prediction of optimum tactics in set situations. Where large databases have been collected (e.g. soccer and squash) models of the games have been created and this has led to predictive assertions of winning tactic (McGarry and Franks, 1995). This has generated some controversy, particularly in soccer, due to misinterpretation of the statistics involved and their range of application. Nevertheless, the function of the systems could well change, particularly as the financial rewards in certain sports are providing such large incentives for success.

Technological advances aside, the real future of notational analysis lies in the growing awareness by coaches, athletes and sports scientists of its potential applications to all sports. Whether the most sophisticated and expensive of systems is being used, or a simple pen and paper analysis, as long as either system produces accurate results that are easy to understand, then coaches, athletes and sports scientists should increase their insights into sport performance. The use of qualitative data together with accurate statistical analyses will contribute to making these systems more relevant to the real world of football.

Summary

Various forms of notation analysis have been used for centuries to assist in the coding of human movements. In contemporary research the technique of motion analysis has been modified to study the movements of soccer players during a game. The advent and development of computers have stimulated sophisticated analysis of patterns of play and sequences of events leading to the creation of goal-scoring opportunities. This type of investigation is complemented by multi-media methods of observing and recording the behaviour of players. Future technological developments are likely to enhance the capture and retrieval of information. These facilities have great potential for assisting coaching staff in making decisions about game tactics. They can also provide performers with insights into many aspects of their own actions.

REFERENCES

Bate, R. (1988) Football chance: tactics and strategy, in *Science and Football* (eds T. Reilly, A. Lees, K. Davids and W. Murphy), E. & F.N. Spon, London, pp. 293–301.

Benesh, J. and Benesh, R. (1956) *Reading Dance – The Birth of Choreology*, Souvenir Press, London.

Downey, J.C. (1973) *The Singles Game*, E.P. Publications, London.

Franks, I.M. and Nagelkerke, P. (1988) The use of computer interactive video technology in sport analysis. *Ergonomics*, **31,** 99, 1593–603.

Harris, S. and Reilly, T. (1988) Space, team work and attacking success in soccer, in *Science and Football* (eds T. Reilly, A. Lees, K. Davids and W. Murphy), E. & F.N. Spon, London, pp. 322–9.

Hughes, M.D. (1993) Notation analysis in football, in *Science and Football II* (eds T. Reilly, A. Stibbe and J. Clarys) E. & F.N. Spon, London, pp 151–9.

Hughes, M.D., Robertson, K. and Nicholson, A. (1988) An analysis of 1984 World Cup of association football, in *Science and Football* (eds T. Reilly, A. Lees, K. Davids and W. Murphy), E. & F.N. Spon, London, pp. 363–7.

Laban, R. (1975) *Laban's Principles of Dance and Music Notation*, McDonald and Evans, London.

McGarry, T. and Franks, I.M. (1995) Winning squash: predicting championship performance from a priori observation, in *Science and Racket Sports* (eds T. Reilly, M. Hughes and A. Lees), E. & F.N. Spon, London.

Mayhew, S.R. and Wenger, H.A. (1985) Time-motion analysis of professional soccer. *Journal of Human Movement Studies*, **11,** 49–52.

Partridge, D. and Franks, I.M. (1989a) A detailed analysis of crossing opportunities from the 1986 World Cup (Part I). *Soccer Journal*, May–June, 47–50.

Partridge, D. and Franks, I.M. (1989b) A detailed analysis of crossing opportunities from the 1986 World Cup (Part II). *Soccer Journal*, June–July, 45–8.

Reep, C. and Benjamin, B. (1968) Skill and chance in association football. *Journal of Royal Statistical Society*, Series A, **131,** 581–5.

Reilly, T. (1990) Football, in *Physiology of Sports* (eds T. Reilly, N. Secher, P. Snell, and C. Williams), E. & F.N. Spon, London, pp. 371–425.

Reilly, T. and Thomas, V. (1976) A motion analysis of work-rate in different positional roles in professional football match-play. *Journal of Human Movement Studies*, **2,** 87–97.

Reilly, T., Stibbe, A. and Clarys, J. (1993) *Science and Football II*, E. & F.N. Spon, London.

Sanderson, F.H. and Way, K.I.M. (1977) The development of an objective method of game analysis in squash rackets. *British Journal of Sports Medicine*, **11,** 188.

Withers R.T., Maricic, Z., Wasilewski, S. and Kelly, L. (1982) Match analyses of Australian professional soccer players. *Journal of Human Movement Studies*, **8,** 158–76.

Yamanaka, K., Hughes, M. and Lott, M. (1993) An analysis of playing patterns in the 1990 World Cup for association football, in *Science and Football II* (eds T. Reilly, A. Stibbe and J. Clarys), E. & F.N. Spon, London, pp. 206–14.

The science of match analysis

21

Ian M. Franks and Tim McGarry

Introduction

The analysis of athletic behaviour can be undertaken at several different levels. For example, at a physiological level of analysis information may focus upon performance of the respiratory system or the cardiovascular system, or could yield information concerning the efficient transport of calcium across the cell membrane during the same performance. A biochemical perspective might relate to the depletion of substrate used as fuel for energy provision to the active muscles during these same events. Moreover, the biomechanist would be interested in the kinetic and kinematic profiles of skilful acts, while the psychologist investigates the mental preparedness of the athlete. Although each level of analysis differs in the measurement tools being used, the dependent variables that are calculated and the theoretical basis from which their enquiries are generated, one critical element remains invariant. They all converge upon one behavioural outcome that has meaning within the context of the sport performance under examination. Match analysis describes the performance at this behavioural level of analysis, coding the actions of individuals or groups in technical terms that have relevance to players and coaches.

All of these scientific disciplines are concerned with investigating various factors that impact upon the observed and notated behaviour. Analyses of task-relevant actions therefore can act as a catalyst for what is presently a multidisciplinary research approach to the study of athletic behaviour. Vickers (1992) defined multidisciplinary research as being

Science and Soccer. Edited by Thomas Reilly. Published in 1996 by E & FN Spon, London. ISBN 0 419 18880 0.

driven by experts within their own specific disciplines generating knowledge that resides in, and is claimed by, the authority of that particular expertise. The final product of multidisciplinary research does not integrate the disciplinary contributions. Match analysis research has the potential to integrate this multidisciplinary research into a common knowledge base and foster an interdisciplinary approach to the study of sport in general and football in particular. In contrast to the definition of multidisciplinary research, the general (or loose) definition of interdisciplinary research adopted by Vickers (1992) stressed that the scientists involved in this endeavour are trained in one discipline but incorporate elements from other disciplines or other interdisciplinary fields. While these researchers are not as proficient in other disciplinary knowledge, they treat that knowledge respectfully. More recently Burwitz *et al.* (1994) have clearly outlined the benefits of this approach to scientific enquiry in their report to the British Association of Sport and Exercise Sciences. There is a need to approach the study of football from an interdisciplinary perspective and match analysis can be of help in this endeavour.

Match analysis not only stimulates interdisciplinary research from outside its own field but interdisciplinarity is integral to the science of match analysis itself. Throughout this chapter research from various disciplines will be used first of all to outline the rationale for match analysis, secondly to highlight the methods of analysis and finally to underline its utility. It serves two fundamental functions: first to provide the coach and athlete with information about past performances, and secondly to provide data for predictive model development.

2.1 PROVIDING INFORMATION

The primary function of match analysis is to provide the coach with information about team and/or individual performance. This then allows accurate, objective and relevant feedback to be available for players. In order to accomplish this task the coach must be cognizant of what has taken place in a recently completed performance and be able to determine how this performance fits into the overall pattern of accumulated performances over the season.

Information provided to players about their own performance is one of the most important variables affecting the learning and subsequent execution of a motor skill. Knowledge about the proficiency with which athletes perform a particular skill is critical to the learning process, and in certain circumstances failure to provide such knowledge, or the provision of irrelevant feedback may prevent learning from taking place. In addition, the nature and scheduling of this information have been shown to be strong determinants of skilful performance (for a review of these findings see Magill, 1993, p. 316–49). That is, precise and task-relevant (realistic) information about the skill, delivered at the correct time, will yield significantly more benefits for the athlete than information that is imprecise, general in nature and scheduled inconsistently. Match analysis can provide such information to coaches and players. However, there has been some resistance to its use. This resistance has stemmed from the traditional view that experienced coaches can observe a match (without any aids to the observation process) and report accurately to the players on the critical elements that have determined the outcome.

Several lines of research in sport and everyday settings have clearly indicated that such observations are not only unreliable but also inaccurate. Franks and Miller (1986) considered the coach as an eyewitness of the sport competition and, using methodology gained from applied memory research, showed that international level soccer coaches could only recollect 30% of the key elements that determined successful soccer performance observed during one half of a televised game. Additional evidence was gained from a different sport setting where the action lasted less than two seconds. The perceptual abilities of novice and experienced gymnastic coaches were compared in a recent study (Franks, 1993) using a forced choice recognition paradigm. The experienced coaches were not significantly better than novice coaches in detecting differences in two performances of the front handspring. However, these experts did produce many more false positives (detecting a difference when none existed) than their novice counterparts and were also very confident in their decisions, even when wrong. This led to speculation that the experience and training that coaches had been exposed to did not necessarily improve the accuracy of observation, but rather predisposed them to seek out and report performance errors even when none existed.

If training and experience have such a profound sensitizing effect upon the perceptual abilities of expert observers, would it be possible to modify coaching education methods to enhance observer accuracy and reliability? Franks and Miller (1991) examined various methods of training the observational skills of soccer coaches. This training involved coaches in predicting highly probable game events. Orienting activities, in the form of video excerpts, were used in this training programme to provide the coaches with a structured framework upon which to observe and remember. The coaches were encouraged to develop a 'top-down' processing observational system that would prepare them to perceive highly probable forthcoming events. The general finding of this study was that coaches who used a structured set of predictive features in order to direct their perceptions were more accurate in their observations than coaches who used non-directed observations (a data-driven or 'bottom-up' processing system). Despite the fact that this type of training regime significantly improved the coaches' ability to observe and remember, the overall recall of these coaches, even after training, was still less than 50% accurate.

Evidence from these studies combined with many others from the field of applied psychology (see Neisser, 1982) strongly suggests that the accuracy of memories of episodic real world events is greatly influenced by many factors. These factors range from environmental variation of the initial observation to the motives and beliefs of the observer. It is important to remember that coaches are active observers and not passive perceivers of information. The perception and remembering of game events is not a copying process but rather a selective and constructive one. Therefore, a problem arises when the coach is required to provide an objective, unbiased accounting of game-related events. The solution is to collect relevant details of performance during the event and then recall these details at the termination of that event. Hughes (1993) detailed several methods of collecting and displaying such information. In soccer these methods have been used for many years. However, it is only recently that the advent of computer and video technology has enabled the coach to elaborate fully upon the component of the coaching process that involves observation of competition.

The flow diagram in Figure 21.1 represents the process of coaching using computer video technology. Observation begins at 'competition' with a time-locked synchronization of computer notation and video tape. The data that are collected, stored and transformed by the computer are subsequently compared to a model of optimum performance that was generated from past data (this will be discussed in greater detail later). Having isolated the priority problems that need to be remedied, the corresponding video excerpts are 'searched' and 'played'. These then form the basis of the organization required for the coaching practice session. Practical solutions along with viewing of video highlights for feedback and modelling purposes are used to prepare the athlete for the next competition. Providing such feedback to teams about tactical performance and to individuals about technical performance within the game significantly modifies playing behaviour (see Partridge and Franks, 1993) toward a predefined model of performance.

21.2 MODEL DEVELOPMENT

Charles Reep has been the architect of quantitative analysis of football since the 1950s. Data on passing movements from individual matches at a professional level were compiled over a 15-year period, the statistical analysis of which revealed compelling evidence for adherence to a well-established mathematical (negative binomial) function and remarkable consistency in the descriptive probabilities of certain behaviours, such as shots on goal and goals scored (Reep and Benjamin, 1968). The general finding from their data was that random chance plays a significant role in determining match outcome. This leads to the reasoned conclusion that those patterns of play which yield favourable probabilities of a desired return, for example a strike on goal, should be maximized to the exclusion of less profitable actions. Similarly, those properties which lead to likely adverse team behaviour should be minimized. The adoption of these recommendations by some football managers has been responsible for what has come to be known in English football as the 'direct style' of play, or more unkindly 'kick and rush'. Not surprisingly, these results have been met with resistance. This is despite 25 years of data which forcefully support the original findings of Reep and Benjamin. The noted stability of these data through to the present day is impressive and cannot easily be discounted.

The patterns of play which are evidenced in the data (Reep and Benjamin, 1968) have a direct implication for optimizing match strategy (for a detailed tactical interpretation of the implications to soccer, see Franks, 1988) The finding that the goal/shot ratio consistently approximates 1:10, for instance, suggests that increasing the number of shots will necessarily lead to an increased goal average per match and, since most shots on goal routinely arise from few playing possessions (passes) into and around the 'scoring area' a direct style of play seems appropriate, thereby enlarging the population of shots from which goals ultimately stem. Data from the 1986 World Cup suggest that goals arise from team play which is even more direct than that which leads to shots on goal (Franks, 1988). This indicates that goals may actually arise from a sub-population within the population of shots initially reported by Reep and Benjamin (1968). A more startling discovery, however, is not simply that goals arise through few possessions, but rather that the attacking team consistently has to lose and regain possession before proceeding to shoot on goal. Reep and Benjamin (1968) reported that between 1953 and 1967, in excess of 50% of shooting opportunities arose from regained possession in the attacking half of the field. Reep (1989) reported these data to still hold fast 40 years after the initial observation, a view reinforced by Franks (1988) who recommended the following:

> Lose as many possessions in the attacking third of the field as possible and reduce (to zero if possible) the number of possessions lost in 'free play' in one's own defending third of the field. Winning teams have been recorded as losing over sixty percent of all possessions in the attacking third of the field. The emphasis changes from *who* has possession of the ball to *where* the ball is in relation to the goal and the attacking team members. (p. 40)

Support for this bold assertion is offered from later analyses of the 1990 World Cup. Partridge and Franks (1991) found that the tournament champions (West Germany) lost the highest percentage of possessions in the attacking third of the field (61%) and, notably, along with Ireland, the lowest percentage of possessions in the defending third of the field (7%). Time has shown these data are consistent and convincing.

The impetus for a sport database and its instructive derivation of generic optimal match strategies is apparent from earlier studies (Reep and Benjamin, 1968; Hughes *et al.*, 1988; Partridge and Franks, 1991). Furthermore, the utility of collecting and analysing quantitative match data provides for an objective accounting and assessment of past sport performance. While the process of analysis is essentially descriptive in nature, the data are invariably used by the coach in a prescriptive fashion to suggest remedial action or to reinforce desired athletic behaviour (see Figure 21.1).

The rationale that an objective evaluation of sport performance more ably directs future decision-making within the coaching process assumes that the data reliably transfer from one sport environment to another. The notion that data, or more specifically the information derived therefrom, are applicable to a future setting may not seem objectionable and, indeed, the traditional and widespread use of scouting in sport proceeds on this very basis. Information regarding the next opponent's expected playing behaviour, often deduced from observation against a different opponent, is used in advance to prepare an appropriate athletic response. The frequent development and subsequent employment of match strategies based on *a priori* observation provide a fitting example.

The inference from prior observation of tactical strategy for application to a future setting necessitates prediction based on inductive reasoning. That is, given a repeat past observation under a given set of conditions, a similar future response is expected for the same set of conditions. This may not be unreasonable: champions seemingly exhibit consistency of performance, at least to the extent that they invariably win. The question is whether champions achieve sporting dominance through successfully imposing their pattern of play (athletic behaviour) on their opponents, and so showing stable behaviour, or whether champions are more variant in their athletic response, and thus adaptable to their particular opponent. If consistent behaviour exists then future sport performance can be predicted from past data using **stochastic** analysis. Indeed, not only can likely outcome be projected through competing **playing profiles** (habitual behavioural responses), but the profiles can be optimized to promote successful performance. This is indeed a powerful application, and moves the role of sport analysis from descriptive analysis to a prescriptive function. How then can competitive sport be modelled using a stochastic process?

Since sports usually consist of two competing players or teams (where a team comprises a number of players) the discussion is limited to this particular situation. The outcome of a contest (win, draw or lose) is usually precipitated by either the scoring system or the time-base. Examples of sports which are **score-dependent** include squash, tennis and volleyball, and examples of sports which are **time-dependent** include soccer, rugby and hockey. The primary distinction is that the score-dependent sports produce a contest result (win or lose) in unspecified time through determining the first competitor to reach a predetermined match score, and time-dependent sports produce a contest

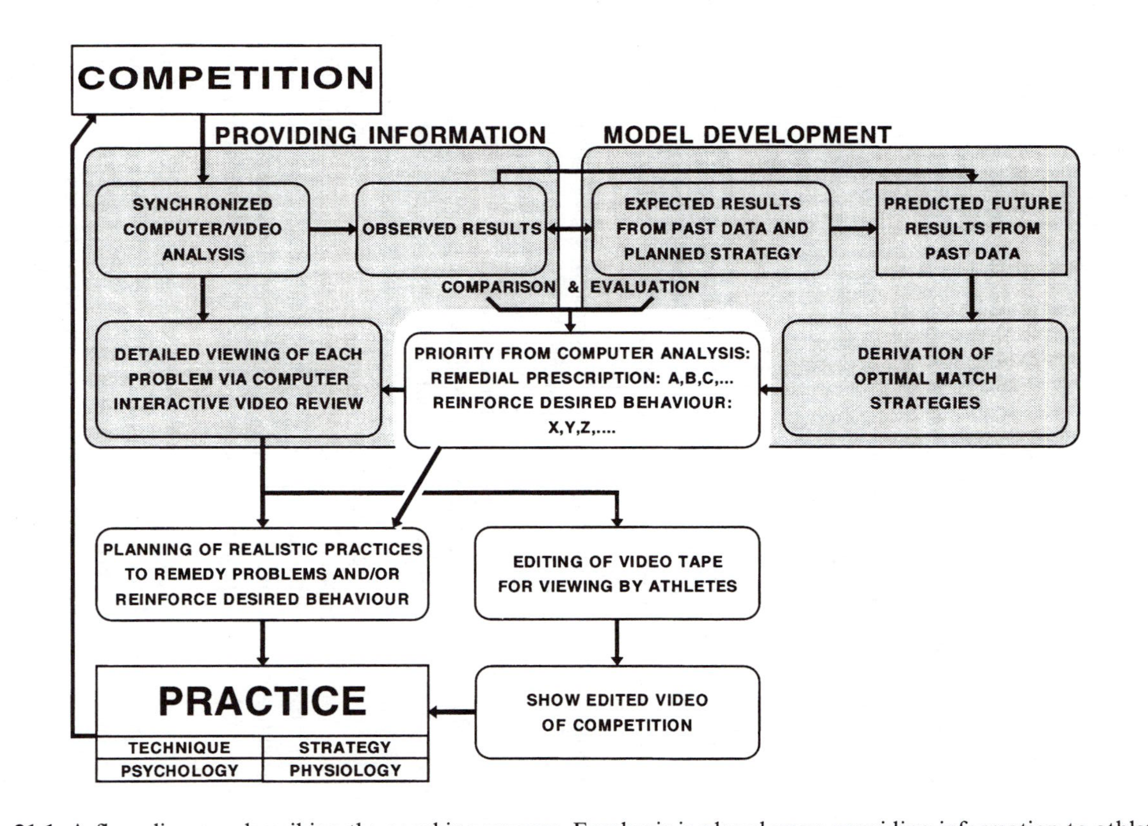

Figure 21.1 A flow diagram describing the coaching process. Emphasis is placed upon providing information to athletes and coaches, and predictive model development.

result (win and lose or draw) in a pre-ordained allotted period of time. In the former, the scoring system determines the contest result, while in the latter the elapsed time period is the determining feature. This distinction suggests a different analytic approach for modelling sport depending on its primary classification.

The score-dependent sports tend to be characterized by a structured sequence of discrete events, where the present event necessarily invokes the next discrete event, usually by the opponent. An example from squash is the serve which, being the first discrete event of the rally, invokes a subsequent discrete event response from the opponent. This state in turn invokes the next event response and so forth, until the end state of the rally is reached (winner, error or let). A rally itself can, of course, be considered a discrete event, and invokes the next rally until a game is reached. The discrete events, which are ascribed to each player, arise in serial and alternate fashion, and, moreover, are explicitly dependent on the preceding event. The relationship between discrete events is thus necessarily related to the opponent since an event specifically cannot arise without its antagonistic predecessor. The exception is the rally-end state which is accessed through a preceding event state (shot) assigned to the player.

Sports that are time-dependent are usually interactive and tend to comprise a chain of discrete events which, while implicitly related to their antecedents, can generally be considered to be primarily contingent on a temporal structure. Winning possession of play from, or losing possession of play to, the opposition offers an example of this phenomenon. Possession of play is not temporally divided in equal portion to each opponent. Nor is it necessary that the discrete events are alternately matched with complementary discrete responses by the opponent because of the sport's interactive character. (While it is true that the number of possessions won by one team will be matched by an equal number of possessions lost by the opponent, the discrete events themselves, for example, pass, dribble and shot, do not necessarily match because an unspecified string of discrete events may be assigned to a team before possession is lost.) Rather, a disparate set of discrete events are ascribed to each team (player) and, together with a corresponding absolute time code for each event, can be thought to interleave on a temporal basis to produce the final contest outcome. The relationship between discrete events in such sports is primarily temporal.

The structure of the particular sport would suggest the most suitable modelling procedure. Sports determined by the scoring system are best simulated using discrete event models and sports governed by time are probably best represented using discrete time models. In score-dependent sports, time is unimportant and need not be modelled. In time-dependent sports, however, the transition to the next event state is dependent on both the present event and the time measure. Discrete event or discrete time models are both appropriate modelling procedures in this instance. The former must model for time on every transition, the latter must, on a unit time stimulus, assign the next event state from the present state or wait until the next unit time event. It is likely that a Poisson process may suitably model for time between discrete events in soccer although a Poisson fit has yet to be demonstrated.

In both discrete event and discrete time models, the transition of a process from the present discrete event state to the next event within the Markov chain is dependent upon the rule set and, importantly, a stimulus, or trigger. In discrete event models the stimulus is the preceding discrete event itself and the process resides in a terminal state when a predetermined match score is reached; in discrete time models the trigger is the time event (or unit) and the process terminates when a fixed period of time is elapsed. In discrete time models the process need not necessarily advance to the next event state simply at the onset of a unit time stimulus.

Another variable to account for in the modelling of sport is the number of players involved in the competition which of course varies between individual and team sports.

An increase in variability is expected in those sports that include a greater number of participants, since each individual behaviour contributes to the observed team behaviour. It is perhaps prudent then, if we are to use the sport model, to project likely future athletic behaviour from empirical data, to model individual sports in the first instance, in an attempt to minimize the variability which naturally occurs within any sport setting. McGarry and Franks (1994), using a discrete event stochastic Markov model of championship squash match-play, tried to predict prospective future behaviour from *a priori* observation. Validation of the stochastic models, which are not detailed here, required that the assumption of stability of a player's profile across different opponents within a short time period be formally tested. A playing profile was defined to comprise: (*a*) the probable shot response of that player to the preceding shot by the opponent; and (*b*) the probable outcome response of the player to his/her shot response. While consistency in athletic response when competing against the same opponent over a period of time was supported from the available data, the profiles were found to be inconsistent when competing against different opponents. An interpretation is that a player's athletic response is dependent on the particular player–player interaction and thus not generalizable to other opponents. It remains unclear at present whether this finding is due to genuine variant athletic behaviour, or whether the result is a function of the derived playing profile itself. It may be, for instance, that the preceding condition of the opponent's shot simply failed to capture the necessary behavioural conditions which elicit a reliable athletic response.

The findings of inconsistent athletic response to preceding conditions do not necessarily imply the analysis has no prescriptive utility whatsoever to future sport performance. Indeed, McGarry and Franks (1995) discovered a seemingly reliable idiosyncratic athletic response pattern of the eventual champion and, retrospectively, tested a derived match strategy for use by earlier opponents. The subsequent data showed that the suggested strategy would have been expected to benefit the opponent and, moreover, that this strategy could have been inferred from the quarter-final data and applied prospectively to the semi-final and final. While some reliable behavioural response(s) of the champion could be identified from a careful perusal of the playing profiles, these responses were not captured through statistical analysis of the playing profiles. The general finding of inconsistency when competing against different opponents counsels caution, especially if match strategies are sought from *a priori* observation and applied in future competition against a different opponent. If invariant behaviour exists in athletic response to preceding conditions then the likely outcome of a sport contest can be predicted using a stochastic approach. If, on the other hand, variant behaviour exists, then the familiar role of scouting may prove less useful than traditionally assumed.

The modelling of competitive sport is an informative analytic technique because it directs the attention of the modeller to the critical aspects of data which delineate successful performance. The modeller searches for an underlying signature of sport performance which is a reliable predictor of future sport behaviour. Stochastic models have not yet, to our knowledge, been used further to investigate sport at the behavioural level of analysis. However, the modelling procedure is readily applicable to other sports and could lead to useful and interesting results. If reliable athletic behaviours can be found which characterize future sport behaviour then the reward is attractive. Imagine the advantage of reliably inferring, and therefore evaluating, the likely contributory effect of an individual's play within a team contest. If, however, reliable athletic behaviours cannot be found, the interpretation is that the findings from an earlier sport setting are not directly transferable to a future sport setting. If this is so, the coach may be best advised to optimize athletic behaviour (maximize the chance of team success and, simultaneously, minimize the chance of success for the opponent) in the **context** of that particular situation, with no expectation of a recurrence of that particular behaviour outside that context. This clearly reduces any emphasis on *a priori* tactical preparation for imminent sport competition and may impugn somewhat present orthodox coaching practice. Further investigation of the traditional assumptions of stable athletic responses which underpin the coaching process is not only warranted, but essential.

Summary

In this chapter we have outlined the contribution that can be made by match analysis to the interdisciplinary study of sport in general and soccer in particular. The notated behaviour of an athlete is the product of many complex processes that comprise the human motor system. The goal of various disciplines within human movement science is to understand how these processes interact to produce skilled human movement. The predominant approach of this scientific enquiry has been either mono- or multidisciplinary. We have made a case for an interdisciplinary approach to the study of sport behaviour using match analysis as the catalyst for this study. One consistent finding from research has been that human actions are environment-specific; that is to say they are task-dependent. The human shows a remarkable ability to adapt to the particular task at hand, whether it be in industry or sport. In our attempts to understand the human in action, consideration must be given to the behaviour that is produced and the environment within which it is produced. Match analysis allows the scientist to understand this specific behaviour and its contextual relevance. The following glimpse into future integrated match analyses should help illustrate this point.

> The analysis from the match has been tabulated and after reviewing a brief summary of game statistics the coaching staff are concerned that late in the game crosses from the right side of the team's attack were being delivered behind the defenders and close to the opposing team's goalkeeper. The result was that the front strikers were not able to contact the ball despite making the correct approach runs (information also gained from the match summary). The coaching staff call for videodisc (immediate recovery) excerpts of each crossing opportunity from the right side of the field in the last 15 minutes of play. Along with this visual information, the computer retrieves other on-line information that is presented in the inset of the large projected video image. This information relates to the physiological condition of the player(s) under review leading up to the crossing opportunity. In addition, a biomechanical 3-D analysis of the crossing technique is presented as each cross is viewed. One player has been responsible for these crosses. Upon advice from the consulting exercise physiologist the coaching staff have concerns about the telemetred respiration and heart rate of the player. A time–motion analysis of the player's movements in the second half of the game is called for, as well as a profile of the player's fitness level and physiotherapy report prior to the game. These are also retrieved from the same videodisc. After considering the information the coaching staff record their

recommendations for team and individual improvement and move on to the next problem provided by a comparison with the predicted data and the real data. A computer program running in the background is busy compiling instances of good performance (successful crosses) and poor performance that will make up an educational modelling programme for the individual player to view. Also, the expert system of coaching practice is being queried about the most appropriate practice for remedial treatment of crossing in this specific setting. An individual fitness programme is prescribed when another expert system is queried. The final question that is asked of the mathematical model is 'given these changes are successfully implemented, what is the likelihood that the number of crosses in the final 15 minutes from the right side of the field will be more successful against our next opponent and what is their expected effect on match outcome?'

All aspects of the above scenario are either in place or are under investigation in notational analysis laboratories throughout the world.

Acknowledgements

This chapter was prepared with the aid of a grant from the Social Sciences and Humanities Research Council of Canada awarded to the first author.

REFERENCES

Burwitz, L., Moore, P. and Wilkinson, D. (1994) Future directions for performance related sports science research: an interdisciplinary approach. *Journal of Sports Sciences*, **12,** 93–109.

Franks, I.M. (1988) Analysis of Association Football. *Soccer Journal*, September/October, 35–43.

Franks, I.M. (1993) The effects of experience on the detection and location of performance differences in a gymnastic technique. *Research Quarterly for Exercise and Sport*, **64,** 227–31.

Franks, I.M. and Miller, G. (1986) Eyewitness testimony in sport. *Journal of Sport Behavior*, **9,** 38–45.

Franks, I.M. and Miller, G. (1991) Training coaches to observe and remember. *Journal of Sports Sciences*, **9,** 285–97.

Hughes, M. (1993) Notational analysis in football, in *Science and Football II* (eds T. Reilly, A. Stibbe and J. Clarys), E. & F.N. Spon, London, pp. 151–9.

Hughes, M., Robertson, K. and Nicholson, A. (1988) Comparison of patterns of play of successful and unsuccessful teams in the 1986 World Cup for soccer, in *Science and Football* (eds T. Reilly, A. Lees, K. Davids and W. Murphy), E. & F.N. Spon, London, pp. 363–7.

McGarry, T. and Franks, I.M. (1994) A stochastic approach to predicting competition squash match-play. *Journal of Sports Sciences*, **12,** 573–84.

McGarry, T. and Franks, I.M. (1995) Modelling competitive squash performance from quantitative analysis. *Human Performance*, **8,** 113–29.

Magill, R.A. (1993) *Motor Learning: Concepts and Applications*, 4th edn, W.C.B. Brown & Benchmark, Madison, WI.

Neisser, U. (1982) *Memory Observed*, W.H. Freeman, San Francisco, CA.

Partridge, D. and Franks, I.M. (1991) Comparative analysis of technical performance: USA and West Germany in the 1990 World Cup Finals. *Soccer Journal*, 57–62.

Partridge, D. and Franks, I.M. (1993) Computer-aided analysis of sport performance: an example from soccer. *The Physical Educator*, **50,** 208–15.

Reep, C. (1989) Charles Reep. *The Punter* (The Scottish Football Association), September, 31–7.

Reep, C. and Benjamin, B. (1968) Skill and chance in Association Football. *Journal of the Royal Statistical Society*, Series A, **131,** 581–5.

Vickers, J. (1992) Where is the discipline in interdisciplinarity? *Interdisciplinarity* (Publication of the Association for Canadian Studies), September, 32–9.

Information technology 22

Tony Shelton

Introduction

In our own lifetimes, information and computer technologies have significantly transformed the world as we know it. Information technology (IT) systems have made our world smaller, safer and ever more efficient by improving communications, by improving education and health care and by opening up multiple information sources to individuals and to many different private and public sector organizations. Information technology systems have helped to make life easier and more pleasant for millions of ordinary people by eliminating or automating mundane tasks and everyday chores and by freeing time for the more innovative and creative side of human nature. There has been a continuing evolution of Western nations from the earliest pre-industrial agricultural societies to more recent industrial societies and to today's 'information society' based on the efficient provision, sharing and management of information.

This major technological evolution has had profound effects on contemporary sport and leisure (see, for example, the review of computers in sport by Lees, 1985). To illustrate these developments for sport in general, and more specifically for soccer, this chapter is focused on:

1. the history and development of computer hardware and software;
2. modern information technologies in soccer, both the potential and the reality.

Science and Soccer. Edited by Thomas Reilly. Published in 1996 by E & FN Spon, London. ISBN 0 419 18880 0.

The intention is to show how computers and other information technologies can gather, analyse and display large sets of notation data to assist individual sports performers, teams, soccer selectors and coaches. A further intention is to show how these have also revolutionized the study and simulation of human movements.

22.1 BACKGROUND

22.1.1 The history

Hull (1992) in his book on the history of IT described how, in 1943, development work started on the Electronic Numerical Integrator and Computer (ENIAC) for the US Army and how in the same year Alan Turing and his colleagues at the British Code and Cipher School were working on the Colossus machines. Turing and his research team, by 1948, were working at Manchester University and were building the world's first electronic, digital, programmable computer.

The earliest computers in the 1940s and 1950s required very large rooms and specialized air conditioned and dust free environments. These machines were used for 'number crunching' of large amounts of arithmetical data. They were slow, expensive and often unreliable. By the late 1950s, these machines, despite their problems, were in widespread use in many commercial organizations for everyday applications such as accounts, pay roll and stock control as well as more sophisticated applications such as forecasting and statistical analyses.

Early time-sharing techniques in which a single computer could be accessed from many remote but integrated terminals (later computer networks) were introduced during the 1960s. This fusion of communication and computing technologies led not only to the development of modern 'information technologies' but also to early attempts at computer system 'hacking' (i.e. gaining unauthorized access to the system) and the development of computer games such as the early versions of Space Invaders (Cornwall, 1986).

22.1.2 The 1990s

Since the 1960s intercontinental television and telecommunications have been achieved by using satellite signals. Satellite and cable television have made networks possible in which many forms of home information and entertainment (films, sports, etc.) are generally available on subscription. In addition, in many countries viewdata systems, in which information is stored on computer databases, can offer a wide range of TV services (e.g. sports results and general news).

Also in the 1990s, while the data processing speed, storage capacity and the accuracy of small computers has been constantly increasing, their cost and size has been constantly decreasing. Stern and Stern (1993) have noted that personal computers (PC) are currently found in 40% of American homes and European domestic sales are not dissimilar. Analogue and digital mobile phones are now common and portable computers with built-in cellular phones are becoming available. Notebook computers of A4 size or smaller were available by the early 1990s and powerful computers which people can wear on their wrists or hang around their necks are predicted for the near future.

22.2 THE REALITY?

Soccer and other football games are without doubt amongst the world's most popular sports and modern IT has had a considerable impact upon them. To examine the ways in which it has done so, let us consider a few recent examples of the impact of television or radio alone on modern top-level soccer.

The UK press has reported soccer supporters' criticism of the timing of the BSkyB's live matches. The satellite TV contract with BSkyB means that for the 1993/94 season some UK Premier League soccer matches started at 1600 hours on a Sunday or at 2000 hours on a Monday night. These times were in sharp contrast to the more traditional Saturday afternoons or mid-week evening game times. As a direct result of the BSkyB contract, UK Premier League matches were played on Saturdays, Sundays, Mondays, Tuesdays and Wednesdays throughout the 1993/94 soccer season.

It is claimed in the UK press that UK TV authorities criticized the 1995 Rugby World Cup organizers for scheduling matches without consulting them and that they might not be able to find programme slots for more than half the matches.

In 1993 UK professional soccer club Portsmouth sent video evidence to the FA to support a plea for leniency for their striker Lee Chapman who had been sent off. Similar video recordings have been used since then in other disciplinary hearings.

Paul Elliot (ex Glasgow Celtic player) has described, in the UK media, the effects of racist crowd taunts and abuse on black soccer players. Elliot claimed that his appearances on Scottish TV and radio helped to raise the level of awareness amongst soccer supporters and that black players are now more readily accepted by soccer supporters.

In the new multi-functional Millwall FC stadium, supporters are entertained before and after the game by 56 TV monitors. It is suggested, however, that 24 of these monitors contain closed circuit surveillance monitors for security purposes.

It is reported in the media that when 1993 French soccer league champions Marseilles were relegated as a punishment for financial irregularities, Monaco (who were not the league runners-up that year) were selected and agreed to take part in the 1993/4 European Cup as replacements as a result of persuasion by TFI, the French national television company. Paris Saint-Germain (the runner-up team) were sponsored by Canal-Plus, the French cable television company who held the exclusive TV rights to the Paris club's matches in the Cup Winners' Cup and so the Paris Saint-Germain organization turned down the opportunity to play in the Champions' Cup.

Many other similar examples could be given which would indicate the pervasive influence of these modern technologies not only on the playing, officiating and spectating of soccer, but on the very structures of the game and its competitions.

22.3 THE POTENTIAL?

22.3.1 Organizational IT and soccer

In major contemporary soccer organizations, business IT systems are now used for diverse purposes which include financial accounting and pay roll, purchasing and investment, marketing and promotion, scheduling and quality control, recruitment and human resource management purposes. Such business IT systems are economical, accurate and fast and improve the flow of information for improved efficiency and more creative management and decision-making. In modern soccer organizations, such systems have (or could have) increased efficiency in all these various administrative areas. Also contemporary communication systems (e.g. fax systems and E-mail) have facilitated the rapid arranging of transport, hotel accommodation, individual dietary requirements and the booking of training facilities and playing venues for overseas soccer tours.

22.3.2 Coaching and notation

Modern IT systems can also assist soccer coaching in several ways. Modern computer-based sports notation systems (see for example Franks and Nagelkerke, 1988) allow the complete analysis both of individual performances and of the efficiency of a team's playing patterns and tactics. It is, for example, possible to note via computer input devices (during the match or from audio or video recordings) each pass and the passing patterns of a team's forwards or midfield players or the defence (Figure 22.1).

Graphic displays of analyses of passing (Figure 22.2) and movement of the ball (Figure 22.3) can demonstrate quite clearly the patterns of play.

Indeed the performances and playing tactics of opposing teams as well as the performances of the match officials can also be studied in great detail and with considerable accuracy. These notation data can be quickly analysed after each match to give rapid feedback on both the positive and negative aspects of a team's performance. When entered in databases, the data for each match

Figure 22.1 Computerized analysis of a video recording of match-play.

could form a complete record of the team's performances during the season and be available for various forms of statistical analysis.

22.3.3 Selection and Artificial Intelligence

It is clearly possible that these performance records could be made available to national team coaches. These computer-based records would allow an in-depth history of each player to be made available for detailed comparisons of the player's abilities for team selection and development. These systems would allow such comparisons to be made in a highly efficient way and could, for example, identify a player's most effective playing position and the effect of differing environmental factors (weather, altitude and so on) on their performance.

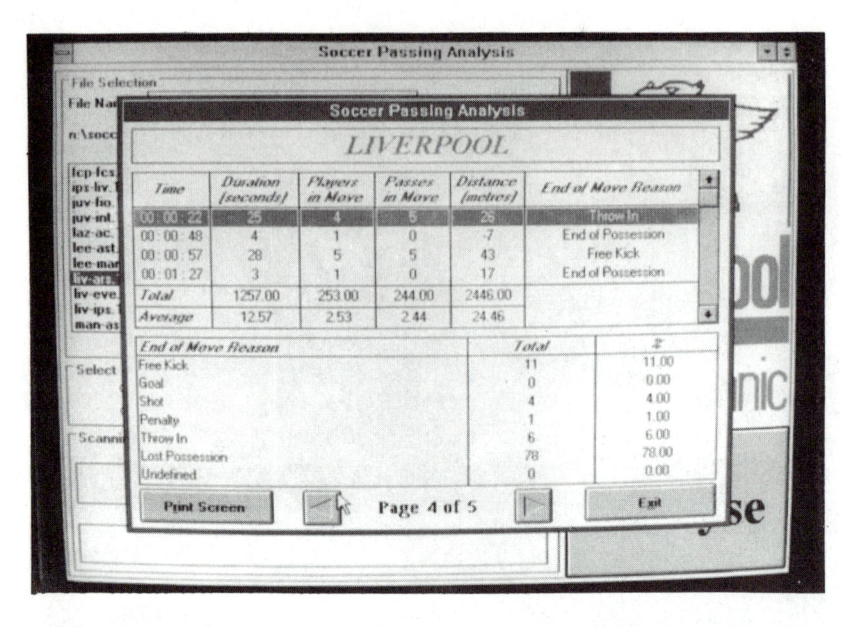

Figure 22.2 Computer display of passing moves in a soccer game.

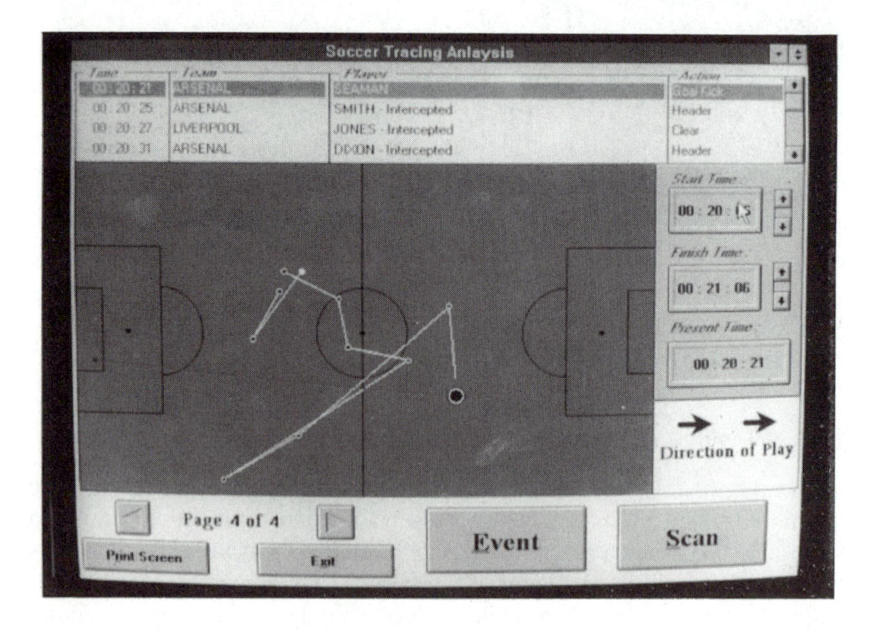

Figure 22.3 The position, the player and the action involved can be monitored and displayed in a computer-based analysis of play.

Modern Artificial Intelligence (AI) computer-based expert systems, which can easily disseminate expert advice, could also aid and accelerate this selection process by allowing the very best soccer selectors' knowledge and reasoning abilities to be readily available in a soccer club's computer systems. Expert systems shells (initially 'empty' AI systems which can be 'filled' with all types of expert knowledge), when coupled with the latest multi-media technologies, also offer the real prospect of developing interactive, self-paced, intelligent systems for motor skills training in soccer.

22.3.4 Video

In the past, many soccer coaches have been totally dependent on their own subjective (and undoubtedly biased) observations of the team's and of individual players' previous performances for evaluation and coaching purposes. In addition to the more sophisticated notation systems detailed above, a team's coaches can now replay the whole match on video tape in slow motion with full freeze frame facilities and even with computer control (see, for example, Lyons, 1988). These useful play-back facilities allow accurate and objective coaching advice to be more widely available. In addition, modern video technologies obviously allow coaching information to be presented in a more complete and dynamic format rather than by just offering the players verbal feedback.

Video recordings, when edited, can also be made available either for resale or for publicity, advertising/sponsorship or for instructional purposes. The use of video also allows the detailed investigation of players' individual techniques (and any variations in their performance) and abilities, and permits scientific, biomechanical or kinematic analysis. The use of video is ideal in coaching, it is versatile and inexpensive in use and tapes can be edited and updated as necessary.

22.3.5 Databases

Modern computer-based information and decision support systems allow players' personal, medical and playing details to be made directly available to managers and coaches and in case of injury or illness to be immediately accessible to doctors or to other medical staff. Database systems could also contain an individual player's personal, attendance, fitness and medical records for monitoring purposes. Such systems would aid the administration of the team by improving both the efficiency of the organization's record-keeping and by improving the organization's internal communications. Modern spreadsheet packages facilitate the analysis of all forms of data and will allow the results to be presented and reported in various tabular or graphic formats for improved managerial decision-making.

Clearly, database systems could also help many soccer clubs and organizations to maintain accurate details either of their supporters or their members. In this way they would offer the supporters improved services and facilities which are more customized to their requirements. The analysis of these records would give the clubs accurate information for the provision and promotion of their facilities and for marketing purposes. In addition, modern computerized graphical software packages and desktop publishing packages now offer the real possibility of rapidly producing publicity material, posters, newsletters and fanzines.

22.3.6 Multi-media systems

In addition, interactive multi-media systems, including video or CD-ROM (compact disk read only memory) technologies, are now being developed for the delivery of entertainment, education and training, information dissemination, or for the marketing or promotion of products or services. Keys known as 'Smart Cards' are used for gaining individual access to hotels, leisure centres and many other areas in which personal or organizational security is required. Systems for the administration of various soccer membership and attendance schemes should improve ground security. It is apparent that the use of debit and/or credit cards and bar code readers at point-of-sale terminals has dramatically improved the efficiency of modern supermarkets and other sales organizations and could have similar applications for the marketing of soccer.

22.4 TOMORROW OR TODAY?

Modern software applications such as word processing (e.g. Word Perfect), spreadsheets (e.g. Lotus 1–2–3), databases (e.g. dBASE III +) and analytical or statistical software (e.g. SPSS), desktop publishing (e.g. Pagemaker) and expert system shells are in standard, everyday use by major organizations. In the near future AI systems, multi-media and the development of virtual reality systems with real-time, full motion soccer action offer outstanding possibilities for highly interactive coaching, training and home entertainment. Information technology systems for soccer are not just science fiction hype for our future but are becoming available as today's reality.

Summary

Information technology (IT) is used for the gathering, analysis and display of large sets of data on soccer play and performance and for the simulation of human movement. In the future, AI expert systems could aid human soccer selection and training. Modern TV and radio has had major effects on the game and its supporters. While video offers solutions to many of the traditional problems of coaching, databases and other commercial software could also relieve many of the organizational, administrative and security problems of modern soccer.

REFERENCES

Cornwall, H. (1986) *The New Hacker's Handbook*, Century Hutchinson, London.
Franks, I.M. and Nagelkerke, P. (1988) The use of computer interactive video in sport analysis. *Ergonomics*, **31,** 1593–603.
Hull, R. (1992) *In Praise of Wimps: a Social History of Computer Programming*, Alice Publications, Pecket Well, West Yorkshire.
Lees, A. (1985) Computers in sport. *Applied Ergonomics*, **16,** 3–10.
Lyons, K. (1988) *Using Video in Sport*, National Coaching Foundation, Leeds.
Stern, N. and Stern, R.A. (1993) *Computing in the Information Age*, John Wiley, New York.

Index